SURFACE ENHANCED RAMAN SCATTERING

SURFACE ENHANCED RAMAN SCATTERING

Edited by

Richard K. Chang

Yale University
New Haven, Connecticut

and

Thomas E. Furtak

Rensselaer Polytechnic Institute
Troy, New York

PLENUM PRESS • NEW YORK AND LONDON

Library of Congress Cataloging in Publication Data

Main entry under title:

Surface enhanced Raman scattering.

Bibliography: p.
Includes index.
1. Raman effect, Surface enhanced. I. Chang, Richard K. (Richard Kounai),
1940 –   . II. Furtak, Thomas E. 1949 –
QC454.R36S87   535.8'46   81-22739
                                                                         AACR2
ISBN-13: 978-1-4615-9259-4   e-ISBN-13: 978-1-4615-9257-0
DOI: 10.1007/978-1-4615-9257-0

PREFACE

     In the course of the development of surface science, advances
have been identified with the introduction of new diagnostic probes
for analytical characterization of the adsorbates and microscopic
structure of surfaces and interfaces. Among the most recently developed techniques, and one around which a storm of controversy has
developed, is what has now been earmarked as surface enhanced Raman
scattering (SERS).

     Within this phenomenon, molecules adsorbed onto metal surfaces
under certain conditions exhibit an anomalously large interaction
cross section for the Raman effect. This makes it possible to
observe the detailed vibrational signature of the adsorbate in the
ambient phase with an energy resolution much higher than that which
is presently available in electron energy loss spectroscopy and when
the surface is in contact with a much larger amount of material than
that which can be tolerated in infrared absorption experiments. The
ability to perform vibrational spectroscopy under these conditions
would lead to a new understanding about the chemical identity, geometry, and bonding of adsorbed material at a level previously unacces-
sible. It is for these reasons that the last few years have brought
an explosion of activity surrounding the exploitation of SERS. The
search for the origin(s) of the anomalous enhancement has given rise
to a research sub-activity of its own. Efforts to explain the enhancement have led to an increased understanding of the whole range
of phenomena associated with the interaction of photons with adsorbates and metal surfaces.

     There is so much information in separate articles at present
that it is difficult for those interested in using SERS to acquire
a comprehensive perspective. The field has produced its share of
confusing and sometimes contradictory experimental results, a characteristic of rapid and non-standardized scientific activity. Many
theoretical models have been proposed, often through simultaneous
developments by geographically separated authors. Activity surrounding the enhancement mechanism itself and the applications of SERS to
surface science will continue for some time.

SERS has been associated with several characteristics which have formed the basis for most of the experimental and theoretical efforts to explain the enhancement. The effect is largest when observed on substrates of silver, copper, and gold; when the substrate is roughened; and when the excitation energy is below the interband threshold for the substrate. Acceptance of these observations, along with supporting non-Raman evidence, has led to a growing consensus that some type of "geometrically defined surface optical resonance" is partially responsible for the enhancement mechanism in SERS. A large body of experimental data and quantitative theory on the long range "electromagnetic" effect has been published along these lines. This is reflected in the emphasis chosen for the various chapters of this book.

There is also significant evidence which indicates that the enhancement depends on short range "chemical" communication between the adsorbate and the surface, i.e., the molecules demonstrate the highest apparent cross section when associated with an "active site." This conclusion is also represented herein. It is likely that the largest enhancement factors are achieved only when both mechanisms are allowed to operate.

The future will bring increased application of SERS, particularly under conditions where there is no competing technique: in electrochemistry, catalysis, combustion, and in other material science of interfaces. Further, it will find its place in vacuum surface characterization because of its advantages over other diagnostic probes. We are also bound to see the results of increased understanding of surface electromagnetic phenomena in such areas as those involving surface photochemical effects and surface nonlinear optical phenomena.

The publication of "Surface Enhanced Raman Scattering" represents a reference point identified by the accumulation of significant insight about this topic and, for the first time, collects authoritative contributions by experts active in the development of the SERS field. While it was impossible to have each individual who has worked with SERS contribute a chapter, the authors were encouraged to refer extensively to the results of others. The first chapters in the book emphasize the theoretical and model-related aspects of SERS; the latter chapters describe experimental results to date. We hope that this volume serves its intended purpose of providing a self-contained reference through which progress in interfacial science can be made through the understanding and exploitation of Raman scattering and related optical processes.

We thank S. Macomber, K. U. von Raben, and especially S. Engelman for their expert assistance in the preparation of this book.

Richard K. Chang
Thomas E. Furtak
(1981)

CONTENTS

THEORY

EXPERIMENT

A SURVEY OF RECENT THEORETICAL WORK

Horia Metiu

Department of Chemistry
University of California
Santa Barbara, California 93106

I. THE BASIC THEORETICAL SCHEME

You marry the person who is available when you are most vulnerable.

K. Berwick

§1. We organize this theoretical discussion around a scheme having three elements: the local field acting on each molecule, the induced molecular dipole and its radiation.

§2. The local field $\vec{E}_\ell(\vec{r},\omega)$ acting on a molecule located at $\vec{r}$ has a variety of sources. The most important is the laser electric field as modified by the presence of the solid surface. To this we must add the "image field" which accounts for the polarization of the metal by the dipole induced at $\vec{r}$, and the field caused by the dipoles induced on the other molecules. These "dipolar" effects modify the absorption spectra (both electronic[1] and infrared[2]), the fluorescence life-time and intensity[3], and the resonant Raman cross section[4] of absorbed molecules. They do not affect substantially normal (*i.e.*, non-resonant) Raman intensity, and will be neglected here.

For normal Raman scattering we are interested in the "static" molecular polarizability only,[5] since the laser is intentionally operated at frequencies well below the excited states of the molecule. In particular we need to know the Raman-Stokes component $\vec{\mu}_{RS} = \overset{\leftrightarrow}{\alpha}_{RS} \cdot \vec{E}_\ell(\vec{r};\omega) \cdot \exp[-i(\omega-\omega_v)t]$. Here $\overset{\leftrightarrow}{\alpha}_{RS}$ is the Raman-Stokes polarizability, $\vec{E}_\ell(\vec{r},\omega)$ is the local field intensity (with the image and other dipolar fields not included), ω is the incident

1

laser frequency and ω_v that of the molecular vibration. If resonant processes are of interest a Drude-Lorentz equation could be used[3,4,5] to compute the frequency dependent polarizability.

The induced molecular dipole $\vec{\mu}_{RS}$ emits Raman shifted (Stokes) photons. The emission intensity can be substantially affected by the presence of the solid surface.

§3. Given the scheme outlined above it is useful to consider the Raman intensity as a product of molecular and electromagnetic effects. The first enters into the theory through the molecular polarizability $\overleftrightarrow{\alpha}_{RS}$. The second through the local field and the dipole emission. We can imagine a very simple, limiting case (the purely electro-dynamic model (PEDM)) in which the polarizability is affected by chemisorption only in a minor way. Then almost all of the enhancement comes from the electromagnetic effects. The local field can be substantially enhanced when the laser excites the electromagnetic resonances of the solid or when the surface has a high curvature. The emission is enhanced if the radiating Raman-Stokes dipole drives an electromagnetic resonance of the solid. The local field and the emission enhancement are related through a reciprocity theorem which in many cases makes them be of comparable magnitude.[6]

Another extreme case is the purely molecular model (PMM) which assumes that the electromagnetic enhancement is minor and most of the SERS intensity comes from profound modifications of the molecular polarizability by the metal. This must be the case if, for example, the existence of a large enhancement ($\sim 10^4$) is documented for molecules adsorbed on flat surfaces.

§4. Those acquainted with the propensity of surface molecules to exhibit perverse behavior might guess that in most situations the molecular and the electromagnetic effects are mixed in proportions depending on the type of molecule, surface, binding and environment. Simpler behavior will be somewhat out of character, though it may appear on some systems, just for the purpose of causing confusion.

One of the important aims of the interaction between theory and experiment is to find ways of determining how much of the observed intensity is due to electromagnetic effects and how much to the molecular ones. Knowledge of the first might suggest ways of manipulating surface shape and properties to tailor the enhancement to our desires. The presence of substantial polarizability modifications may help to diagnose the existence of a peculiar binding or profound modifications in the electronic structure of the adsorbed molecule.

II. PURELY ELECTROMAGNETIC EFFECTS

> *Everything should be made as simple as possible,*
> *but not simpler.*
>
> A. Einstein

II.1. Introductory Remarks

§5. The electromagnetic effects, *i.e.*, the local field and the
molecular emission, can be computed if: (a) the polarization of
the molecule is represented by a point dipole (or a spatially
extended, oscillating, polarization charge); (b) the phenomenolo-
gical equation $\vec{D}(\vec{r},\omega) = \varepsilon(\omega)\vec{E}(\vec{r};\omega)$ can be used even though we are
interested in fields and sources located as close as one angstrom
from the surface; (c) the surface has a simple shape.

So far most experiments have been done on surfaces which
differ substantially from the theoretical models, thus preventing
direct quantitative tests of the theory. For this reason our
strategy is bimodal. We try to isolate those theoretical
predictions that seem to be common to various models (*e.g.*, long
range enhancement) and assume that they also apply to the
experimental situation. The presence of the predicted effects in
the data is circumstantial evidence for the existence of the
electromagnetic enhancement. Second, we pay special attention to
those experiments which work, when done with outmost care, with
systems on which exact electrodynamic calculations can be performed
(*e.g.*, colloidal systems).

II.2. Electromagnetic Enhancement on Flat Surfaces

§6. The theory predicts that the electromagnetic enhancement at
a flat surface is "minor".[7] The local field is given by the
Fresnel equations and it has, roughly, twice the value of the
incident field.[7] The emitted field, at the detector, is about
twice that emitted by the same dipole in vacuum. This gives an
enhancement factor of four (two from the local field multiplied by
two from emission) in the detected field, which amounts to 16 in
intensity.

The Raman intensity is further affected by the fact that
chemisorption clamps the molecule down and prevents it from
tumbling. Since the local field is almost perpendicular to the
surface some interesting "selection rules" appear. If the
chemisorptive bond keeps the "most polarizable direction" of the
molecule perpendicular to the surface, the Raman intensity will be
increased with resepct to the gas phase value. If the most
polarizable direction is parallel to the surface the orientation
effects could lower the intensity substantially. Such "selection
rules" are customary in specular EELS and IR reflection measurements

of vibrational frequencies. They may be very useful, for example, in detecting phase transitions in which aromatic molecules lying flat on the surface rearrange to an edgewise position. Such a transition should be accompanied by a sudden increase in Raman intensity at the transition temperature (or coverage).

§7. It is difficult, but possible,[8] to detect the Raman spectrum of a monolayer on a flat surface in the absence of any polarizability enhancement. Udagawa *et al.*[8] carried out such an experiment in UHV on a "flat" Ag(100) surface and found an enhancement factor of 4.4×10^2 above the cross section of gaseous or liquid Py. Schultz *et al.*[9] worked with an electrochemical cell and reported that the enhancement factor for a "flat" Ag surface that was not anodized is 10^4. Subsequent anodization increased the enhancement factor to a total of 10^6. This would imply the existence of a molecular enhancement mechanism since the electromagnetic enhancement on a flat surface is much smaller than the observed one. Unfortunately, these experiments cannot exclude the possibility that a very small number of "boulders" (of a radius less than 250 Å for Ref. 9) might cause part or even all of the observed enhancement. Recent detection of SERS for molecules adsorbed on a liquid mercury surface[10] (enhancement factor 10^4-10^5) is not subject of such doubts. Thermal excitation of Rayleigh surface waves seems to be the only possible source of roughness but the wave amplitude is much too small to allow an electrodynamic explanation of the effect.

Since the electrodynamic effects at flat surfaces are well understood such systems are ideal for the study of molecular enhancement effects. We hope that they will be vigorously pursued.

II.3. Use of Surface Plasmons

§8. A flat metallic surface has an electromagnetic resonance, namely the surface plasmon. Due to the requirements of momentum conservation it cannot be excited optically even though its existence can be seen in electron energy loss spectra. Surface plasmon excitation becomes possible if the parallel momentum of the incident photon is modified to match that of the plasmon. This can be done by using a prism with carefully chosen refractive index[11] (Attenuated Total Reflection ≡ ATR), or by constructing a grating on the surface[12] to change the photon parallel momentum by diffraction.

II.3.A. The ATR technique. §9. Chen *et al.*[13] seem to have been the first to point out the possibility of using the large evanescent field created by the ATR configuration, to enhance the normal Raman and the coherent anti-Stokes spectrum of monolayers adsorbed on a silver surface. This was well before people were aware of the existence of SERS. The predicted[13] enhancement factor depends on the experimental arrangement, and under the most favorable conditions (on Ag) it can be of order 150-200. The theory assumes a perfectly flat surface while under experimental conditions some unavoidable surface roughness scatters the plasmon and broadens it, lowering the expected enhancement.

§10. It is easy to confirm the existence of an ATR induced enhancement since the Raman intensity must have a peak at the incidence angle for which the reflectivity dips. This has been observed experimentally.[14] The Raman intensity at this resonance angle is estimated to be an order of magnitude larger than the off-resonance value. The s-polarized light does not cause such a resonant enhancement, and this is also in agreement with the theory.

   II.3.B. Raman spectroscopy on gratings. §11. The local field near a grating can be computed by a method originated by Rayleigh and improved by a number of contemporary authors.[12] The application to Raman scattering by molecules adsorbed on the surface of a grating was made by Jha *et al.*[15]

   The computation of the local field follows closely the derivation of the Fresnel equations with the difference that a refracted wave is added to the incident, reflected and transmitted ones. Perturbation theory, with $(\xi\omega/c)$ as a small parameter (ξ is the mean grating height), is used to compute the amplitudes of the refracted waves. These have several interesting properties.
   (a) The diffracted wave momentum has a component perpendicular to the surface, given by

$$\Gamma = (\omega/c)\{\cos^2\theta - 2\sin\theta\sin\phi(2\pi nc/L\omega) - (2\pi c/L\omega^2\}^{\frac{1}{2}} . \qquad (1)$$

Here L is the wavelength of the grating, θ and ϕ are the polar and the azimuthal angles of the direction of incidence and $n=\pm1,\pm2$, *etc.* The grooves of the grating are directed along the x-coordinate. If the expression in the curly bracket is negative Γ becomes imaginary and the refracted wave is evanescent. Its amplitude decays exponentially as a function of the distance to the surface. The decay length is given by Γ^{-1} and it can be as long as 1000 Å.
   (b) The amplitude of the evanescent wave has a maximum value when

$$(\omega/c)^2\sin^2\theta+2(\omega/c)\sin\theta\sin\phi(2\pi n/L)+(2\pi n/L)^2 = (\omega/c)^2\frac{\text{Re }\epsilon(\omega)}{1+\text{Re }\epsilon(\omega)} .$$

$$(2)$$

The r.h.s. of this equation is the square of surface plasmon parallel momentum. The l.h.s. is the square of the parallel momentum of the diffracted photon. Once the laser frequency, ω, the material and the grating wavelength, L, are chosen, one can satisfy Eq. (2) by working at the appropriate angles of incidence

θ and ϕ. At these angles Raman intensity is enhanced as a result of a local field enhancement caused by plasmon excitation.

(c) The maximum value of the local field is proportional to Re $\varepsilon(\omega_R)/Im\ \varepsilon(\omega_R)$. Again, a small imaginary part of the dielectric constant at the resonance frequency ω_R makes the metal a good enhancer. Silver will thus be an excellent material.

(d) The local field is proportional to the mean amplitude ξ of the grating. Since the theory is valid only for $(\xi\omega/c) \ll 1$ one cannot increase the local field indefinitely by making gratings with larger and larger ξ. Since the plasmon is scattered by the grating (this effect is not contained in this simple theory) the increase of the grating amplitude ξ will increase the plasmon line width[16] making it of the form $a + b\xi^2$. The square of the local field will then vary as $\xi^2(a + b\xi^2)^{-1}$ and saturation[17] occurs as ξ is increased.

(e) If the incident laser beam is directed perpendicular to the grooves ($\phi = 90°$) only p-polarized light can excite the plasmon (hence enhance the Raman lines). For $\phi \neq 90°$ both s- and p-polarized light can cause enhancement. The peaks will be seen at the same angle for both s and p components.

§12.   The presence of a grating can involve the surface plasmon in enhancing the emission of the molecular Raman-Stokes dipole. This effect has been considered by Aravind *et al.*[18]

The physics of the process is rather simple. The near field of the oscillating dipole is spatially inhomogeneous and because of this it can excite a surface plasmon even if the surface is flat.[3,4] This does not lead to any radiative gain since on a flat surface the plasmon cannot emit photons. The presence of the grating makes photon emission possible. The conservation of the parallel momentum requires that the angle of this emission is very precisely defined. The grating acts as a device that removes energy from the near field of the dipole and turns it into precisely directed photons. This is energy that would not be radiated if the surface was absent or flat; therefore, it leads to a Raman signal enhancement which appears as an "angular resonance" as the detection angles are varied. Detailed calculations[18] show that the efficiency of this process varies substantially with oscillator frequency, the angles of detection and the properties of the grating. The largest intensity found numerically[14] is 500 times above that of the dipole emitting from the neighbourhood of a flat surface. This takes place at a frequency of 2.4 eV, for a silver grating height (peak to bottom) of 1148 Å, a grating wavelength of 820 nm and an azimuthal angle of 85°. The resonance is at a polar angle of 25°. Lowering the azimuthal angle diminishes the intensity and shifts the resonance to higher polar angles. At low azimuthal angles double angular resonances appear, with peaks at two values of θ. One can in fact compute the value of ϕ at which resonance splitting occurs.

The two peaks correspond to photons whose wave vector is diffracted by plus or minus the grating wavevector (n =±1).

§13. Both the calculation of the local field[15] and enhanced emission[18] use a perturbation theory that requires $(\xi\omega/c) \ll 1$. The theory does not "renormalize" the plasmon to take into account changes caused by the presence of the grating. Therefore the plasmon of the flat surface appears in the theory, while experimentally the plasmon of the grating is excited. The presence of a molecular layer will also modify the properties of the plasmon. Further complications can appear if the metal film is so thin that the surface plasmons of the two faces of the film interact with each other. All these effects can be taken into account and it is probably just a matter of time before improvements are made. As it stands, both Jha *et al.*[15] and Aravind *et al.*[18] overestimate the enhancement and the theory is semi-quantitative.

§14. The experimental work with gratings[19-21] has established that the Raman intensity has a peak at the angle of incidence for which the reflectivity has a dip caused by plasmon excitation. A scan of the emission angle shows a peak which can be attributed to plasmon emission driven at Raman-Stokes frequency, as suggested by Aravind *et al.*[18] The enhancement casued by the plasmon local field at the resonant angle of incidence is 25 times larger than the off-resonance signal.[19] Other estimates[20] give a signal of 50-80 times larger than that on flat films (this is the total enhancement for a 10 nm polystyrene film). The Raman signal when the detector is located at the emission resonance angle is five times larger than the off-resonance signal. This points out that, as expected, the computations over-estimate the magnitude of the plasmon local field[15] and of the dipole driven plasmon emission.[18] The latter seems to be in error to larger extent. This is somewhat expected since the perturbation theory is not as good for dipole fields as it is for planar waves.[18]

The only work using s-polarized light is that of Girlando *et al.*[16] They find that while the reflectivity is quite flat the Raman signal is not zero. Some residual-non-plasmon-enhancement is therefore present. This might be caused by the presence of a metal "boulder" on the grating, or some form of small scale roughness, or molecular enhancement.

The long range distance dependence of the enhancement factor has been verified by Sanda *et al.*[21] who monitored the change of the Raman intensity with the number of layers deposited on the surface. The results agree with the prediction that the intensity falls off exponentially with the distance and the decay length is that of the surface plasmon. The plasmon enhancement computed (or extrapolated from the large distance data) for the first adsorbed layer turns out to be smaller than the detected one. The

discrepancy seems to be confined to the first layer. Therefore, we
might consider this to be an indication that the effect is not
caused by additional roughness of the grating surface. If such
roughness was present, its effect would extend beyond the first
layer, and this is not the case. The intensity estimated from a
local theory by Jha *et al.*[15] is in general agreement with the
observations for the first layer.[21]

II.4.   Electrodynamic Enhancement by Using Isolated Solid Particles

   II.4.A.   Introductory remarks.   §15.   A simple way of using
electromagnetic resonances spectroscopically is to work with
molecules adsorbed around small solid particles. This can be done
in liquid (using colloids)[22] or gas phase (using "molecular beams"
of particles[23]); or the particles can be trapped in a matrix[24] or
deposited on a supporting surface.[25] If electron microscopy,
detailed light scattering and spectroscopic measurements are
jointly used, such systems can play an important role in our understanding of the enhancement mechanism, since the electrodynamic
effects can be computed in detail.

   II.4.B.   The use of the electrodynamic resonances of small
spheres.   §16.   If the diameter of the sphere is smaller than the
wavelength of light, the local field caused by the incident laser is
easily computed.[26]   To the laser field $\vec{E}_i$ we must add the field
caused by the dipole

$$\vec{p} = [(\varepsilon_1(\omega) - \varepsilon_2)/(\varepsilon_1(\omega) + 2\varepsilon_2)] \, a^3\vec{E}_i \equiv \beta(\omega)\vec{E}_i, \qquad (3)$$

located at the center of the sphere. This enhances the local field.

§17. The sphere can also enhance the intensity of the radiation of
the molecular Raman-Stokes dipole $\vec{\mu}_{RS}$. This happens because the
field $\vec{E}_\mu$, caused by $\vec{\mu}_{RS}$ polarizes the sphere inducing a point
dipole $\vec{p}' = \beta(\omega-\omega_v)\vec{E}_\mu$ at its center. This radiates at the Raman-
Stokes frequency adding to the total Raman signal.

§18. Detailed calculations[27] support the qualitative description
given above and lead to several interesting conclusions.
   (a) The Raman excitation spectrum [divided by $(\omega-\omega_v)^4$] is
proportional to $|\beta(\omega-\omega_v)\beta(\omega)|^2$.   The polarization of the sphere
$\beta(\omega)$, peaks at the resonance frequency ω_R satisfying the equation
Re $\varepsilon_1(\omega_R) + 2\varepsilon_2 = 0$. Since the vibrational frequency ω_v is
generally smaller than the width of the resonance, both $\beta(\omega-\omega_v)$ and
$\beta(\omega)$ can be simultaneously on resonance.
   (b) At the resonance frequency the enhancement factor is
proportional to $Im \, \varepsilon_1(\omega_R)^{-4}$. Materials, such as Ag for which
$Im \, \varepsilon_1(\omega_R)$ is small are obviously better enhancers. The total
enhancement caused by small Ag spheres can be of order 10^6.

(c) If the enhancement is purely electromagnetic the frequency
dependence of the enhancement factor is determined by the nature of
the metal and the vibrational frequency of the line of interest.
The latter enters only through the factor $|\beta(\omega-\omega_v)|^2$. These
predictions can be tested by studying various lines, from various
molecules co-adsorbed on the same sphere.

(d) The enhancement factor for one molecule, located at a distance d from a sphere of radius a, depends on $[a/(a+d)]$ (see Ref. 12).
This is a slowly varying function of d; for example, if a = 400 Å,
the enhancement factor for a molecule located at d = 30 Å is 0.42
times that of a molecule located at the surface.

II.4.C. Use of electromagnetic resonances of large spheres.
§19. There are two detailed calculations[27b,d] related to the enhanced
Raman spectrum on large spheres. The theory used to compute the
local field is that developed by Mie. The calculation of the dipole
emission is more complicated, but feasible.[27b] We summarize here
the results.

(a) The enhancement produced by large spheres is always below
that obtained in the small radius limit. This is due to the
decrease of both the local field[27b,d] and the emission.[27b] For
example, for Ag spheres the results are as follows. If the radius
is 5000 Å the enhancement is at most 100. For 500 Å it can go as
high as 10^4. For 50 Å it could be 10^6.

(b) The excitation spectrum for large particles is very
complicated[27b,d] since a photon can excite not only just the dipole
of the sphere but a large number of multipolar Mie resonances.

(c) The dependence of the Raman intensity on the radius of the
sphere is very complex[27b] and it oscillates in a way that should
be detectable experimentally.

II.4.D. Use of the electrodynamic resonances of ellipsoids.
§20. There are a number of calculations[28] that study the Raman
spectroscopy of molecules located near ellipsoids smaller than
the wavelength of light. We review here the results of Gersten and
Nitzan[28b] who took the electric field of the incident laser parallel
to the principal axis and considered a molecule located on the
principal axis. This simplifies the computation and displays the
basic concepts more clearly, but has the shortcoming of diminished
realism.

The local field is composed of the incident laser field plus
that caused by a dipole induced in the ellipsoid. This can be
computed since the polarizability of the ellipsoid is known.[28b]
Since the incident field was chosen along the principal axis the
induced dipole has the same direction. Therefore only one of the
components of the polarizability tensor is put to work in this
calculation.

The molecular emission is enhanced, like in the case of the
sphere, because the molecular dipole drives the ellipsoid inducing
in it a dipole which radiates at Raman-Stokes frequency.

§21. Detailed calculations yield an enhancement factor proportional to

$$\left| 1 + \frac{(1-\varepsilon)\xi_o Q_1' \ (\xi_1)}{\varepsilon Q_1 (\xi_o) - \xi_o Q_1'(\xi_o)} \right|^4 . \tag{4}$$

We dropped here the contribution of the image effect since we
assume that, unless resonant or pre-resonant Raman scattering is of
interest, it has a negligible influence on the intensity. Here $\varepsilon(\omega)$
is the dielectric constant of the metal. The geometric parameters
are $\xi_o = a/(a^2-b^2)^{\frac{1}{2}}$ and $\xi_1 = (a+H)/(a^2-b^2)^{\frac{1}{2}}$, where a and b are the
principal axes, $a \geqslant b$, and H is the distance of the molecule (*i.e.*,
of the point dipole) to the tip of the spheroid. Q_1 is the Legendre
polynomial of second kind and Q_1' is its derivative.
    (a) The enhancement factor can be resonantly increased if the
incident laser frequency has the value ω_R satisfying (see the
denominator of Eq. (4)) $Re[\varepsilon(\omega_R)Q_1(\xi_o) - \xi_o Q_1'(\xi_o)] = 0$. Compared
to the resonance condition $Re \ \varepsilon(\omega_R) + 2 = 0$ valid for a small sphere
in vacuum this equation introduces a new element. For a given
metal the resonance frequency depends on the ratio a/b, through
ξ_o, while for the sphere the resonance frequency is fixed. Varying
a/b from 1 to 7 can change the resonance[28b] frequency for a Ag
ellipsoid from 3.5 eV (a/b=1) to 1.6 eV. Large changes are also
obtained[28b] for Cu and Au.
    (b) The magnitude of the enhancement caused by the electromag-
netic resonance is again controlled by $(Im \ \varepsilon(\omega))^{-4}$. The leading
term in the enhancement factor is

$$\left[1 - \varepsilon(\omega_R) \right]^4 \left[\frac{\xi_o Q_1'(\xi_1)}{Im \ \varepsilon(\omega_R)Q_1(\xi_o)} \right]^4 . \tag{5}$$

Like in the case of the sphere, a large enhancement is expected for
materials, such as Ag, which have small values of $Im \ \varepsilon(\omega_R)$.
    (c) We have mentioned previously that one could think that the
total enhancement is caused by the excitations of the electrodynamic
resonance and by the existence of a large surface curvature. Since
the shape influences strongly the properties of the resonance, one
cannot strictly separate the two effects. The simplest classifica-
tion seems to be obtained by taking the limit $\varepsilon(\omega) \to -\infty$ corresponding to a perfect conductor. Since no electromagnetic resonances
are possible for perfect conductors we propose to regard the
enhancement obtained in such a case as a shape (curvature) effect.
The remainder is then the resonant contribution.

    For ellipsoids, the perfect metal limit[28b] gives an enhancement
factor proportional to

$$\left| 1 - \frac{\xi_o Q_1'(\xi_1)}{Q_1(\xi_o)} \right|^4 .$$ (6)

Since the metal was assumed to be a perfect conductor neither its
properties nor the incident frequency appear in this equation.
Only the aspect ratio a/b and the distance of the molecule to the
surface are important. Variations in the enhancement factor as
large as two orders of magnitude can be obtained by changing these
quantities.

Note that in Eq. (4) the term $\xi_o Q_1'(\xi_1)/Q_1(\xi_o)$ which controls the
magnitude of the "shape enhancement" is multiplied by $(1/Im\,\epsilon(\omega_R))$
which controls the effectiveness of the resonance enhancement. The
two effects multiply each other. The combined effect of shape and
resonant enhancement can give an enhancement factor as large[28b] as
10^{11}, for Ag at $\omega \simeq 2.03$ eV, and a/b = 5, for a molecule-spheroid
distance H = 5 Å. At this distance image effects are unimportant.
Such a large value may require consideration of non-linear effects
since the local field is rather large. It also suggests that the
observed enhancement of 10^6 (this is the maximum value suggested
experimentally) may come from a fraction of 10^{-5} of the total
number of molecules, which happen to be located on a sharp tip of
the surface. These are very interesting predictions, not yet
substantiated experimentally.

(d) The change of the enhancement factor with the distance of
the molecule to the surface of the spheroid shows the same "long
range" behavior as in the case of the sphere. For example,[28b] the
enhancement factor for a Ag spheroid with a/b = 2, a molecule with
polarizability 10 $Å^3$ and a laser frequency of 3 eV (very close to
resonance) is 10^9 when the molecule is near the surface and $2 \cdot 10^8$
if the molecule is 25 Å away. The same behavior is seen at other
frequencies. This long range effect is a general feature of the
electromagnetic enhancement which, if observed experimentally, cannot be attributed to other mechanisms.

II.4.E. Brillouin scattering from spheroids. §22. A very
interesting suggestion, that may aid testing the predictions of the
electromagnetic model, has been made by Gersten and the Exxon
group.[29] It is well known, from the theory of elasticity, that
small objects such as spheres or ellipsoids exhibit "eigen"
vibrations. They represent oscillations of the object's shape and
size and their frequency $\tilde{\omega}$ is small enough so they can be thermally
excited. Since the polarizability of the ellipsoid depends on its
shape and size, it is modulated by the "eigen" vibrations. As a
result if we place an ellipsoid in a laser field of frequency ω,
the mechanical eigen vibrations will induce dipoles of frequencies
$\omega \pm \tilde{\omega}$. The effect may be regarded as inelastic Brillouin scattering by the ellipsoid, with the sound waves, appearing in the
ordinary Brillouin scattering, replaced by the "eigen" vibrations.
This idea seems to explain the presence of a low frequency Raman

line[30] (which was also attributed to bending molecular
vibrations).[31]

We feel that this effect should be kept in mind and looked for
in future experiments. If observed, it will allow to test whether
the Brillouin line position is consistent with the existing information concerning the shape and the dielectric properties of the
particles. In aerosol research this phenomenon may be useful in
investigations of particle shapes as an addition to the customarily
employed elastic scattering.

II.4.F. <u>The interaction between resonances.</u> §23. The local
electromagnetic field of a sphere, driven at resonance by a laser,
extends far outside the sphere. For this reason two spheres can
interact electromagnetically even if separated from each other by a
large distance. One expects that the two degenerate resonances of
the isolated spheres are split by this interaction and the excitation spectrum of the two sphere system will have two peaks. The
same feature must appear in the excitation spectrum of SERS which
follows that of the electromagnetic resonance of the two sphere
system.

This simple idea is confirmed by recent calculations[32] which
in addition show that the electromagnetic energy of these double
resonances is concentrated in the space between the spheres. For
this reason, for Ag, the value of $\vec{E}_\ell \cdot \vec{E}_\ell$ (which is proportional to
the local field intensity) in the region between the spheres is ten
times higher than what it would be if the spheres did not interact.
Since qualitatively one expects the gain in emission to be comparable[6] to that due to the local field, a molecule located between
spheres may have a Raman enhancement factor a hundred times larger
than what it would be if the spheres did not interact. This number
is given for orientation only; it depends on the distance between
the spheres, their radius and the angle between the incident field
and the line joining the centers of the spheres.

§24. As a side remark we note that the electromagnetic interaction
between spheres may serve as a sensitive tool for the study of the
dynamics of the Brownian motion in colloidal solutions. The
polarizability of small isolated spheres is, like that of atoms,
isotropic. For this reason the light scattered at ninety degrees
with respect to the incident beam is not depolarized. The
polarizability of two spheres that are close enough to interact
electromagnetically, is like that of a diatomic molecule. There
are now two important directions, one along the axis of the centers
and the other along the polarization of the incident electric field
This anisotropy causes depolarization of the ninety degree scattering.[33]   The properties of the depolarized intensity depend on
the relative position of the pair of spheres. It is thus a direct

measure of their Brownian motion and if it is monitored experimentally and carefully interpreted, it should allow an interesting test
of the dynamic theories of Brownian motion. Note the analogy of
this process to the collision induced depolarization in light
scattering by monoatomic fluids.

II.4.G. A brief look at experiments. §25. Even though the
electrodynamic model makes detailed predictions concerning the
dependence of the enhancement on the particle size and incident
frequency, no definitive experimental test of these predictions is
yet available. The reason is mainly technical. A conclusive
experiment must satisfy certain conditions. The shape and size of
the particles must be known with precision. This requires electron
microscopy as well as light scattering (and extinction) studies.
The electromagnetic interaction between the solid particles must
be prevented. This requires dilution control and avoidance of
coagulation. The incident frequency must span a wide range, to
cover both on and off resonance situations.

There is no experiment which satisfies all these conditions.
The measurements of Creighton *et al.*,[22a] Kerker *et al.*[22b] and
Wetzel and Gerischer[22c] are either in contradiction with the theory
or are strongly influenced by coagulation. For example, Creighton's
measurements[22a] show that the absorbance has initially a sharp peak
corresponding to extinction by the small sphere resonance. A second
peak grows rather rapidly at a lower frequency in a region where our
calculations[32] predict the presence of a second resonance, turned
on by the electromagnetic interaction between spheres. The Raman
excitation spectrum peaks at the frequency corresponding to this
second resonance. We believe that colloid coagulation causes
electromagnetic interactions and possibly the molecules located in
the space between spheres may give most of the enhanced Raman
signal.

The effects of colloid coagulation can be prevented by working
with matrix isolated metal spheres. Abe *et al.*[24] have demonstrated
the possibility of such measurements. Unfortunately, the limited
number of incident frequencies used makes this very interesting
paper inconclusive as far as the confirmation of PEDM is concerned.

Recent measurements of von Raben *et al.*[22e] seem to lead to an
interesting and perhaps unexpected conclusion: a large enhanced
Raman signal is obtained in a frequency region where PEMT predicts
practically no enhancement. Since this work monitors carefully the
aggregation and works with large colloid dilution, it seems to
provide rather strong evidence for a non-electrodynamic mechanism.
It would be interesting to know whether this observation is peculiar
to the $Au(CN)_2^-$ -Au system used in Ref. 22e or if it is more general.

§26. Obviously, this is still an open ended story. We feel that
a detailed and convincing experimental proof of the existence of

the electromagnetic enhancement by small spheroidal particles (or by small particles of any shape) is still to come. I find it hard to believe that the predictions made by the electrodynamic calculations will turn out to be incorrect. Problems may appear when working with particles smaller than the electron mean free path ($\sim$300 Å), in which electron scattering by the surface can shorten the relaxation time making thus $Im\ \varepsilon(\omega)$ larger than in the bulk metal. Since the calculations use bulk dielectric constants, they will overemphasize the enhancement factor (which is proportional to $Im\ \varepsilon(\omega)^{-4}$). For larger particles this problem does not appear. Furthermore, calculations very similar to those needed for SERS have been carried out to predict the emission from fluorescent molecular layers painted on dielectric spheres[34] or "infinite" cylinders.[35] The agreement with the experiment is very impressive,[36] leaving little doubt that barring the existence of size effects, the electromagnetic calculations are reasonably reliable. The real question seems to be not whether the electromagnetic effects are present but to what extent they are accompanied by molecular ones.

II.5. Particles on Surfaces

§29. Moskovits[37] seems to be the first to formulate a model based on the idea that the rough surface participates in SERS through its electrodynamic resonances. He regarded the protuberances of the surface as metal islands seated on an "indifferent" support. Using an argument that is clever, but rather hard to evaluate, Moskovits predicted that the excitation spectrum of the enhanced Raman cross section follows the absorption profile of the electromagnetic resonance. His basic idea has been refined by Burstein *et al.*[25] and confirmed by subsequent work[28b],[38] which computed the local field and the emission enhancement for models simple enough to permit the solution of Maxwell equations. Gersten and Nitzan[28b] (G-N) considered a half-ellipsoid of arbitrary dielectric constant imbedded in a perfectly conducting flat surface. Aravind and Metiu[38] (A-M) looked at a sphere located near the flat surface, all solids being of arbitrary dielectric constant.

The aim of such calculations is to mimic the behavior of the systems used in Bell Labs experiments,[39] which consist of metallic "boulders" of all sizes (up to thousands of Å) which are randomly spread on a more or less flat surface. The A-M system[38] could in principle be prepared by making metallic spheres through evaporation in a carrier gas[23],[24] and co-freezing them on a flat metallic surface, together with the molecules whose Raman spectrum is desired.

Unless prepared in a special way the customary rough surfaces are not like those discussed by theorists. The shape of the "boulders" differs and this will modify the properties of their electromagnetic resonance. A further modification is introduced by the electromagnetic interactions between boulders. Calculations for two spheres[32] show that the interaction is long ranged and modifies to a large extent the absorption spectrum and the local field.

Therefore it seems that the correct strategy in trying to verify to what extent the electrodynamic contributions are important in SERS on rough surfaces is to try to identify--by concerted analysis of all the models--what are the features most likely to be transferable from the oversimplified models to the real thing and simultaneously to decide which features are so model specific that ought to be disregarded when analyzing data taken on real surfaces.

§30. Gersten and Nitzan[28b] consider a half-ellipsoid of arbitrary dielectric constant imbedded in a flat perfect conductor. One semi-axis is perpendicular to the flat surface and the induced molecular dipole is located on and along that axis. The incident electric field is perpendicular to the flat surface. It is assumed that the perfect conductor is grounded and therefore its potential is zero.

One can immediately guess that the local field for this system is closely related to that near an isolated ellipsoid. Consider a free ellipsoid whose half is identical to the one considered by G-N. Since the dipole and the external field are directed along the principal semi-axis they excite a resonance which has a zero potential at the mid-plane of the free ellipsoid (the mid-plane is perpendicular to the external field). Therefore cutting the ellipsoid along the mid-plane and replacing half of it by a flat, perfectly conducting body does not affect the resonance at all. The image theorem assures that the field outside the system consisting of a half-ellipsoid and a perfect conductor is equal to that of the isolated ellipsoid, as if the missing half of the ellipsoid is still there; the same image theorem puts a molecular dipole inside the perfect conductor symmetrically with respect to the surface plane. The total field of the composite system is that of a free ellipsoid driven by two dipoles located on the principal axis symmetrically with respect to the mid-plane.

The equations obtained by a detailed calculation are consistent with this picture.[28b] Therefore all the conclusions described when we discussed free ellipsoids (§21) are also valid for a half-ellipsoid imbedded in a perfect conducting half-space. These are: (a) the enhanced local field is long ranged; (b) the enhancement comes from resonance plus curvature effects; (c) the maximum enhancement is proportional to $(Im\ \varepsilon(\omega))^{-4}$; (d) the Raman excitation spectrum follows the absorption spectrum of the electromagnetic resonance.

The question is which of these results are model specific and which will survive even if the geometry is modified? We feel that all of the statement (a) through (d) are survivors. However many of the specifics of the G-N calculation will be altered. For example the excitation spectrum is very model dependent. By taking a perfectly conducting flat surface G-N's model precludes the

excitation of resonances of the flat solid. Furthermore, the
choice of the zero potential flat surface to pass through the
middle of the ellipsoid, and the choice of the field along the
semi-axes perpendicular to the flat surface, and the choice of
the dipole on and along the semi-axes, all of these assure that
the resonance of the half-ellipsoid on a perfectly conducting
plane is identical to that of an isolated ellipsoid; the restrictions of the model do not allow a distortion of the resonance by
the change in geometry. If any of these conditions are changed
the resonance of the protruding ellipsoid will be a (severely?)
distorted version of that of the isolated ellipsoid and the
spatial distribution of the local field and the excitation spectrum
are changed.

§31. The calculations carried out for a system consisting of a
sphere near a flat surface allow us to gain some understanding of
such effects. The case of a perfectly conducting small sphere near
a flat surface with an arbitrary dielectric surface can be solved
analytically. This model is realistic and physically relevant for
an incident field of infrared frequency and a flat material having
a surface polariton (e.g. SiC) or plasmon (e.g. InSb) in that
frequency range.

§32. It is clear from such an illustrative calculation that the
presence of the boulder will always excite a "gap mode" whose
origin is that of a "distorted surface plasmon." This mode is suppressed by any model which takes the flat surface to be that of a
perfectly conducting material.

§33. The calculations of the electromagnetic fields for a sphere
near a flat surface shed some light on the manner in which the two
resonances--that of the sphere and the surface plasmon of the flat
surface--interact with each other and modify the fields and the
spectrum of the system. Consider first the case when the sphere is
located at a large distance from the flat surface. The laser can
excite only the $\ell = 1$ resonance; the other resonances, whose
frequencies are given by $\ell \mathrm{Re}\ \varepsilon_1(\omega) + (\ell+1)\varepsilon_2 = 0$ [here $\ell = 1,2,\ldots,$
$\varepsilon_1(\omega)$ is the dielectric constant of the sphere and ε_2 that of
the material around it] are not optically active. The surface
plasmon is not excited because of restrictions imposed by parallel
momentum conservation. As the sphere is moved closer to the surface,
the dipole corresponding to polarized sphere ($\ell = 1$) can excite the
surface plasmon. The two resonances interact with each other. The
sphere resonance frequency ($\omega_p/\sqrt{3}$, for a Drude model) is pushed down
and the surface plasmon frequency ($\omega_p/\sqrt{2}$, for a Drude model) is pushed
up.[38] At even smaller sphere-plane distances the dipole induced by
the laser at the center of the sphere has an image with respect to
the flat surface. The image dipole acts on the sphere exciting the
modes with $\ell = 2,3$, etc., which otherwise are not optically acces-
sible. These, in turn, interact with and excite the surface plasmon.

As a result a large number a "gap modes" appear, roughly bunched around the two initial resonances ($\omega_p/\sqrt{2}$ and $\omega_p/\sqrt{3}$) of the system.

Through their interaction the sphere and the plane assist each other in becoming optically active. The sphere breaks the translational symmetry so photons can excite the surface plasmon. The plane breaks rotational symmetry so the photons can excite resonances with high "angular momentum" $\ell > 1$. The spectrum thus becomes much richer than in the case when the two objects are isolated and none of these resonances are optically detectable.

§34. Computer calculations for the sphere-planar surface system[38,40] confirm these qualitative results. Several features are worth reporting.

(α) In most cases the frequency spacing between the gap modes is less than their width. Only two imhomogeneously broadened resonances are observed, one below the Rayleigh resonance of the sphere and one above the surface plasmon. For a metal sphere on a SiC flat surface the surface polariton of SiC is so narrow that some of the gap modes can be resolved.[40] This however seems to be an exception. The surface plasmons, of metals (e.g. Ag) or semiconductors (InSb), the surface polaritons of ionic crystals (e.g. NaCl) and the surface excitons of molecular crystals (e.g. naphalene) all are shifted by a metal sphere into a broad, inhomogeneously broadened gap mode.[40]

(β) The electromagnetic field between the two objects is enhanced to a greater extent than the maximum resonant enhancement given by either object kept in isolation. We found this to be true when two spheres[32] interact with each other or when a sphere is located near a plane, for metallic objects as well as on systems like metal sphere-flat SiC.[38,40] For example, for a sphere of radius $a = 100$ Å at $d = 5$ Å from an Ag flat surface there are two resonances $\omega_- = 3.17$ eV and $\omega_+ = 3.77$ eV. The magnitude of the square of the local field, $\vec{E}_\ell \cdot \vec{E}_\ell$, at a point on the flat surface located under the sphere is 1.3×10^4 times larger than that of the incident laser, when $\omega = \omega_-$. The isolated sphere gives roughly a factor of 100. The Ag sphere does not aid much the excitation of the surface plasmon of the flat Ag surface and the value of $\vec{E}_\ell \cdot \vec{E}_\ell$ at $\omega = \omega_+$ is 1.7 times larger than that of the incident laser. This is less than what one gets in ATR geometry ($\sim$100) or with a grating. It is interesting to note that at some frequencies $\vec{E}_\ell \cdot \vec{E}_\ell$ is lower than $\vec{E}_i \cdot \vec{E}_i$ for the incident field. If we put a metal sphere on flat SiC we can excite a surface polariton at $\omega \sim 700$ cm^{-1}. The value of $\vec{E}_\ell \cdot \vec{E}_\ell$ is extremely large, roughly 10^6 times that of the incident laser.[40]

(γ) The enhanced field is maximum between objects and falls-off slowly as one moves away from that region. The long range effects predicted by other models seem to be always present even though the distance dependence differs from case to case.

§35. The models used so far have one "boulder" on a flat surface.
All the calculations carried out for systems that have two interacting resonances show a substantial modification of their optical
properties. Real surfaces have thousands of "boulders" randomly
seated on a planar surface. The question is, how do these interact
and what is this interaction doing to their electrodynamic properties? Do they become better enhancers than the isolated boulders?
Is there an optimum boulder concentration? These are certainly all
important questions and I do not believe that as yet we have the
answers.

       Consider two extreme limits, that of pairwise interaction and
that of the closest packing. For the first we have enough model
calculations[32,38,40] to suspect that the local field between the
interacting objects is larger than when they are isolated. Increased enhancement might be possible. The other extreme limit can be
understood by carrying out a "thought experiment". Assume a surface
which looks like a flat terrain with high rise buildings on it. At
very low density these behave almost as isolated ellipsoids on a
flat surface. Large electromagnetic enhancement around them is expected. Now imagine a contractor gone mad, who filled up the space
with buildings as tightly as possible. The high-rise buildings--all
of identical size--can be fitted to construct, with their roofs, a
perfectly flat surface. There is now no electromagnetic enhancement
besides the minor one discussed for flat surfaces. Similarly, closely packed half ellipsoids and spheres can build a grating, thus
giving a smaller enhancement than isolated half spheres. No doubt
these are extreme situations, but there is a hint here that if the
boulders are too crowded the enhancement may go down, while a low
concentration--which however permits electromagnetic interactions--
can give a larger enhancement than in the case of completely isolated
boulders.

§36. Obviously, comparing theory and experiment for rough surfaces
is not an easy task. The experiments at Bell Labs[39] are however
among the most convincing indications that the electromagnetic
enhancement is present. No other theory can explain the long range
effect demonstrated by them.[39]

       It is not yet clear whether the electrodynamic enhancement
induced by roughness is the dominant effect or if it provides just
a part of the total intensity. Recent experiments, detecting SERS
on a liquid mercury surface,[10] on a flat (?) surface in an electrochemical cell[9] or in UHV[8] may indicate that molecular effects
are comparable or maybe more important than the roughness induced
ones. The absence of the enhanced spectrum of water in colloidal
solutions and electrochemical experiments should also be an
indication that the long range electrodynamic enhancement for rough
surfaces is smaller than some model calculations, using isolated
"boulders", would predict. The recent discovery[41] that an increased
ionic concentration in solution leads to the appearance of a SERS

signal for water is difficult to explain with a theory in which the electrodynamic enhancement is the dominant effect.

II.6. Small Random Roughness

§37. The limit in which the rough features of the surface are very crowded can be treated by a model[42] which assumes that the surface has a random shape. In other words the height $z = \xi(x,y)$ of the surface (z is perpendicular to the surface) at the point (x,y) is not known, but we know the probability that ξ takes a specific value. The electrodynamic calculations for such a system are greatly simplified if it is assumed that the mean square height is much smaller than the wave vector of the radiation.[42] This allows the use of perturbation theory. Calculations for reflectivity have been carried out[42] under the assumption that the probability for ξ is Gaussian. In this case the surface roughness enters into the calculation through two parameters: the mean square height and the correlation length of surface height in a direction parallel to the surface. Comparison with reflectivity measurements for metal films deposited on CaF_2[43] (this allows roughness variation) are in general agreement with the theory.[44]

There are few applications to surface spectroscopy. Dipole emission is slightly affected[45] and the fluorescence lifetime is modified.[46] Little systematic experimental work has been done. The fact that CaF_2 roughening of a tunneling junction affects its SERS has been demonstrated by Tsang *et al.*[19a] and the change of the fluorescence lifetime by Adams *et al.*[47]

II.7. Can We Use Phenomenological Maxwell Equations?

§38. The theoretical scheme discussed above assumes the validity of the phenomenological material equations. These relate the displacement vector $\vec{D}(\vec{k},\omega)$ to the electric field $\vec{E}(\vec{k},\omega)$ inside the metal through the relation $\vec{D}(\vec{k},\omega)=\varepsilon(\omega)\vec{E}(\vec{k},\omega)$. It is assumed that the presence of the surface causes a sudden jump in the dielectric constant; because of this Maxwell's equations impose certain boundary conditions on the values of $\vec{D}$ and $\vec{E}$ at the metal and vacuum side of the surface. There are reasons to believe that the material equations must be changed if the external field is spatially inhomogeneous (such as the field of a dipole located at 1 Å from the surface) or if we are interested in the magnitude of the field at points very close to the surface. The question is, how large might the errors be if such changes are not made and the phenomenological procedure is employed?

Maniv and Metiu[48-50] have attempted to provide an answer to this question by using methodology developed by Newns[51] and Feibelman[52] (a simpler, less accurate, but very useful method was developed by Fuchs and Kliewer[53]). Their model uses jellium to describe the nuclear charges and an electron gas to compute (with RPA) the dielectric susceptibility of the metal. The presence of a surface, where jellium terminates abruptly, puts the electrons under two kinds of fields: the external one, which might be of long wavelength, and the one exerted by jellium termination, which has high k components. Furthermore, the translational symmetry is broken and only parallel momentum is conserved. For these reasons, the material equation becomes:

$$\vec{D}(\vec{k}_\shortmid, z; \omega) = \int_0^{+\infty} dz' \epsilon(\vec{k}_\shortmid, z, z'; \omega) \cdot \vec{E}(\vec{k}_\shortmid, z'; \omega).$$

If both points z and z' are deep enough inside the metal, this equation goes into $\vec{D}(\vec{k}, \omega) = \epsilon(\vec{k}; \omega) \cdot \vec{E}(\vec{k}; \omega)$. This still differs from the phenomenological one since the dielectric constant depends on the wave vector $\vec{k}$ of the incident field (spatial dispersion). If the incident wave vector is small, $\vec{k}$ can be taken to be zero and we recover $\vec{D}(\vec{k}, \omega) = \epsilon(\omega) \cdot \vec{E}(\vec{k}; \omega)$.

It turns out that the field caused by an incident laser at one or two Å outside the metal is roughly the one given by the phenomenological theory (Fresnel equations).[49] The field caused by an oscillating dipole located near the surface is a different story. Here the phenomenological theory (image theorem) clearly breaks down since the field becomes infinite as the dipole approaches the surface. Taking spatial dispersion into account "heals" this divergence.[50] The rate of energy transfer and the level shifts caused by the presence of the metal are much smaller in the microscopic model than those predicted phenomenologically.[50]

Several papers[54] have attempted to estimate the errors made by using point dipoles to describe the spectroscopic properties of the adsorbed molecules. The interaction of the molecule with the laser is not altered by this assumption since the smallness of the molecular size, as compared to the laser wavelength, makes it quite correct. Problems appear when a point dipole is used to compute the interaction of the polarized molecule with the electrons in the metal. The latter are close to the molecule and see the entire polarization charge density $\rho(\vec{r})$ not just its dipole (i.e., $\vec{\mu} = \int \vec{r} \rho(\vec{r}) d\vec{r}$). Concern with the use of point dipoles has also been expressed in the literature on tunneling junctions[55] and infrared spectroscopy of adsorbed molecules.[2b] There is no doubt that errors are caused.

Perhaps a better question is to what extent, if at all, we can describe the polarization of a chemisorbed molecule by modeling it with an oscillating, classical charge density? We might have a long wait before a reliable answer to this question is provided.

II.8.   Underline: Conclusions and Prospects

> *Good judgement is the result of experience, but*
> *experience is often the result of bad judgement.*
>
>   R. Lovett

§39. The weight of the theoretical "evidence" is such that one can
hardly doubt the existence of a curvature induced enhancement of the
local field and a resonant enhancement of both local field and dipole
emission. The magnitude of such effects will still be hotly debated
for a while. There are two reasons for this. First, the models
have basic shortcomings: the use of phenomenological Maxwell
equations at very close distances to the surface, the treatment of
molecules as classical point dipoles, the use of bulk dielectric
constants for the small metallic particles composing the "roughness"
and finally the approximations (perturbation theory) sometimes
involved in solving the equations (*e.g.* for gratings and small
roughness). Second, the experimental systems differ from those
discussed by theorists (except for the case of isolated spheres or
ellipsoids) to the extent that unequivocal quantitative comparison
with the theory is not yet possible.

Under these (temporary) conditions one might want to defer a
detailed quantitative analysis of the data and concentrate at the
present and in the nearest future on finding the foot prints of
the electromagnetic enhancement. There are many things to look for.

(a) If the enhancement is purely electromagnetic the normal
Raman excitation spectrum is a property of the surface. Various
molecules and modes on the same surface, should have the same
excitation spectrum. Such a behavior has been noticed by
Creighton[56] and Allen *et al.*[57]

(b) The electromagnetic enhancement extends beyond the first
monolayer and, depending on the shape and the properties of the
surface, it may be present at many tens of angstroms away from
the surface, No molecular mechanism can explain such long range
effects. Rowe *et al.*[39a] have observed this for "boulder" roughness,
Murray *et al.*[39b] for rough tunneling junctions and Sanda *et al.*[21]
for gratings.

(c) The nature of the metal enters into the electromagnetic
theory only through its dielectric constant. Studies of carefully
chosen alloys whose dielectric constant can be changed in a
controlled way, may be used to verify this statement. Work in this
direction has just started.[58]

(d) The polarization of the Raman scattering and its angular
properties are controlled by the tensor and directional properties

of Maxwell's equation solutions. These effects deserve more attention from both theory and experiment.

(e) Perhaps the most neglected prediction of the electromagnetic theory is the fact that the same effects invoked to explain Raman enhancement must participate in a host of other spectroscopic processes. (e1) Silver spheres deposited on solid surfaces may be used to enhance the photoelectron yield[59] for solids whose work function is below the Rayleigh resonance of the sphere. The dependence of the excess photoelectron yield (this is the yield in the presence of the spheres minus the yield in their absence) on the frequency of the incident laser can be correlated with the expected excitation spectrum of the SERS of molecules deposited on the same system.[59] (e2) Tunneling junctions offer a similar opportunity. Roughening the surface (with CaF_2 or by depositing metal spheres) permits[60] light emission from the electromagnetic resonances of the junction, driven by fluctuations in tunneling currents.[17,61] If molecules are then deposited on the junction's rough surface, the induced molecular dipoles will drive the same electromagnetic resonances. The emission caused by tunneling currents ought to be related to Raman emission. Such correlations have been in fact observed.[19a]

(e3) Clearly other processes--such as electron scattering-- that cause the rough surface to emit photons can be used to sort out the electromagnetic properties of the surface. In the case of electron scattering, the measurement of the electron energy loss in conjunction with photon emission by the rough surface will be useful. One can learn about the excitation process of the electromagnetic resonances on a given rough surface and the "photon yield" of such resonances, before using the surface for surface enhanced Raman scattering. The light-emission-electron-energy-loss experiments can be correlated with the SERS spectrum on the surface, especially if coincidence measurements of forward scattered electron and the emitted photon could be made.

(e4) Finally, whenever possible the SERS measurements on a given system should be combined with resonance Raman, electronic absorption and fluorescence studies. In doing this the frequency dependence of the molecular polarizability is probed without changing the electromagnetic properties of the surface. Much can be learned from analysing such experiments.

(f) If surface roughness can indeed produce such high local fields, non-linear spectroscopy can be easily performed to further test the electromagnetic model. Such work is in progress[62] and will undoubtedly play a very important role in the future development of the field.

III. MOLECULAR ENHANCEMENT MECHANISMS

Any critic can establish a wonderful batting
average by just rejecting every new idea.

J.D. Williams

III.1. Introduction

§40. We have already mentioned in the previous pages that there are
experimental reasons to believe that part of the enhancement is due
to a chemisorption induced increase of the molecular polarization
(§7, §25 and §36). There are many theories[4,7,24,63-70] that suggest possible mechanisms for such an enhancement, but unfortunately
most models are such that their quantitative predictions are not
very reliable and the qualitative ones are not singular enough to
lead to easy experimental identification.

We review only a limited number of these papers here. Other
recent reviews,[71,72] as well as other chapters of this book cover the
material quite satisfactorily.

III.2. Two Level Molecules Interacting with a Metal

§41. There is a large number of models which consider the Raman
cross section of a two level molecule bound to a metal. The
methods are diverse, but they are more closely related than a
superficial examination would suggest. The earliest and simplest
is the image dipole model. The Drude-Lorentz equation describes
the polarization of a two level molecule and gives the same results
as a two level quantum model.[4,63] The polarized molecule--represent-
ed as a point dipole--interacts with its image and as a result the
position and the width of the upper molecular level is changed.
This affects the frequency dependent polarizability. There are
slightly different ways of interpreting how the model should be
used to analyse SERS results. We feel that the model is relevant
to SERS only if the joint action of chemisorption and image effect
brings an upper electronic state of the molecule in resonance with
the laser. In this case resonance Raman takes place at an unexpected frequency. Furthermore, the vibration of the molecule with
respect to the surface plane and/or the existence of a distribution
("static disorder") of molecule surface distances, modifies this
"surface induced resonant Raman"[4] practically beyond recognition.
The Raman intensity resembles[4,73] the data on SERS. This theory was
proposed[4,63] at a time when only very few molecules were known to
give SERS. Now the number of such molecules is so large that it is
very difficult to believe that all of them have the required
electronic state in the right frequency range. There is however
some evidence that Py on Ag might have such a state.[74] The
image model fits some data   for this system but it is not clear

whether this is due to the existence of an electronic state or it is an accident.

As we have already discussed in §38 the model has shortcomings that have been partly removed by using an electron-gas model to include spatial dispersion[50,64] and the gradual change of the electron density at the surface.[50] The most important conclusion of such calculations is that both the width and the shift of the upper level, caused by the polarization of the metal, are smaller than those predicted by the image model[4,63] and as a result the Raman excitation spectrum is sharper than the one given by the image formula.

§42. An alternative treatment uses the Fano-Anderson model[75] to compute the change in upper level width and position.[66] In principle this has advantages: charge transfer is allowed by the theory (the dipole model precludes it) and the electronic charge of the molecule is spatially distributed on a molecular orbital (not a point dipole). Unfortunately numerical computations with the model are very difficult and their reliability is not known.

III.3. Vibrational Modulation of the Electrons of the Metal

§43. The simplest classical description views the Raman effect as an electronic process. The electric field of light drives the electrons of the molecule and polarizes them. The same electrons are also driven by the electric field caused by the oscillating nuclei. If the electron polarization, caused by the joint effect of these two forces, is computed to second order there will be an induced dipole component with the frequency $\omega - \omega_v$, which radiates Raman photons. The quantum description consists of a two-step process: the absorption of a virtual photon by the electrons, followed by emission of a photon with simultaneous excitation of a vibrational mode.

The models discussed below are based on the idea that if a molecule is chemisorbed the electrons of the metal can also participate in the Raman process, exactly in the manner in which the molecular ones do. Of course only a small fraction of metal electrons can be involved, namely those that are close enough to interact with the oscillating nuclei. However, since these electrons are highly polarizable they may add substantially to the Raman cross section.

III.3.A. Charge transfer model. §44. McCall and Platzman[67] managed to produce a simple formulation of the idea that part of the contribution of the metal's electrons to the Raman cross section is due to the fact that the nuclear oscillations are accompanied by charge transfer from the metal into the molecule.

If one denotes by V the volume of the metal that could be involved in charge transfer and by χ the metal's susceptibility, the polarizability of the metal is $\alpha_M = V\chi$. The change δr in the position of the molecular ion core bound to the metal causes a charge transfer $\delta q/\delta r$, and this causes a polarizability change $(\partial \alpha_M/\partial r) \cdot \delta r \equiv [\delta(V\chi)/\delta q][\delta q/\delta r]\delta r$. The susceptibility depends on the dielectric constant through $\chi = (\varepsilon-1)/4\pi$, and $\varepsilon = 1-(\omega_p/\omega)^2$ (Drude) depends on the charge density $n=q/V$ through $\omega_p^2 = (4\pi n e^2)/m$. A change in q causes a change $\delta\omega_p/\delta q = (4\pi e^2)/mV$ in ω_p. These equations allow the computation of $(\partial\alpha_M/\partial r) \cdot \delta r$ and give the result $(\partial\alpha_M/\partial r)\cdot\delta r = (e/m\omega^2)(\delta q/\delta r)\cdot\delta r$. It is straight-forward[67] to get from this the corresponding Raman cross section $(d\sigma/d\Omega) = (1/3)(e/mc^2)^2(\delta q/\delta r)^2(\Delta r)^2$; here $(\Delta r)^2$ is the square of the matrix element $<0|\delta r|1>$ with $|0>$ and $|1>$ being the ground and excited vibrational states. Using $(\Delta r)^2 = 0.01$ Å^2 and $\partial q/\partial r = 0.2$ electrons/Å, McCall and Platzmann[67] get $d\sigma/d\Omega = 10^{-29}$ cm^2/sr. This is 20 times larger than the Raman cross section of N_2 (5×10^{-31} cm^2/sr). If $dq/dr = 0.1$ el/Å the enhancement is 5 and $dq/dr = 0.05$ el/Å gives 0.125, which is negligible.

This is obviously a crude estimate which is not easy to improve upon. It makes some specific predictions which might be studied experimentally. (a) The Raman Stokes cross section provided by this mechanism cancels the well known ω^4 frequency dependence appearing in the intensity formula. The only frequency dependence left in intensity is that due to the electromagnetic factors (local field and dipole emission). (b) Only vibrational modes that cause a displacement of the nucleus bound to the metal are expected to cause significant reflectivity modulation. The displacement δr of the atom is a linear combination $\delta r = \sum a_i \delta Q_i$ of the normal mode amplitudes δQ_i. The Raman intensity of a given mode i is proportional to the extent that the mode participates in δr.

A model similar, in many respects, to the one discussed here has been studied by Abe *et al*.[24]

III.3.B. <u>Reflectance modulation by the nuclear vibrations</u>
§45. Another mechanism through which the electrons of the metal can be involved in Raman scattering has been suggested by Otto[68] and developed by Maniv and Metiu.[69] This exploits the microscopic theory of the reflectivity of the metal surface.[48,49] Within the random phase approximation the microscopic processes corresponding to photon reflection consist in photon absorption by the metal, with formation of an electron-hole pair, followed by electron-hole pair recombination with photon emission. The ion cores of the chemisorbed molecule interact--through coulomb forces--with the electrons of the metal and can get involved in the reflection process in the following way. The photon is absorbed to form an

electron-hole pair; then the electron (or the hole) can interact
with the molecular ion cores and excite them vibrationally; after
that, the electron and the hole recombine to emit a photon. The
frequency of this photon is $\omega-\omega_v$ since the process leaves the
molecule vibrationally excited, with the excitation energy $\hbar\omega_v$.

The calculation of the cross section for this process is very
tedious and complex. One must compute properly the local field
inside the metal (which excites the electron-hole pair), the screening of the coulomb interaction between the ion cores and the metal
electrons (or holes), and finally the photon emission (through the
metal) after electron-hole recombination.

We find that if the charge oscillating at the vibrational
frequency is very close to the jellium edge, polarizability
enhancement as large as 10^4 can be obtained. However, this would
happen only in exceptional cases when the oscillation of the
nuclear charges might cause some of the electrons located near
jellium to follow them. This electronic charge, oscillating at the
vibrational frequency ω_v, can then get involved in the Raman process.
If the oscillating charge is an ion core, the enhancement is one or
two orders of magnitude depending on frequency, metal electron
density and ion-jellium distance.

III.3.C. <u>Plasmon modulation by the nuclear vibration</u>. §46.
Another model has been developed[15]   on the assumption that the
molecular ion core charges modulate the surface charge density
induced by the laser and make it emit at Raman-Stokes frequency.
The theory has been constructed for the case when the surface is
a sinusoidal grating, but it can be extended to other cases.

The surface charge density $\rho(\vec{r},\omega)$ can be obtained from
Poisson equation $\nabla\cdot\vec{E} = -4\pi\rho$. This yields ρ since the electric
field $\vec{E}$ inside and outside the metal grating can be computed by
perturbation theory.[12]   The charge density thus obtained is multiplied by a step function at the grating surface, appearing because
the dielectric constant has a step function. Jha et $al.$[15] have
replaced it by a smooth, exponentially decaying function whose form
is phenomenologically chosen to simulate the electron density
profile present because the electrons tunnel through the surface
barrier. The function depends on the position and charge of the
ion cores and the work function of the metal. Through these
assumptions the ion core position $\vec{R}_i$ enters into the expression of
the charge density $(\rho(\vec{r},\vec{R}_i;\omega)$. The dipole moment of the surface
is given by the usual formula $\vec{\mu}(\vec{R}_i;\omega) = \int d\vec{r} \ \vec{r}\rho(\vec{r},\vec{R}_i;\omega)$, and the
component $(\partial\vec{\mu}/\partial\vec{R}_i)\cdot\delta\vec{R}_i$, where $\delta\vec{R}_i$ is the vibrational amplitude,
emits Raman-Stokes photons.

A number of interesting predictions are made. (a) The dipole
is proportional to the electric field which has a resonant behavior

when the surface plasmon is excited by the laser (see §11). Therefore, the surface dipole has the same resonant behavior. (b) The
Raman intensity caused by this mechanism depends on the work function of the surface. (c) A static electric field will change the
surface dipole, affecting thus the intensity. The formula of
Jha *et al.*[15] is in general agreement with the measurements of
Sanda *et al.*[21]   However, we feel that further tests are needed.

IV. CONCLUSIONS

> *Jumping to conclusions seldom leads to
> happy landings.*
>
>   S. Siporin

     It seems to me that the only unqualified general conclusion I
can subscribe to, at this time, is that we need further theoretical
and experimental work; the story of SERS is not yet concluded.

ACKNOWLEDGMENTS

     I am grateful to NSF, Sloan Foundation and Camille and Henry
Dreyfus Foundation for the support of my work. Special thanks are
given to my Santa Barbara colleagues, P.K. Aravind, R. Petschek,
Jose Arias, Eric Hood, Greg Korzeniewski, Paul Hansma, T.K. Lee,
Walter Kohn, Bob Schrieffer, Doug Scalapino and Art Hubbard for many
useful and stimulating conversations. Finally, I would like to
apologize to all my colleagues (especially to Andreas Otto) whose
work has not been given, in this review, the attention it deserves.

REFERENCES

1. A. Bagchi, R.G. Barrera, and B.B. Dasgupta, Classical local
       field effect on an adsorbed overlayer, Bull. Am. Phys. Soc.
       25:259 (1980).
2. (a) R.A. Shigeshi and D.A. King, Surface structure and surface
       lattice constant of (001) vapor deposited Au films using
       high resolution transmission electron microscopy, Surf. Sci.
       58:484 (1976); G.D. Mahan and A.A. Lucas, Collective vibra-
       tional modes of adsorbed CO, J. Chem. Phys. 69:5126 (1978);
       M. Moskovits and J.E. Hulse, Frequency shifts in the spectra
       of molecules adsorbed on metals with emphasis on the infra-
       red spectrum of adsorbed CO, Surf. Sci. 78:397 (1978); M.
       Scheffler, The influence of lateral interactions on the
       vibrational spectrum of adsorbed CO, Surf. Sci. 81:562.
    (b) S. Efrima and H. Metiu, Vibrational frequencies of a
       chemisorbed molecule: The role of the electrodynamic inter-
       actions, Surf. Sci. 92:433 (1980); S. Efrima and H. Metiu,

Surf. Sci., to be published.

3. R.R. Chance, A. Prock, and R. Sibley, Molecular fluorescence and energy transfer near interfaces, Adv. Chem. Phys. 37:1 (1978).

4. (a) S. Efrima and H. Metiu, Resonance Raman scattering by adsorbed molecules, J. Chem. Phys. 70:1939 (1979).
    (b) S. Efrima and H. Metiu, Surface induced resonance Raman scattering, Surf. Sci. 92:417 (1980); S. Efrima and H. Metiu, Raman scattering from adsorbed molecules in electrochemical systems, Israel J. Chem. 18:17 (1979).

5. J. Behringer, in: "Molecular Spectroscopy," Vol. 2, The Chemical Society, London (1974).

6. H. Metiu, to be published.

7. S. Efrima and H. Metiu, Classical theory of light scattering by an adsorbed molecule, J. Chem. Phys., 70:1602 (1979); Light scattering by a molecule near a solid surface, J Chem. Phys. 70:2297 (1979).

8. M. Udagawa, C-C. Chou, J.C. Hemminger, and S. Ushioda, preprint, Raman scattering cross sections of an adsorbed pyridine molecule on a smooth silver surface.

9. S.G. Schultz, M. Janik-Czachor, and R.P. Van Duyne, Surface enhanced Raman spectroscopy: a re-examination of the role of surface roughness and electrochemical anodization, Surf. Sci. 104:419 (1981).

10. R. Naaman, S.J. Buelow, O. Cheschnovsky, and D. R. Herschbach, Surface enhanced Raman scattering from molecules adsorbed on mercury, J. Phys. Chem. 84:2692 (1980); L.A. Sanchez, R.L. Birke, and J.R. Lombardi, The surface enhanced Raman spectrum from pyridine on mercury, Chem. Phys. Lett. 79:219 (1981); R.P. Van Duyne, private communication.

11. (a) A. Otto, Excitation of nonradiative surface plasma waves in silver by the method of frustrated total reflection, Z. Physik 216:398 (1968); E. Kretschmann, Die bestimmung optischer konstanten von metallen durch anregung von oberflächenplasmo-schwingungen, Z. Physik, 241:313 (1971).
    (b) M.R. Philpott, in: "Topics in Surface Chemistry,", E. Kay and P. S. Bagus, eds., Plenum, New York (1979), p. 329.
    (c) A. Otto in: "Optical Properties of Solids-New Developments," B. O. Seraphin, ed., North Holland, Amsterdam (1976), p. 677.

12. (a) V. Celli, A. Marvin, and F. Toigo, Light scattering from rough surfaces, general; angle and polarization, Phys. Rev. B 11:2777 (1975).
    (b) A.A. Maradudin and D.L. Mills, Scattering and adsorption of electromagnetic radiation by a semi-infinite medium in the presence of surface roughness, Phys. Rev. B 11:1392 (1975).

13. W.P. Chen, G. Ritchie, and E. Burstein, Excitation of surface electromagnetic waves in attenuated total-reflection prism configurations, Phys. Rev. Lett. 37:993 (1976).

14. (a) B. Pettinger, A. Tadjeddine, and D.M. Kolb, Enhancement in
     Raman intensity by use of surface plasmons, Chem. Phys. Lett.
     66:544 (1979); (b) R. Dornhaus, R.E. Benner, R.K. Chang, and
     I. Chabay, Surface plasmon contribution to SERS, Surf. Sci.
     101:367 (1980).

15. S.S. Jha, J.R. Kirtley, and J.C. Tsang, Intensity of Raman scat-
     tering from molecules adsorbed on a metallic grating, Phys.
     Rev. B 22:3973 (1980).

16. D.L. Mills, Attenuation of surface polaritons by surface rough-
     ness, Phys. Rev. B 12:4036 (1975); A.A. Maradudin and W.
     Zierau, Effects of surface roughness on the surface polari-
     ton dispersion relation, Phys. Rev. B 14:484 (1976); E.
     Kroeger and E. Kretschmann, Surface plasmon and polariton
     dispersion at rough boundaries, Phys. Stat. Sol. 76b:515
     (1976).

17. B. Laks and D.L. Mills, Roughness and the mean free path of
     surface polaritons in tunnel junction structures, Phys. Rev.
     B 20:4962 (1979).

18. P.K. Aravind, E. Hood, and H. Metiu, Angular resonances in the
     emission from a dipole located near a grating, Surf. Sci.
     (in press).

19. (a) J.C. Tsang, J.R. Kirtley, and J.A. Bradley, Surface en-
     hanced Raman scattering and surface plasmons, Phys. Rev.
     Lett. 43:772 (1979); (b) J.C. Tsang, J.R. Kirtley, and T.N,
     Theis, Surface plasmon polariton contributions to Stokes
     emission from molecular monolayers on periodic silver
     surfaces, Solid State Commun. 35:667 (1980).

20. A. Girlando, M.R. Philpott, D. Heitman, J.D. Swalen, and R.
     Santo, Raman spectra of a thin organic film enhanced by
     plasmon surface polaritons on holographic metal gratings,
     J. Chem. Phys. 72:5187 (1980).

21. P.N. Sanda, J.M. Warlaumont, J.E. Demuth, J.C. Tsang, K.
     Christman, and J.A. Bradley, Surface enhanced Raman scatter-
     ing from pyridine on Ag (111), Phys. Rev. Lett. 45:1519
     (1980).

22. (a) J.S. Creighton, C.G. Blatchford, and M.G. Albrecht, Plasmon
     resonance enhancement of Raman scattering by pryidine ad-
     sorbed on silver or gold particles of size comparable to
     the excitation wavelength, J. Chem. Soc. Faraday Trans. II,
     75:790 (1979) (b) M. Kerker, O. Siiman, L.A. Bumm, and D.-S.
     Wang, Surface enhanced Raman scattering of citrate ions
     adsorbed on colloidal silver, Appl. Opt. 19:3253 (1980),
     (c) H. Wetzel and H. Gerischer, Surface enhanced Raman scat-
     tering from pyridine and halide ions adsorbed on silver and
     gold sol particles., Chem. Phys. Lett. 76:460 (1980).
     (d) M.E. Lippitsch, Observation of surface enhanced Raman
     spectra by adsorption to silver colloids, Chem. Phys. Lett.
     74:125 (1980). (e) K.U. von Raben, R.K. Chang and B.L.
     Laube, Surface enhanced Raman scattering of $Au(CN)_2^-$ ions
     adsorbed on gold colloids, Chem. Phys. Lett. 6:272 (1981).

23. J.D. Eversole and H.P. Broida, Electron microscopy of size dis-
    tribution and growth of small zinc crystals formed by homo-
    geneous nucleation in a flowing inert gas system, J. Appl.
    Phys. 45:596 (1974).

24. H. Abe, K. Manzel, W. Schulze, M. Moskovits, and D.P. DiLella,
    Surface enchanced Raman spectroscopy of CO adsorbed on
    colloidal silver particles, J. Chem. Phys. 74:792 (1981).

25. C.Y. Chen, E. Burstein, and S. Lundquist, Giant Raman scatter-
    ing by pyridine and CN⁻ adsorbed on silver, Solid State
    Commun. 32:63 (1979); E. Burstein and C.Y. Chen, Raman scat-
    tering by molecules adsorbed at metal surfaces. The role of
    surface roughness, in: "Proceedings of the VIIth Inter-
    national Raman Conference," W.F. Murphy, ed., North Holland,
    New York (1980), p. 346; C.Y. Chen and E. Burstein, Giant
    Raman scattering by molecules at metal island films, Phys.
    Rev. Lett. 45:1287 (1980); C.Y. Chen, I. Davoli, G. Ritchie,
    and E. Burstein, Giant Raman scattering by molecules adsorbed
    on Ag and Au metal island films, Surf. Sci. 101: 363 (1980).

26. J.A. Stratton, "Electromagnetic Theory," McGraw Hill, New York,
    (1941).

27. (a) D.-S. Wang, H. Chew, and M. Kerker, Enhanced Raman scatter-
    ing at the surface of a spherical particle, Appl. Opt. 19:2256
    (1980). (b) M. Kerker, D.-S. Wang, and H. Chew, Surface
    enhanced Raman scattering by molecules adsorbed at spherical
    particles, Appl. Opt. 19:3373 (1980). (c) S.L. McCall, P.M.
    Platzman, and P.A. Wolff, Surface enhanced Raman scattering,
    Phys. Lett. 77A:381 (1980). (d) B.J. Messinger, K. Ulrich
    von Raben, R.K. Chang, and P.W. Barber, Local fields at the
    surface of noble metal microspheres, Phys. Rev. B (in press).

28. (a) J.I. Gersten, The effect of surface roughness on surface
    enhanced Raman scattering, J. Chem. Phys. 72:5779 (1980).
    (b) J.I. Gersten and A. Nitzan, Electromagnetic theory of
    enhanced Raman scattering by molecules adsorbed on rough
    surfaces, J. Chem. Phys. 73:3023 (1980). (c) F.J. Adrian,
    Surface enhanced Raman scattering by surface plasmon enhance-
    ment of electromagnetic fields near spheroidal particles on
    a roughened metal surface, Chem. Phys. Lett. 78:45 (1981).
    (d) M. Kerker, Phys. Rev. B (in press).

29. J.I. Gersten, D.A. Weitz, T.J. Gramila, and A.Z. Genack, Inelas-
    tic Mie scattering from rough metal surfaces: theory and
    experiment, Phys. Rev. B 22:4562 (1980); D. Weitz, A.Z.
    Genack, T.J. Gramila, and J.I. Gersten, Anomalmous low fre-
    quency Raman scattering from rough metal surface and the
    origin of surface enhanced Raman scattering, Phys. Rev.
    Lett. 45:355 (1980).

30. A.Z. Genack, D.A. Weitz, and T.J. Gramila, Very low frequency
    surface enhanced Raman scattering, Surf. Sci. 101:381 (1980).

31. H. Morawitz and T.R. Koehler, A model for Raman active vibra-
    tional modes on a metal surface: pyridine and CN⁻ on silver,
    Chem. Phys. Lett. 71:64 (1980).

32. P.K. Aravind, A. Nitzan, and H. Metiu, The interaction between electromagnetic resonances and its role in spectroscopic studies of molecules adsorbed on colloidal particles or metal spheres, Surf. Sci. (to be published).

33. P.K. Aravind and H. Metiu (unpublished).

34. H. Chew, M. Kerker, and P.J. McNulty, Raman and fluorescence scattering by molecules embedded in concentric spheres, J. Opt. Soc. Am. 66:440 (1976).

35. H. Chew, D.D. Cook, and M. Kerker, Raman and fluorescence scattering by molecules embedded in dielectric cylinders, Appl. Opt. 19:44 (1980).

36. E.H. Lee, R.E. Benner, J.B. Fenn, and R.K. Chang, Angular distribution of fluorescence from monodispersed particles, Appl. Opt. 17:1980 (1978); J.P. Kratohvil, M.-P. Lee, and M. Kerker, Angular distribution of fluorescence from small particles, Appl. Opt. 17:1978 (1978).

37. M. Moskovits, Surface roughness and the enhanced intensity of Raman scattering by molecules adsorbed on metals, J. Chem. Phys. 69:4159 (1978); Solid State Commun. 32:59 (1979).

38. P.K. Aravind and H. Metiu (to be published).

39. (a) J.E. Rowe, C.V. Shank, D.A. Zwemer, and C.A. Murray, Ultra high vacuum studies of enhanced Raman scattering from pyridine on Ag surfaces, Phys. Rev. Lett. 44:1770 (1980); D.A. Zwemer, C.V. Shank and J.E. Rowe, Surface enhanced Raman scattering as a function of molecule surface separation, Chem. Phys. Lett. 73:201 (1980); (b) C.A. Murray, D.L. Allara, and M. Rhinewine, Silver-molecule separation dependence of surface-enhanced Raman scattering, Phys. Rev. Lett. 46:57 (1981).

40. P.K. Aravind, R. Rendell, and H. Metiu (to be published).

41. B. Pettinger, M.R. Philpott, and J.G. Gordon, Contribution of specifically adsorbed ions, water and impurities to the surface enhanced Raman spectroscopy of Ag electrodes, J. Chem. Phys. (to be published).

42. A.A. Maradudin and D.L. Mills, Scattering and adsorption of electromagnetic radiation by a semi-infinite medium in the presence of surface roughness, Phys. Rev. B 11:1392 (1975).

43. J.G. Endriz and W.E. Spicer, Surface plasmons: 1. Electron decay and its observation in photoemission, Phys. Rev. Lett. 24:64 (1970); Study of aluminum films: 2. Optical studies of reflectance drops and surface oscillations on controlled roughness films. Phys. Rev. B 4:4144 (1971).

44. D.K. Cohen, S.O. Sari, and K.D. Scherkoske, Interfacial scattering and electrodynamics due to roughness at silver surfaces, Surf. Sci. 101:355 (1980).

45. P.K. Aravind and H. Metiu, The enhancement of Raman and fluorescence intensity by small surface roughness, changes in dipole emission, Chem. Phys. Lett. 74:301 (1980).

46. P.K. Aravind, J. Arias, and H. Metiu, Chem. Phys. Lett. (submitted).

47. A. Adams, R.W. Rendell, W.P. West, H.P. Broida, P.K. Hansma, and H. Metiu, Luminescence and nonradiative energy transfer to surfaces, Phys. Rev. B 21:5565 (1980).

48. T. Maniv and H. Metiu, Electron gas effects in the spectroscopy of molecules chemisorbed at a metal surface: 1. Theory, J. Chem. Phys. 72:1996 (1980); Electrodynamics at a metal surface with applications to the spectroscopy of adsorbed molecules: 1. General theory, Phys. Rev. B 22:4731 (1980).

49. T. Maniv and H. Metiu, J. Chem. Phys. (to be published, 1982).

50. G. Korzeniewski, T. Maniv, and H. Metiu, The interaction between an oscillating dipole and a metal surface described by a jellium model and the random phase approximation, Chem. Phys. Lett. 73: 212 (1980); J. Chem. Phys. (to be published, 1982).

51. D.M. Newns, Dielectric response of a semi-infinite degenerate electron gas, Phys. Rev. B 1:3304 (1970); D.E. Beck and V. Celli, Linear response of a metal to an external charge distribution, Phys. Rev. B 2:2955 (1970).

52. P.J. Feibelman, Microscopic calculation of electromagnetic fields in refraction at a jellium vacuum interface, Phys. Rev. Lett. 34:1092 (1975); Surface electronic structure information from surface plasmon photoexcitation in free electron metal films, Phys. Rev. B 12:1319, 4282 (1975).

53. K. L. Kliewer and R. Fuchs, Theory of dynamical properties of dielectric surfaces, Adv. Chem. Phys. 27:301 (1978).

54. W.H. Weber and G. W. Ford, Optical electric field enhancement at a metal surface arising from surface plasmon excitation, (to be published); P.R. Hilton and D.W. Oxtoby, Surface enhanced Raman spectra: A critical review of the image dipole description, J. Chem. Phys. 72:6346 (1980); W.E. Palke, The effect of image charges on molecular adsorption, Surf. Sci. 97:L331 (1980).

55. J. Kirtley, D.J. Scalapino, and P.K. Hansma, Theory of vibrational mode intensities in inelastic electron tunneling spectroscopy, Phys. Rev. B 14:3177 (1976).

56. J.A. Creighton, private communication.

57. C.S. Allen, G.C. Schatz, and R.P. Van Duyne, Tunable laser excitation profiles of surface enhanced Raman scattering from pyridine adsorbed on a copper electrode, Chem. Phys. Lett. (to be published).

58. T. Furtak and J. Kester, Do metal alloys work as substrates for surface enhanced Raman spectroscopy?, Phys. Rev. Lett. 45:1652 (1980)

59. P.K. Aravind, J. Arias, and H. Metiu (unpublished).

60. P.K. Hansma and H. P. Broida, Light emission from gold particles excited by electron tunneling, Appl. Phys. Lett. 32:545 (1978); A. Adams, J.C. Wyss, and P.K. Hansma, Possible observation of local plasmon modes excited by electrons tunneling through junctions, Phys. Rev. Lett. 42:912 (1979).

61. D. Hone, B. Mühlschlegel, and D.J. Scalapino, Theory of light
     emission from small particle tunnel junctions, Appl. Phys.
     Lett.33:203 (1978); R.W. Rendell, D.J. Scalapino, and B.
     Mühlschlegel, Role of local plasmon models in light emission
     from small particle tunnel junctions, Phys. Rev. Lett.
     41:1746 (1978).

62. (a) C.K. Chen, A.R.B. de Castro, and Y.R. Shen, Surface enhanced
     second harmonic generation, Phys. Rev. Lett. (to be published);
     (b) J.P. Heritage, J.G. Bergman, A. Pinczuk, and J.M. Worlock,
     Surface picosecond Raman gain spectroscopy of a cyanide mono-
     layer on silver, Chem. Phys. Lett. 67:229 (1979).

63. (a) F.W. King, R.P. Van Duyne and G.C. Schatz, Theory of Raman
     scattering by molecules adsorbed on electrode surfaces, J.
     Chem. Phys. 69:4472 (1978); (b) G.L. Eesley, and J.R. Smith,
     Enhanced Ramam scattering on metal surfaces, Solid State
     Commun. 31:815 (1979); (c) R. Loudon, "The Quantum Theory
     of Light," Oxford University Press, London (1973), Ch. IV.

64. W.H. Weber and G.W. Ford, Enhanced Raman scattering by adsor-
     bates including the nonlocal response of the metal and the
     excitation of nonradiative modes, Phys. Rev. Lett. 44:1774
     (1980); R. Fuchs and R.G. Barrera, Phys. Rev. B (to be
     published).

65. E. Burstein, Y.J. Chen, C.Y. Chen, S. Lundquist, and E. Tosatti,
     Giant Raman scattering by adsorbed molecules on metal
     surfaces, Solid State Commun. 29:567 (1979).

66. (a) J.I. Gersten, R.L. Birke, and J.R. Lombardi, Theory of
     enhanced light scattering from molecules adsorbed at the
     metal-solution interface, Phys. Rev. Lett. 43:147 (1979);
     (b) G.W. Robinson, Surface enhanced Raman effect, Chem. Phys.
     Lett. 76:191 (1980); (c) T.K. Lee and J.L. Birman, Molecules
     adsorbed on plane metal surfaces: Coupled system eigenstates,
     Quantum theory of enhanced Raman scattering by molecules on
     metals: Surface plasmon mechanism for plane metal surfaces,
     Phys. Rev. B 22:5953,5961 (1980); (d) H. Ueba, Effective
     resonant light scattering from adsorbed molecules, J. Chem.
     Phys. 73:725 (1980); H. Ueba and S. Ichimura, Raman scatter-
     ing of adsorbed molecules, J. Chem. Phys. 74:3070 (1981);
     (e) M. Philpott, Effect of surface plasmons on transitions
     in molecules, J. Chem. Phys. 62:1812 (1975).

67. S.L. McCall and P.M. Platzman, Raman scattering from chemi-
     sorbed molecules at surfaces, Phys. Rev. B 22:1660 (1980).

68. A. Otto, J. Timper, J. Billmann, G. Kovacs, and I. Pockrand,
     Surface roughness induced electrode Raman scattering,
     Surf. Sci. 92:L55 (1980).

69. T. Maniv and H. Metiu, Some comments concerning the microscopic
     theory of Raman scattering by adsorbed molecules, Surf. Sci.
     101:399 (1980); R. Maniv and H. Metiu, Raman reflection:
     A possible mechanism for enhancement of Raman scattering by
     an adsorbed molecule, Chem. Phys. Lett. (in press).

70. A. Otto, J. Timper, J. Billmann, and I. Pockrand, Enhanced in-
     elastic light scattering from metal electrodes caused by
     adatoms, Phys. Rev. Lett. 45:46 (1980); A. Otto, Raman scat-
     tering from adsorbates on silver, Surf. Sci. 92:145 (1980);
     I. Pockrand and A. Otto, Coverage dependence of Raman scat-
     tering from pyridine adsorbed to silver - vacuum interfaces,
     Solid State Commun. 35:861 (1980); A. Otto, Surface Enhanced
     Raman scattering. What do we know?, Surf. Sci. 6:309 (1980);
     J. Billmann and A. Otto, Experimental evidence for a local
     mechanism of surface enhanced Raman scattering, Appl. Surf.
     Sci. 6:356 (1980); I. Pockrand and A. Otto, Raman scattering
     from silver/vacuum interfaces, Appl. Surf. Sci. 6:362 (1980).
71. T.E. Furtak and J. Reyes, A critical review of theoretical
     models of surface enhanced Raman scattering, Surf. Sci.
     93:351 (1980).
72. H. Metiu, Prog. Surf. Sci. (to be published).
73. G. Schatz and R.P. Van Duyne, Image field theory of enhanced
     Raman scattering by molecules adsorbed on metal surfaces:
     Detailed comparison with experimental results, Surf. Sci.
     101:425 (1980).
74. B. Pettinger, U. Wenning, and D. M. Kolb, Raman and reflectance
     spectroscopy of pyridine adsorbed on single crystal Ag
     electrodes, Ber. Bunsenges. Physik. Chem. 82:1326 (1978);
     J. E. Demuth and P.N. Sanda, Observation of charged transfer
     states for pyridine chemisorbed on Ag (111), preprint.
75. P.W. Anderson, Localized magnetic states in metals, Phys. Rev.
     124:41 (1961); U. Fano, Effects of configuration interaction
     on intensities and phase shifts, Phys. Rev. 124:1866 (1961).

THE IMAGE FIELD EFFECT: HOW IMPORTANT IS IT?

George C. Schatz

Department of Chemistry
Northwestern University
Evanston, Illinois 60201

I. INTRODUCTION

Shortly after the discovery of surface enhanced Raman scattering (SERS),[1] several mechanisms were proposed to explain the observed factor of 10^6 enhancement in Raman intensity seen for molecules adsorbed on Ag surfaces. (Reference 2 contains a review.) One of these mechanisms,[3,4] now known as the image field effect (IFE) mechanism, considered the interaction between the oscillating dipole moment induced in the adsorbed molecule and its image in the metal. Under the appropriate circumstances and assumptions, the interaction between the adsorbed molecule and its image was found to be large enough to cause a substantial enhancement in Raman intensity, and from this it was concluded that the IFE might at least partially be responsible for the SERS enhancement. Since these first studies, a number of papers have appeared in which the various approximations and assumptions of the IFE model have been studied and tested.[5-16] At present, the conclusions of these studies are the subject of significant disagreement and, because of this, the IFE model has been both praised and condemned. In this paper I review these efforts to test the IFE model of SERS and I will summarize the current status of the comparison between IFE model predictions and experiment. With this review, I will attempt to assess just where the image model sits at this point in its application to SERS, and I will point out some approaches to the further testing of the validity of the IFE model.

Surface enhanced Raman scattering is certainly not the first phenomenon where the application of image theory has been the subject of controversy. The current debate about the role of image effects in producing the coverage dependent frequency shifts

observed in the infrared spectroscopy of CO on metals is very
active.[17] Image and related theories have, however, been very successful in applications to a number of other surface phenomena, thus
giving at least some support to their use in certain physical situations. The image theory of fluorescence lifetimes of molecules as
a function of distance from metal surfaces is quite successful (with
some improvements and modifications).[18] Recent spacer experiments
indicate that the R^3 dependence of lifetime on molecule-surface
separation remains valid at separations as small as 7 Å.[19] Image
models have also been used in theories of chemisorption[20] and adsorption induced work function changes[21] where more accurate density
functional calculations indicate the usefulness of these models for
distances as close as 2 Å. Image theories have also seen application in studies of photoemission,[22] of electron and ion scattering
from surfaces,[23] and other phenomena, often with quantitative results if carefully applied.

Of course the application of image theories to surface phenomena
becomes increasingly suspect as the separation between molecule and
surface becomes small. This is perhaps at the heart of the controversy with respect to the IFE interpretation of SERS. Density functional calculations[21] indicate that at small molecule-surface separations, the point charge image formula can be accurate provided that
distances are measured relative to an appropriately defined image
plane. The location of this plane relative to the "true surface"
(usually taken as the positive background edge in jellium calculations) is roughly independent of molecule-surface separation for
distances down to 2 Å but thereafter varies rapidly with distance.[20]
The incorporation of an effective distance into image model calculations provides one method of accounting for the effects of screening and the absence of a sharp discontinuity in the electron density
at the surface.

Still another complication in the use of image models at short
range is the problem of finite molecular size. For adsorbed atoms
or symmetrical molecules, it is common to locate the point charge
or dipole representing the atom or molecule at its center of symmetry.[3,21] Other prescriptions have been used,[22] but only a few
studies have attempted to model finite size effects through the use
of distributions of charges.[24]

For studies of frequency dependent properties such as Raman intensities, an additional difficulty in the use of image models con-
cerns the proper modelling of the metal dielectric response in
determining the magnitude of image charges and dipoles. Most of the
simple image models use bulk metal optical dielectric constants to
model this frequency dependence. This ignores the fact that the
surface dielectric constants differ from bulk constants, and that
they exhibit independent frequency and wavevector dependence. Only
very recently have studies of the wavevector dependence of image

models been made,[10,12,13] but these show that such effects are quite important.

Still another problem with the use of image models at short molecule-surface distances arises from chemisorption effects. Since the image model treats the Coulomb interactions between the adsorbed molecule and surface, it is important that the molecular charge distribution is not strongly altered by adsorption; otherwise the "bare" molecule properties will not be representative of ad-molecule interactions. To date the only studies of chemisorption effects are the density functional calculations mentioned above,[21] and most of these refer to H atom adsorption (where the bare molecule is assumed to be a proton).

In spite of these numerous problems associated with image model applications to molecules close to metal surfaces, the simplest point dipole IFE model of SERS has been surprisingly successful as a phenomenological theory. The current status of comparisons between IFE predictions and experiment will be reviewed in Section II, and there we will find that such features of SERS intensities as their dependence on frequency, on metal, and on adsorbate are properly described qualitatively by the IFE model. Indeed, several of the IFE model predictions concerning the magnitude of enhancements on Cu, Au and Hg were made well in advance of observations on these metals and were at least partially responsible for motivating the observations. Thus, in discussing the IFE model we are faced by an apparent contradiction in that the model's predictions are qualitatively correct; yet the assumptions underlying the model are suspect and certainly cannot be quantitative.

As mentioned previously, there have now been several papers published in which certain of the approximations of the simple IFE model have been tested. These studies, which will be reviewed in Section III, include for the wavevector dependent dielectric response of the metal,[8-10,12-13] for the smooth variation of the electron density across the interface[8-10,12] and for finite molecular size.[11,16] It should be emphasized that none of these calculations represent a quantitative microscopic treatment of the complete SERS problem, but they do provide insight concerning the importance of certain assumptions made in the IFE model. In the conclusion (Section IV) will be described some very recent ab initio calculations which apparently do contain the necessary elements of a truly microscopic theory of SERS.

II. CURRENT STATUS OF PREDICTIONS OF THE SIMPLE POINT DIPOLE IFE MODEL

The point dipole IFE model was first introduced by King et al.[3] and has been greatly expanded upon by Efrima and

Metiu.[4-6] A closely related model has also been presented by Eesley and Smith[7] and a less closely related model by Morawitz and Koehler.[14] In all of these studies, a point dipole is used to represent the oscillating dipole moment induced by the applied electromagnetic field in the molecule. In calculating the image field associated with this induced dipole, all difficulties associated with dispersion and chemisorption described above are neglected. The overall expression for the SERS enhancement factor ε arising from this model is given by[15]

$$\varepsilon = G \left| 1 - \frac{\gamma \alpha_0}{4R^3} \right|^{-4} \qquad \qquad (1)$$

where

$$\gamma = (\varepsilon_M - \varepsilon_A) / (\varepsilon_M + \varepsilon_A) \, \varepsilon_A . \qquad \qquad (2)$$

In this expression, ε_M is the metal dielectric constant, ε_A the adsorbate dielectric constant, α_0 the unperturbed molecular polarizability (the component associated with the axis perpendicular to the surface) and R is the molecule to image plane separation. G is a geometrical factor which has been discussed previously for the case of smooth surfaces[4-6,15] where its value is roughly 10. For scattering from rough surfaces, G incorporates the effect of roughness induced surface electromagnetic field enhancements and its value is correspondingly larger.

Recently, the electrodynamics of random distributions of hemispheroidal metal bosses on flat perfectly conducting metal surfaces has been studied for several metals.[25] These distributions were chosen to simulate the types of rough surfaces that are prepared by anodization in electrochemical cells. For Ag and other strong SERS enhancing metals (Cu, Au, Hg), maximum roughness induced enhancements of 10^2 were found, along with relatively flat dependences of enhancement on excitation frequency below the flat surface plasmon frequency. This estimate of roughness induced enhancement is in good agreement with recent experimental estimates for electrochemical systems.[26] Combining the factor of 10^2 with the factor of 10 which arises from flat surface contributions to local field enhancements[15] leads to an estimate of $G = 10^3$ for SERS scattering from rough Ag surfaces.

The factor $\left| 1 - \gamma \alpha_0 / 4R^3 \right|^{-4}$ in Eq. (1) arises from image field induced enhancement of the molecular polarizability derivative. The parameters γ and α_0 are not difficult to estimate in evaluating this enhancement for any given system, but the choice of R is not at all straightforward. This is because neither the location of the image plane relative to the surface nor the position of the point dipole which represents the ad-molecule is well defined. While previous estimates do indicate that "reasonable" prescriptions for choosing R lead to large image enhancements, it is also true that $|1 - \gamma\alpha_o/4R^3|^{-4}$ varies rapidly with R for small molecule-surface separations. This makes any simple prescription which happens to give large enhancements suspect. To circumvent this problem, R can be chosen by requiring ε to have approximately the correct value for some metal (such as Ag) at some frequency (say 500 nm). In addition, it is desirable to pre-average the factor $|1 - \gamma\alpha_o/4R^3|^{-4}$ over a Gaussian distribution of R's with standard deviation σ (typically 0.1 Å) to simulate the possibility of a distribution of distances from the surface. Thus if G = 10 (for smooth Ag surfaces), R is chosen so that $|1 - \gamma\alpha_o/4R^3|^{-4}$ averages to 10^5 at 500 nm in order to make $\varepsilon = 10^6$.

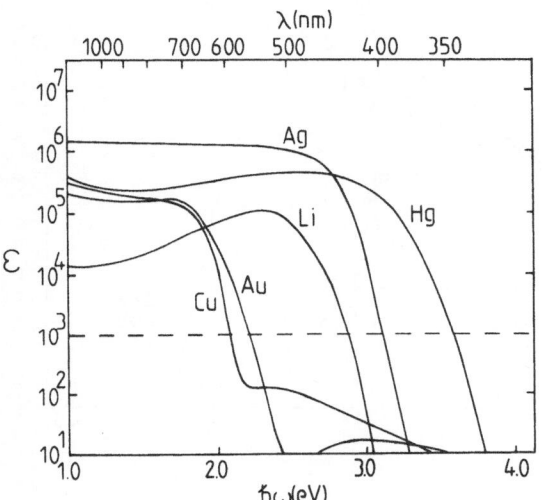

Fig. 1. Smooth surface IFE-SERS enhancement factor ε versus photon energy $\hbar\omega$ (in eV) and wavelength λ (in nm) for Ag, Hg, Li, Cu and Au. The parameters in this calculation apply to adsorbed pyridine and are taken from Ref. 15 with the exception of R. R is chosen to make $\varepsilon = 10^6$ at 500 nm for Ag and is shifted in value for other metals as discussed in the text. The dashed line at $\varepsilon = 10^3$ indicates the approximate value of ε required for experimental observation to be possible in electrochemical systems.

While it is clear that the assumptions used to evaluate Eq. (1) are somewhat ad hoc, it is interesting to see what predictions come from this phenomenological model for different metals and frequencies. Figure 1 summarizes the results for pyridine adsorbed on Ag, Au, Cu, Hg and Li using pyridine polarizabilities discussed previously[15] and literature values of the metal dielectric constants.[27-29] This graph incorporates the smooth surface enhancement factor of Ref. 15 in determining G and uses an R value for Ag of 1.55 Å. For metals other than Ag, a small shift in R is made in order to describe shifts in image plane location with changing metal properties (screening and surface electron density profiles). Using shifts obtained from jellium calculations,[20] the R values are 1.51 for Cu, 1.55 for Au, 1.51 for Hg and 1.58 Å for Li.

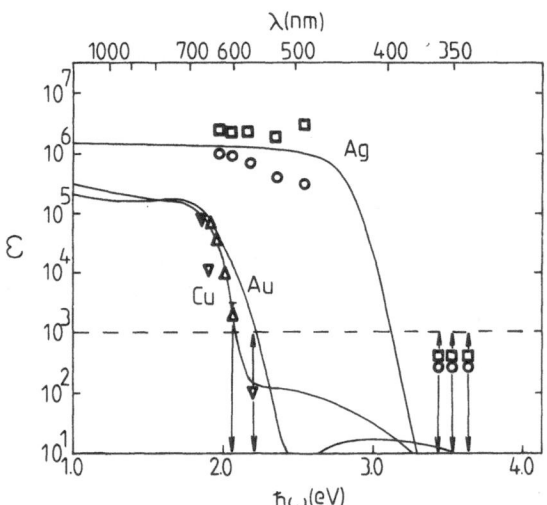

Fig. 2.   IFE and experimental SERS enhancement factor for pyridine
          adsorbed on Ag, Au, Cu. The IFE curves are identical to
          those plotted in Fig. 1. The experimental results are
          from Refs. 26 and 30 for Ag [the ε's for Raman enhancements
          at 1008 (circles) and 1215 cm^{-1} (squares) are plotted],
          Ref. 26 for Cu [1015 cm^{-1} is plotted (triangles Δ)], and
          Refs. 31 and 32 for Au (1015 cm^{-1} is plotted, with open
          triangles ∇ for Ref. 31 and filled triangles $\blacktriangledown$ for Ref. 32).
          All measured ε values at or below 10^3 are upper bounds due
          to instrument limitations.

Figure 2 compares the IFE results from Fig. 1 with experimental ε values for Ag from Refs. 26 and 30, Cu from Ref. 26 and Au from Refs. 31 and 32. The two experimental reports of enhancements for Hg[32,33] are not very quantitative, but they appear to indicate that ε is roughly $10^4 - 10^6$ at 515 nm. Neither the frequency dependence of ε on Hg nor the experimental enhancements on Li has been reported.

Comparison of the simple point dipole IFE model results with experiment in Figs. 1 and 2 indicates the following:

1. Both IFE and experimental enhancements on Ag are relatively flat up to $\hbar\omega$ = 2.6 eV, then drop to below detection threshold ($\varepsilon = 10^3$) at higher ω.

2. Enhancements on Cu and Au are large below $\hbar\omega$ = 2 eV and drop off at higher ω. The experimental and theoretical enhancements for Cu and Au at 1.9 eV are quite similar in magnitude, with values lower than the analogous Ag enhancements by factors of 10–100.

3. The IFE enhancements on Hg are quite large throughout the visible though not as large as on Ag except in the near UV. IFE enhancements on Li are one to two orders of magnitude smaller than on Ag.

It is especially noteworthy that the IFE enhancements for Au and Cu drop suddenly by over two orders of magnitude near 2 eV. This indicates a strong sensitivity to metal dielectric properties which is in good agreement with experiment. Although the rough surface electrodynamic models[25] also predict a drop in enhancement near 2 eV, the magnitude of the predicted drop is over an order of magnitude smaller than is seen experimentally.

The overall agreement between theory and experiment in Fig. 2 is amazingly good in view of the crudeness of the theory and the fact that the one adjustable parameter R is fixed for only one metal at one frequency. For this reason, the IFE model has been useful as a predictive tool in guiding experiments on new metals. (Indeed, curves similar to those in Figs. 1 and 2 were originally calculated well before experiments on Cu, Au, and Hg were first done.) At the same time, the calculations described in the next section indicate that more sophisticated image models often give poorer agreement between theory and experiment for many properties.

The simple IFE model can also be used to predict other optical properties (depolarization ratios, second-harmonic generation enhancements, etc.) some of which have been discussed previously in Ref. 15. Because the IFE and roughness enhancement mechanisms work in parallel, many properties seem to reflect that mechanism which has the most rapidly varying dependence on measurable parameters.

For example, the angular dependence of scattering on weakly roughened surfaces is dominated by the geometrically defined conditions which optimize surface plasmon excitation.[34] Since the IFE mechanism shows a weak variation of scattered intensity with incident or outgoing angle,[15] measurements of angular distributions are primarily sensitive to the enhancement which arises from surface plasmon excitation. Related statements can be made concerning depolarization ratios, relative mode intensities and other geometrically based information. In addition, both the roughness and IFE mechanisms predict enhancements in second-harmonic generation (SHG) cross sections. Existing SHG experiments[35] have demonstrated the importance of the roughness mechanism, but estimates of adsorbate induced smooth surface SHG enhancements have not yet been made.

From the above comments, it is clear that the process of differentiating the contributions of different enhancement mechanisms to SERS will be difficult without more quantitative intensity calculations. This makes the improved IFE models to be discussed below especially important.

III. REVIEW OF MORE SOPHISTICATED IFE MODELS

Table 1 summarizes those studies of SERS which have used or tested the image field model. Refs. 3-8 and 14,15 can all be categorized as simple point dipole models in which the wavevector dependence of the dielectric constant has been ignored and a sharp surface boundary is assumed. The remaining papers[9-13,16] represent attempts at removing one or more of these assumptions, as will now be described.

The study of Hilton and Oxtoby[11] determined the static polarizability of a hydrogen atom which is next to a flat perfect metal surface. In this model, the image of the hydrogen atom is simply an antihydrogen atom, and the wavefunction describing this system of hydrogen-antihydrogen is determined by a Hartree SCF procedure. This model thus relaxes the point dipole approximation, though it oversimplifies the metal response and ignores any chemical bonding effects. The calculated polarizability as a function of the proton-antiproton separation distance d (d=2R) shows very little change from the infinite separation distance result for d as small as 1 Å. This is followed by a rapid increase with decreasing d for d < 1 Å. The analogous image model polarizability exhibits a similar dependence of polarizability on d, but the rapid increase begins at d = 1.6 Å. Hilton and Oxtoby used their calculated total energies to estimate that the minimum allowed d at room temperature was 2.4 Å. This rather large value is a consequence of the neglect of chemisorption, since this causes the hydrogen-antihydrogen interaction to be repulsive at longer distances than would normally be the case. Since the polarizability enhancement was negligible at 2.4 Å, they

Table 1. Summary of Image Model Calculations

Reference	Point Dipole Approximation?	Include Wave-Vector Dependence of Dielectric Constants?	Include for Continuous Variation of Electron Density?	Estimated SERS Enhancement (Ag, R~1.6Å)
3-8	Yes	No	No	10^3-10^6
14-15	Yes	No	No	10^3-10^6
11	No	No[a]	No[a]	~1
13	Yes	Yes	No	10^3
16	No	Yes	No	10^3
9,10	Yes	Yes	Yes	~10^4
12	Yes	Yes	Yes	~1

[a]Perfect mirror approximation used.

concluded that the IFE mechanism is not important in SERS. An analogous calculation of image induced infrared frequency shifts by Palke[36] leads to a similar conclusion concerning the importance of image effects, though with some reservations because of difficulties with overlap between molecular and image charge clouds.

Perhaps the two main problems with Hilton and Oxtoby's conclusion concerning SERS are: (1) that their calculation actually refers to the static polarizability, and not to the frequency dependent polarizability derivative which is relevant to Raman scattering, and (2) that because of omission of chemisorption, the 2.4 Å distance d at which they located their H atom is much larger than that which characterizes true H atom adsorption on metals. An illustration of the importance of the first problem is found in the recent ab initio calculations of frequency dependent polarizability derivatives for H_2 adsorbed onto Li clusters by Pandey and Schatz[37] (to be described in Section IV). They find that the frequency dependent polarizability derivative enhancements at optical frequencies can be quite large (several orders of magnitude) even when the static polarizability enhancements are close to unity. To understand the significance of the second problem, it is important to realize that molecule to surface distances are not constrained to be larger than the cube root of the molecular polarizability. As an example, consider the case of atomic iodine adsorbed onto Ag in UHV. The SERS spectrum of this has recently been observed, with the AgI stretch mode clearly resolved[38] in submonolayer coverages. This system has also been studied by SEXAFS[39] and the AgI bond distance is found to be 2.87 Å, corresponding to a distance to the surface of 0.92 Å. This distance is certainly smaller than the cube root of the iodine static polarizability and is much smaller than the cube root of the frequency dependent polarizability in the visible region (where I has resonant transitions). All of this indicates that $\alpha_o/4R^3$ can be unity or greater for I on Ag, and thus one cannot rule out the importance of image effects on geometrical grounds alone for this and many other systems.

Perhaps the more useful conclusion to be drawn from Hilton and Oxtoby's calculation is that the calculated dependence of polarizability on d is qualitatively similar though quantitatively different from that obtained from the simple image model. This is presumably one manifestation of finite molecular size effects and illustrates a major difficulty in choosing appropriate values of d (or R) in image theory applications.

A rather different approach to improving the image model has been used by Weber and Ford.[13,16] Using the Kliewer-Fuchs model of metal dielectric response,[40] they have studied the influence of spatial dispersion (wavevector dependent dielectric constants) on the IFE mechanism. Their first paper[13] used the point dipole approximation and found that spatial dispersion reduced the maximum image enhancement by about 10^2 (down to roughly 10^6 if no average over R is included). The dependence of ε on R remains as strong as for the simple image model, however. In a more recent paper,[16] Weber and Ford have included for the effects of finite molecular size in their model by replacing the molecule by a sphere of finite radius. By using multipole polarizabilities to describe the response of the adsorbed molecule to the image field (also including for spatial dispersion), Weber and Ford find that the maximum Raman enhancement is reduced by 10^3 from the point dipole result to an overall value of $\varepsilon = 10^3$. Combining this with a factor of 10^3 which they estimate as due to roughness effects[13] (see also Ref. 41), gives an overall enhancement factor of 10^6. While this result agrees with the estimates we presented in Section II, the Weber-Ford model still contains several approximations which require further study before these conclusions can be considered reliable. These include: (1) the use of a free electron model to describe the metal dielectric response, (2) assumption of a sharp boundary at the metal interface and for the sphere which describes the molecule, (3) omission of chemisorption effects.

A closely related series of papers by Korzeniewski, Maniv, and Metiu[9,10] and by Feibelman[12] have recently tackled the very difficult problem of incorporating the continuous variation of electron density across the interface into the SERS enhancement factor calcu-

lations. These papers also include for the wavevector dependent
dielectric response, but to date all numerical evaluations have
treated the adsorbed molecule as a point dipole. The Korzeniewski
et al. and Feibelman models are physically identical, but each group
has focused attention on somewhat different pieces of information.
Particularly, Feibelman has determined the asymptotic corrections to
the classical image dipole enhancement factor while Korzeniewski
et al. have directly evaluated the short distance image field enhancement. In these models, the response of the metal electrons to
the oscillating induced molecular dipole is treated using linear
response theory, with a polarization propagator which is obtained
for an infinite barrier jellium model using the random phase approximation (RPA).

Using a radius parameter r_s = 5.0, Korzeniewski et al.[9] found
that at R $\sim$ 1.6 Å the polarizability derivative enhancement factor
was reduced by roughly 10^2 in their model compared to the simple
image model result. They attributed the origin of this reduction
as due to screening, and if so then their reduction factor is consistent with that of Weber and Ford.[13] It was also found that the
position averaged enhancement factor exhibited a more resonant frequency dependence than is exhibited in Fig. 1 for the simple image
model.

Feibelman[12] analyzed his results in terms of the leading asymptotic correction to the image plane location. This is analogous to
the corrections found by Lang and Kohn for the static image model[20]
but generalizes these to the frequency dependent case. An important
feature of this generalization is that the image plane location [and
hence R in Eq. (1)] is complex. This causes a reduction in the maximum image enhancement which is inversely proportional to the imaginary part of R. Estimates of this reduction for r_s = 2 jellium lead
Feibelman to conclude that the image enhancement was quite small,
perhaps on the order of unity. Just why this enhancement factor is
so much smaller than that estimated by Korzeniewski et al. for the
same model is not entirely clear at this time. The Korzeniewski
calculation did find that the imaginary part of the image field was
substantially larger than in the simple image model, but the influence of this on image enhancement was not as large as was concluded by Feibelman. Somewhat different r_s parameters were used in
the respective calculations though both groups argued that this was
not important. It is also possible that the asymptotic corrections
of Feibelman underestimated the magnitude of image fields at the
small distances considered by Korzeniewski et al.

Although the calculations of Korzeniewski et al. and of
Feibelman provide a significant improvement in the sophistication
of ad-molecule electrodynamics, they are still far from being truly
realistic for several reasons. The use of a point dipole is an
obvious approximation, as is the use of a jellium model to describe

the metal response. To remove either of these limitations without
simultaneously including for chemisorption effects can lead to in-
consistencies (such as the problem of overlapping charge clouds
described earlier), but to incorporate chemisorption into this treatment seems extremely difficult at present.

IV. CONCLUSION

    We can succinctly summarize the current status of the image
field effect mechanism of SERS by the following four statements:

1. The simple point dipole phenomenological image model [whose
    enhancement factor is given by Eq. (1) and which is plotted
    in Figs. 1 and 2 for several metals] shows remarkably good
    correspondence with experiment.

2. The attempts to improve this model discussed in Section III
    all conclude that the simple image model is quantitatively
    and perhaps even qualitatively inaccurate, and that the
    correct IFE enhancements should be several orders of magni-
    tude lower than is predicted by the simple models.

3. There is currently substantial disagreement concerning how
    large the image effect is for real ad-molecule-metal sys-
    tems. Estimates of the enhancement vary from 1 to 10^4.

4. None of the models in Table 1 have relaxed all of the ap-
    proximations indicated in the Table simultaneously. Even
    if they did, they would still be unrealistic unless chemi-
    sorption effects are also included, and unless both the
    molecule and metal electronic properties are described
    accurately.

    All of this indicates that the current status of the IFE mechan-
ism is uncertain, and that it may remain so unless a truly microscopic description of SERS is obtained. Fortunately it appears that
such a description may soon be available. Recently, Pandey and
Schatz[37] have calculated ab initio time dependent Hartree-Fock frequency dependent polarizability derivatives for H_2 adsorbed onto Li
clusters. Considering the H_2Li_2 system, they found that for frequencies close to the lowest excitation frequency of Li_2 and at
geometries close to equilibrium, the polarizability derivative with
respect to the H_2 stretch was enhanced by $10-10^2$ compared to that in
isolated H_2. This corresponds to a Raman enhancement of 10^2-10^4 not
including local field effects. This enhancement factor was found to
be a strong function of the molecule-metal distance, dropping to a
small value at just a few Å separation. Although the precise magnitude of this enhancement factor depends somewhat on how the widths
of the resonant metal states are modelled, this calculation does

indicate that strong enhancements are obtained from truly microscopic calculations. A major advantage of this approach is that all of the problems associated with finite molecular size, choice of molecule to metal separation, proper metal dielectric response and chemisorption are simultaneously taken care of in the Hartree-Fock description. What is not clear at this point is what is the coupling responsible for the enhancement effect. It should be possible to disentangle the calculation, however, to assess how much of this enhancement is due to image effects and how much to chemical effects. Presumably, at that point a fairly concrete assessment of the IFE mechanism will (at last!) be possible.

ACKNOWLEDGEMENTS

Helpful discussions with U. Laor, P. K. K. Pandey and R. P. Van Duyne are gratefully acknowledged. This research was supported by the Office of Naval Research (Contract N00014-79-C-0794).

REFERENCES

1. D. J. Jeanmaire and R. P. Van Duyne, Surface Raman spectro-electrochemistry. Part I. Heterocyclic, aromatic and aliphatic amines adsorbed on the anodized silver electrode, J. Electroanal. Chem. 84:1 (1977); R. P. Van Duyne, Applications of Raman spectroscopy in electrochemistry, J. Physique (Paris) 38:C5 (1977); M. G. Albrecht and J. A. Creighton, Anomalously intense Raman spectra of pyridine at a silver electrode, J. Am. Chem. Soc. 99:5215 (1977).
2. T. E. Furtak and J. Reyes, A critical analysis of theoretical models for the giant Raman effect from adsorbed molecules, Surf. Sci. 93:351 (1980).
3. F. W. King, R. P. Van Duyne and G. C. Schatz, Theory of Raman scattering by molecules adsorbed at electrode surfaces, J. Chem. Phys. 69:4472 (1978).
4. S. Efrima and H. Metiu, Classical theory of light scattering by a molecule located near a solid surface, Chem. Phys. Lett. 60:59 (1978).
5. S. Efrima and H. Metiu, Classical theory of light scattering by an adsorbed molecule I. Theory, J. Chem. Phys. 70:1602 (1979); S. Efrima and H. Metiu, Resonant Raman scattering by adsorbed molecules, J. Chem. Phys. 70:1930 (1979); S. Efrima and H. Metiu, Light scattering by a molecule near a solid surface. II. Model calculations, J. Chem. Phys. 70:2297 (1979).
6. S. Efrima and H. Metiu, Raman scattering from adsorbed molecules in electrochemical systems, Israel J. Chem. 18:17 (1979).
7. G. L. Eesley and J. R. Smith, Enhanced Raman scattering on metal surfaces, Solid State Commun. 31:815 (1979).
8. S. Efrima and H. Metiu, Surface induced resonant Raman scattering

(SIRRS), Surf. Sci. 92:417 (1980).

9. G. Korzeniewski, T. Maniv and H. Metiu, The interaction between
    an oscillating dipole and a metal surface described by a
    jellium model and the random phase approximation, Chem. Phys.
    Lett. 73:212 (1980).

10. T. Maniv and H. Metiu, Electron gas effects in the spectroscopy
    of molecules chemisorbed at metal surfaces. I. Theory, J.
    Chem. Phys. 72:1996 (1980).

11. P. R. Hilton and D. W. Oxtoby, Surface enhanced Raman spectra:
    A critical review of the image dipole description, J. Chem.
    Phys. 72:6346 (1980).

12. P. J. Feibelman, Local field at an irradiated adatom on jellium-
    exact microscopic results, Phys. Rev. B 22:3654 (1980).

13. W. H. Weber and G. W. Ford, Enhanced Raman scattering by ad-
    sorbates including the nonlocal response of the metal and
    the excitation of nonradiative modes, Phys. Rev. Lett.
    44:1774 (1980).

14. H. Morawitz and R. T. Koehler, A model for Raman active libra-
    tional modes on a metal surface: Pyridine and CN$^-$ on silver,
    Chem. Phys. Lett. 71:64 (1980).

15. G. C. Schatz and R. P. Van Duyne, Image field theory of en-
    hanced Raman scattering by molecules adsorbed on metal sur-
    faces: Detailed comparison with experimental results,
    Surf. Sci. 101:425 (1980).

16. G. W. Ford and W. H. Weber, Electromagnetic effects on a mole-
    cule at a metal surface. I. Effects of nonlocality and
    finite molecular size, preprint.

17. G. D. Mahan and A. A. Lucas, Collective vibrational modes of
    adsorbed CO, J. Chem. Phys. 68:1344 (1978) and references
    therein; G. Blyholder, CNDO Model of carbon monoxide chemi-
    sorbed on nickel, J. Phys. Chem. 79:756 (1975) and references
    therein.

18. R. R. Chance, A. Prock and R. Silbey, Lifetime of an emitting
    molecule near a partially reflecting surface, J. Chem. Phys.
    60:2744 (1974).

19. A. Campion, A. R. Gallo, C. B. Harris, H. J. Robota and P. M.
    Whitmore, Electronic energy transfer to metal surfaces: A
    test of classical image dipole theory at short distances,
    Chem. Phys. Lett. 73:447 (1980).

20. N. D. Lang and W. Kohn, Theory of metal surfaces: Induced
    surface charge and image potential, Phys. Rev. B 7:3541
    (1973); E. Zaremba and W. Kohn, Van der Waals interaction
    between an atom and a solid surface, Phys. Rev. B 13:2270
    (1976); J. A. Appelbaum and D. R. Hamann, Variational calcu-
    lation of the image potential near a solid surface, Phys.
    Rev. B 6:1122 (1972).

21. W. C. Meixner and P. R. Antoniewicz, Effective polarizability
    of polarizable atoms near metal surfaces, Phys. Rev. B
    13:3276 (1976).

22. J. W. Gadzuk, Screening energies in photoelectron spectroscopy

of localized electron levels, Phys. Rev. B 14:2267 (1976).

23. S. Andersson and B. N. J. Persson, Inelastic electron scattering by a collective vibrational mode of adsorbed CO, Phys. Rev. Lett. 45:1421 (1980); J. Kirtley and P. K. Hansma, Vibrational-mode shifts in inelastic electron tunneling spectroscopy: Effects due to superconductivity and surface interactions, Phys. Rev. B 13:2910 (1976).

24. S. Efrima and H. Metiu, Vibrational frequencies of a chemisorbed molecule: The role of the electrodynamic interactions, Surf. Sci. 92:433 (1980).

25. U. Laor and G. C. Schatz, The effect of surface roughness in surface enhanced Raman scattering: The importance of multiple plasmon resonances, submitted to Chem. Phys. Lett.

26. C. S. Allen, G. C. Schatz and R. P. Van Duyne, Tunable laser excitation profile of surface enhanced Raman scattering from pyridine adsorbed on a copper electrode, Chem. Phys. Lett. 75:201 (1980).

27. P. B. Johnson and R. W. Christy, Optical constants of the noble metals, Phys. Rev. B 6:4370 (1972).

28. E. G. Wilson and S. A. Rice, Reflection spectrum of liquid Hg from 2-20 eV, in, "Optical Properties and Electronic Structure of Metals and Alloys," F. Abeles, ed., Wiley, New York (1965), p. 271.

29. J. N. Hodgson, The optical properties and electronic band structure of lithium, in, "Optical Properties and Electronic Structure of Metals and Alloys," F. Abeles, ed., Wiley, New York (1965), p. 60.

30. R. P. Van Duyne, Laser excitation of Raman scattering from adsorbed molecules on electrode surfaces, in, "Chemical and Biochemical Applications of Lasers," Vol. 4, C. B. Moore, ed., Academic Press, New York (1978), p. 101.

31. U. Wenning, B. Pettinger and H. Wetzel, Angular-resolved Raman spectroscopy of pyridine on copper and gold electrodes, Chem. Phys. Lett. 70:49 (1980).

32. R. P. Van Duyne, private communication.

33. R. Naaman, S. J. Buelow, O. Cheshnovsky and D. Herschbach, Enhanced Raman scattering from molecules adsorbed on mercury, J. Phys. Chem. 84:2692 (1980).

34. B. Pettinger, U. Wenning and H. Wetzel, Angular resolved Raman spectra from pyridine adsorbed on silver electrodes, Chem. Phys. Lett. 67:192 (1979); B. Pettinger, U. Wenning and H. Wetzel, Surface plasmon enhanced Raman scattering frequency and angular resonance of Raman scattered light from pyridine on Au, Ag and Cu electrodes, Surf. Sci. 101:409 (1980).

35. C. K. Chen, A. R. B. de Castro and Y. R. Shen, Surface enhanced second harmonic generation, Phys. Rev. Lett. 46:145 (1981); C. K. Chen, T. F. Heinz, D. Ricard and Y. R. Shen, Detection of molecular monolayers by optical second-harmonic generation, Phys. Rev. Lett. 46:1010 (1981).

36. W. E. Palke, The effect of image charges on molecular absorption, _Surf. Sci._ 97:L331 (1980).

37. P. K. K. Pandey and G. C. Schatz, to be published.

38. M. Barr, P. C. Stair and R. P. Van Duyne, unpublished.

39. P. Citrin, P. Eisenberger and R. Hewitt, Extended X-ray adsorption fine structure of surface atoms on single-crystal substrates: Iodine adsorbed on Ag (111), _Phys. Rev. Lett._ 41:309 (1978).

40. K. L. Kliewer and R. Fuchs, Anomalous skin effect for specular electron scattering and optical experiments at non-normal angles of incidence, _Phys. Rev._ 172:607 (1968); R. Fuchs and K. L. Kliewer, Optical properties of an electron gas: Further studies of a nonlocal description, _Phys. Rev._ 185:905 (1969).

41. W. H. Weber and G. W. Ford, Optical electric-field enhancement at a metal surface arising from surface-plasmon excitation, _Opt. Lett._ 6:122 (1981).

COUPLED EXCITATION MODEL AND QUANTUM TEST OF IMAGE FIELD EFFECT

T. K. Lee

Institute for Theoretical Physics
University of California
Santa Barbara, California 93106

J. L. Birman

Department of Physics
City College, CUNY
New York, New York 10031

I. INTRODUCTION

In this paper we present a quantum theory of surface enhanced
Raman scattering based upon coupled eigenstates or coupled excitations of a physiadsorbed molecule and surface plasmons of the metal.
Our theory permits a quantum mechanical test to be made of the
classical image field effect which is one limiting case of our
model. In view of the physical simplicity and ease of calculation
of classical theories based on image fields, such a test is important in deciding how much credence to give such theories, which have
been previously intensively studied[1,2] and are reviewed by G. C.
Schatz in another chapter in this volume.

The image field can produce an "effective resonance" condition
for the laser frequency, giving a maximum enhancement of Raman
scattering at "resonance." We must turn to a quantum theory, however, to clarify the microscopic conditions needed to produce this
image-field based resonance. It is the result of our theory that
this appealing image mechanism is only capable of producing a relatively small enhancement (about 10-100) under realistic approximations appropriate to the canonical pyridine-silver system.

Our conclusion is based on a model in which the physiadsorbed
molecule is a two-level system with discrete dipole-allowed excita-

tions strongly coupled with the surface plasmon elementary excitations of the metal. Our theory considers both flat metal surface
geometry and a rough surface geometry - the latter treated in first
order (weak roughness) perturbation.

Calculation of the Raman scattering cross section proceeds by
first determining a set of coupled eigenstates of the molecule and
metal using Fano's method,[3] and then using these as intermediate
states.[4],[5] The apparent resonant frequency depends sensitively on
the density of states of the coupled eigenstates. The latter depends
on the density of states of the surface plasmon and its polarization
field. These in turn depend on the structure of the volume dielec-
tric function $\varepsilon(\omega,\vec{k})$ of the metal. Of course $\varepsilon(\omega,\vec{k})$ must in general
be taken as a function of both frequency and wave vector. Neglecting
$\vec{k}$ dependence by taking the long wavelength limit of $\varepsilon(\omega,\vec{k})$, and
working in the non-retarded limit ($c \rightarrow \infty$), classical image field
theory is recovered. But in this coupled system the molecule is
only a few angstroms away from the surface; the short wavelength $\vec{k}$
dependence in $\varepsilon(\omega,\vec{k})$ must be considered. The corrections to classical image field results are so important as to vitiate any possi-
bility of applying them to the surface enhanced Raman scattering
problem. Including surface roughness permits a direct photon-
surface plasmon interaction which is absent when the surface is
taken as flat. In the latter case the molecular dipole-excitation
part of the eigenstate acts as the mediator of the photon plasmon
coupling. But the result is essentially unchanged: the image field
mechanism fails to produce the observed 10^4-10^6 enhancement ratio.

Part of the present work is already published in Refs. 4 and 5.
Below we shall refer to them as papers I and II, respectively.

II. THE MODEL HAMILTONIAN

We consider the system as a metal surface with a point molecule
at distance R_z above the surface. The metal occupies the half
space z<0 and has a dielectric function $\varepsilon(\omega,\vec{k})$. We first treat a
flat surface, and then in section V estimate the effect of weak
roughness.

The surface plasmon has a strong polarization field outside
the metal. This field induces the electronic transitions in the
molecule. The molecular transitions can also excite surface plasmons. This is the only interaction to be considered in this model.

The total Hamiltonian consists of three parts $H = H_m+H_{sp}+H_I$.
The molecule is taken as a two level system with states $|0>$ and $|1>$;
its Hamiltonian is

$$H_m = \sum_{i=0,1} \varepsilon_i |i><i| \quad . \tag{1}$$

We shall choose $\varepsilon_0 = 0$, so ε_1 is the molecular excitation energy. The Hamiltonian for the surface plasmon field is

$$H_{sp} = \sum_q \hbar \omega_{\vec{q}}\, a_{\vec{q}}^{\dagger} a_{\vec{q}} \quad . \tag{2}$$

In the dipole approximation the polarization field of the surface plasmon $\vec{E}$ interacts with the molecule through $\sum_i e \vec{r}_i \cdot \vec{E}$, where r_i are the electrons of the molecule. In quantized form, the interaction Hamiltonian is

$$H_I = \sum_q V_{\vec{q}}(a_{\vec{q}} - a_{-\vec{q}}^{\dagger})(|0><1| + |1><0|) \quad . \tag{3}$$

The interaction matrix element $V_{\vec{q}}$ can be easily derived[4,6,7] in the case of neglect of $\vec{k}$ dependence in $\varepsilon(\omega,\vec{k})$. Taking the dielectric function $\varepsilon(\omega) = \varepsilon(\omega,k=0)$ to be of the Drude form $\varepsilon(\omega) = 1-\omega_p^2/\omega^2$, we obtain

$$V_{\vec{q}} = \frac{\omega_q}{c} \left(\frac{4\pi\hbar c}{L^2 P_q}\right)^{1/2} (\vec{\mu}_{\perp} \cdot \vec{q}/q + i\, \frac{q}{\nu}\, \mu_z)\, e^{-\nu R_z} \tag{4}$$

where L^2 is the area of the surface, $\mu_{\perp}$ and μ_z are the dipole moments of the molecule in the direction parallel and perpendicular to the surface. q, ν and ω_q are related by the equations[8,9,10]

$$\nu^2 = q^2 - \omega_q^2/c^2 \tag{5}$$

and

$$\frac{c^2 q^2}{\omega_q^2} = \frac{\varepsilon(\omega_q)}{1 + \varepsilon(\omega_q)} \quad . \tag{6}$$

The quantity P_q in Eq. (4) is given by

$$P_q = \frac{\varepsilon^4(\omega_q) - 1}{\varepsilon^2(\omega_q)[-1 - \varepsilon(\omega_q)]^{1/2}} \quad . \tag{7}$$

The expression of $V_{\vec{q}}$ for a more general form of dielectric function, namely including interband contribution, is given in paper I.

In the nonretarded limit $cq/\omega_q \gg 1$, the coupling constant reduces to the form

$$|v_{\vec{q}}|^2 = \frac{\pi\hbar q\omega_q}{L^2} \, (|\vec{\mu}_\perp \cdot \hat{q}|^2 + \mu_z^2)e^{-2qR_z} \tag{8}$$

This expression was first derived by Gersten and Tzoar.[6]

It should be noticed that the above result is derived by neglecting the wave-vector dependence in the dielectric function. The simplest way to introduce the wave-vector dependence is to use the hydrodynamical form of the dielectric function. Fuchs and Kliewer[11] have shown that the surface-plasmon dispersion relation obtained by using the hydrodynamic form of $\varepsilon(\omega,\vec{k})$ is similar to the result obtained by using the complicated Lindhard dielectric function.

In the hydrodynamical approximation, the longitudinal dielectric function is of the form

$$\varepsilon_\ell(\omega,\vec{k}) = 1 - \frac{\omega_p^2}{\omega^2 - \beta^2 k^2} \quad , \tag{9}$$

where $\beta^2 = \frac{3}{5} v_F^2$, $\mathbf{v}_F$ is the Fermi velocity. The transverse dielectric function $\varepsilon_t(\omega) = \varepsilon_\ell(\omega, k=0)$.

In paper I we have shown that in the nonretarded limit and using the hydrodynamic dielectric function, the coupling constant $V_{\vec{q}}$ is given by[12]

$$|v_{\vec{q}}|^2 = \frac{\pi\hbar q}{L^2} \frac{\omega_p^2}{2\omega_q} \, e^{-2qR_z}(|\vec{\mu}_\perp \cdot \hat{q}|^2 + \mu_z^2) \cdot (\frac{2\omega_p^2}{\omega_p^2 + 2\omega_q^2}) \quad . \tag{10}$$

Now the surface-plasmon dispersion relation is not given by Eq. (6) but is changed to

$$\frac{\beta^2 q^2}{\omega_q^2} = \frac{1}{4} \, (1 + \varepsilon(w_q))^2 \quad . \tag{11}$$

It is interesting to observe that the total Hamiltonian given by Eqs. (1)-(3), $H = H_m + H_{sp} + H_I$, is exactly of the same form used by R. Loudon[13] in the discussion of a two-level atom interacting with radiation.

III. THE COUPLED EIGENSTATES

To diagonalize the Hamiltonian, we shall follow Fano's[3] general theory of discrete-continua coupling. Details of the calculation can be found in paper I. The mixed eigenstates of the coupled Hamiltonians are found to be

$$|\psi_E> = a(E) |1,g> + \sum_{\vec{q}} b_{\vec{q}}(E) |0,\vec{q}> \quad .$$ (12)

We have used the notation $|n,\vec{q}>$ to represent the product of molecular states $n=0,1$ and the plasmon states $\vec{q}$, or g (the ground state, no excited plasmons). The mixing coefficients $a(E)$ and $b_{\vec{q}}(E)$ are

$$|a(E)|^2 |V(E)|^2 = [\pi^2 + Z^2(E)]^{-1}$$ (13)

and

$$b_{\vec{q}}(E) = \left[\frac{P}{E - \hbar\omega_q} + Z(E) \ \delta(E - \hbar\omega_q) \right] V_{\vec{q}}^* \ a(E) \quad .$$ (14)

The function $Z(E)$ is defined by the equation

$$Z(E) |V(E)|^2 = E - \varepsilon_1 - F(E) \quad .$$ (15)

$F(E)$ and $\pi|V(E)|^2$ are real and imaginary parts of the function $\Delta(E)$

$$\Delta(E) \equiv \sum_{\vec{q}} \frac{|V_{\vec{q}}|^2}{E - \hbar\omega_q - i\delta} = F(E) + i\pi|V(E)|^2$$ (16)

where $\delta \to 0+$.

In the derivation of the above result, we have neglected two interactions in H_I; they are $a_{\vec{q}}|0><1|$ and $a_{\vec{q}}^{\ddagger}|1><0|$. These interactions couple states $|0,g>$ and $|1,\vec{q}>$, their energy difference $\varepsilon_1 + \hbar\omega_q$ is much larger than energy difference $\varepsilon_1 - \hbar\omega_q$ between states $|1,g>$ and $|0,\vec{q}>$. The states $|1,\vec{q}>$ also do not contribute to Raman scattering.

Here we wish to emphasize that the eigenstates $|\psi_E>$ of Eq. (14) represent a new set of eigenstates of the entire system. They have properties of both surface-plasmon states and molecular states. They should not be identified as shifted and broadened original molecular states as discussed by Philpott.[7]

So far we have assumed that the surface plasmon has no damping. In practice, the surface plasmon has a finite lifetime due to interaction with surface irregularities, bulk plasmon, single particle excitation, interband transition etc. The damping can be taken into account phenomenologically by introducing Γ in Eq. (16), i.e.,

$$\Delta(E) = \sum_{\vec{q}} \frac{|v_{\vec{q}}|^2}{E - \hbar\omega_q - i\Gamma} \quad .$$ (17)

For the case in which the wave-vector dependence in the dielectric function is neglected, the function $\Delta(E)$ is obtained by substituting Eq. (4) into Eq. (17):

$$\Delta(E) = \frac{e^2\hbar^3}{m\varepsilon_1} \frac{1}{c^3} \int_0^{\omega_{sp}} d\omega\, \omega^3 \left(f_{01}^\perp - \varepsilon(\omega) f_{01}^z\right) \frac{\varepsilon^2(\omega)}{[-1-\varepsilon(\omega)]^{5/2}[1-\varepsilon(\omega)]}$$

$$\times \frac{1}{E-\hbar\omega-i\Gamma} \exp\left(-\frac{2\omega}{c} \frac{R_z}{[-1-\varepsilon(\omega)]^{1/2}}\right) \quad ,$$ (18)

where ω_{sp} is the surface plasmon frequency defined by $\varepsilon(\omega_{sp}) = -1$. The oscillator strengths $f_{01}^\perp$ and f_{01}^z are related to the dipole moments by

$$f_{01}^\perp = \frac{m\varepsilon_1}{e^2\hbar^2} \mu_\perp^2 \quad \text{and} \quad f_{01}^z = \frac{2m\varepsilon_1}{e^2\hbar^2} \mu_z^2 \quad .$$

For $2R_z\omega_{sp}/c \ll 1$, Eq. (18) reduces approximately to the form

$$\Delta(E) \simeq \frac{1}{8R_z^3} (\mu_z^2 + \frac{1}{2} \mu_\perp^2) \frac{\hbar\omega_{sp}}{E-\hbar\omega_{sp}-i\Gamma} \quad .$$ (19)

At $E = \varepsilon_0 = 0$, the real part of $\Delta(E)$ gives the classical expression for the self-energy of a dipole and its image field.

Including the wave-vector dependence in the dielectric function greatly reduces the image field. The expression $\Delta(E)$ now becomes

$$\Delta(E) = \frac{e^2\hbar^2}{m\varepsilon_1} \left(\frac{\omega_p}{c}\right)^3 \frac{1}{8} \int_0^{Q_c} Q^2 dQ (f_{01}^\perp + f_{01}^z) \exp\left(-\frac{2\omega_p R_z}{c} Q\right)$$

$$\times \frac{1}{\Omega} \frac{1}{\frac{E}{\hbar\omega_p} - \Omega - i\frac{\Gamma}{\hbar\omega_p}} \frac{2}{1+2\Omega^2} \quad ,$$ (20)

where $\Omega = \omega/\omega_p$ and $Q = cq/\omega_p$; they are related by the equation

$$\frac{\beta^2}{c^2} Q^2 = \frac{\Omega^2}{4} \left(2 - \frac{1}{\Omega^2}\right)^2 \quad .$$ (21)

We have chosen a cut-off $Q_c = c/2\beta$ in Eq. (21), under the assumption that above the volume plasmon frequency ω_p, essentially no surface plasmon exists.

The real and imaginary parts of the function $\Delta(E)$, namely $F(E)$ and $\pi|V(E)|^2$, are plotted in Fig. 1 as a function of the molecular distance R_z at $E = 2.5$ eV. We have chosen $\varepsilon_1 = 4.31$ eV, $\hbar\omega_p = 5.4$ eV, and $\Gamma = 0.08$ eV.[14] These values are roughly in accordance with the values of pyridine and silver. The oscillator strength f_{01} for the particular pair of electronic levels considered here is not known. Since $\Delta(E)$ is directly proportional to f_{01}, for simplicity we have chosen $f_{01}^{\perp} = 0$ and $f_{01}^{z} = 1$. The solid curves are obtained from Eq. (18) for the case of no spatial dispersion. The dashed curves are calculated from Eq. (20) with spatial dispersion. As expected, at small R_z and neglect of spatial dispersion, the values of $\Delta(E)$ are over estimated by an order of magnitude. Having obtained the function $\Delta(E)$, we can now use the eigenstates $|\psi_E\rangle$ to calculate the ratio of the Raman scattering cross sections for adsorbed and free molecules.

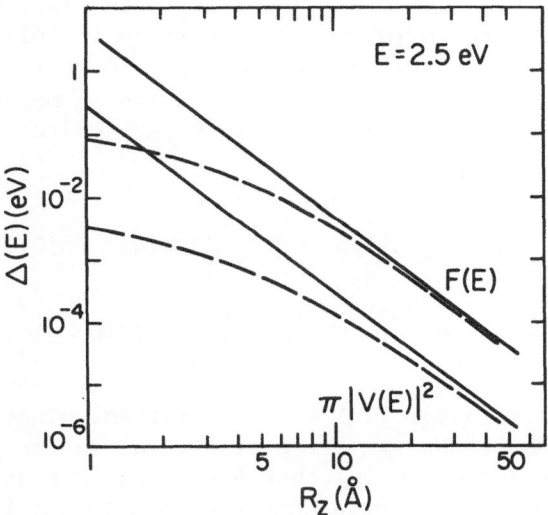

Fig. 1. The real and imaginary parts of the function $\Delta(E)$ as a function of R_z at $E = 2.5$ eV. The dashed curves are calculated with spatial dispersion, and the solid curves neglect spatial dispersion. The parameters $\Gamma = .08$ eV, $\hbar\omega_p = 5.4$ eV, $\varepsilon_1 = 4.31$ eV, $f_{01}^{\perp} = 0$, $f_{01}^{z} = 1$, and $c/\beta = 2.76 \times 10^2$.

IV. RAMAN SCATTERING

The matrix element for Raman scattering between an initial state $|i>$ and a final state $|f>$ is of the form

$$M = \sum_j \left\{ \frac{<f|\hat{e}_s\cdot\vec{p}|j><j|\hat{e}_i\cdot\vec{p}|i>}{E_j - E_i - \hbar\omega_L} + \frac{<f|\hat{e}_i\cdot\vec{p}|j><j|\hat{e}_s\cdot\vec{p}|i>}{E_j - E_f + \hbar\omega_L} \right\} \quad (22)$$

where $|j>$ are intermediate states, E_i, E_j and E_f are energies of the initial, intermediate and final states, $\hat{e}_i$ and $\hat{e}_s$ are the polarization vectors of the incident and scattered light, and ω_L is the incident laser frequency.

For the system of molecules adsorbed on a semi-infinite metal surface, the initial state $|i>$ is taken to be the ground state of the coupled system, i.e. the molecule in the lowest electronic level and vibrational state, and no surface plasmons are excited. The final state $|f>$ has the same electronic configuration as the initial state $|i>$, but the molecule is in the excited vibrational state. Therefore for the Stokes process, $E_f - E_i = \hbar\omega_v$, where ω_v is the vibrational frequency.

It should be remarked that the molecular states are treated in the Born-Oppenheimer approximation. Vibrational state overlap integrals are important in determining the relative probability of transitions between different states. In this work, we shall assume that the molecule has only one vibrational state associated with the excited electronic level, and vibrational-state overlap integrals will be omitted. It is worth pointing out that in principle the coupled eigenstates given by Eq. (12) should include the vibrational state.

The intermediate states $|j>$ are now chosen to be the coupled eigenstates. By using Eq. (12), the first part of the matrix element M of Eq. (22) becomes

$$M = \int \frac{dE}{E-\hbar\omega_L} \left[<0,g|\hat{e}_s\cdot\vec{p}|1,g>a(E) + \sum_{\vec{q}} <0,g|\hat{e}_s\cdot\vec{p}|0,\vec{q}>b_{\vec{q}}(E) \right]$$

$$\times \left[<1,g|\hat{e}_i\cdot\vec{p}|0,g>a^*(E) + \sum_{\vec{q}} <0,\vec{q}|\hat{e}_i\cdot\vec{p}|0,g>b_{\vec{q}}^*(E) \right]. \quad (23)$$

The second term in Eq. (22) will be omitted, since it does not have a resonance denominator. For a flat metal surface light cannot excite surface plasmon directly; thus the matrix element $<0,\vec{q}|\hat{e}_i\cdot p|0,g> = 0$. If the metal surface is not flat but with some roughness or protruding metal shapes then the matrix element is not zero and in some instances could be very important. Fano anti-resonance behavior might occur due to interference between the matrix elements $<1g|\hat{e}_i\cdot p|0g>a(E)$ and $\sum_{\vec{q}} b_{\vec{q}}(E)<0\vec{q}|\hat{e}_i\cdot\vec{p}|0g>$.

We shall consider the matrix element M in two separate cases: a flat metal surface and a surface with weak roughness. The latter case will be discussed in the next section. The flat surface case has been discussed in detail in paper II. Near resonance the ratio of the scattering cross section per molecule for the adsorbed molecule and isolated molecule is given by

$$R \simeq \left| \int \frac{dE}{E - \hbar\omega_L} \left| a(E) \right|^2 \right|^2 \Bigg/ \left| \frac{1}{\varepsilon_1 - \hbar\omega_L} \right| \quad . \tag{24}$$

By using Eqs. (13) and (15), we obtain

$$\left| a(E) \right|^2 = \frac{\left| V(E) \right|^2}{[E - \varepsilon_1 - F(E)]^2 + \pi^2 \left| V(E) \right|^4} \quad . \tag{25}$$

After substitution of Eq. (25) into Eq. (24) we obtain

$$R = \frac{(\varepsilon_1 - \hbar\omega_L)^2}{[\hbar\omega_L - \varepsilon_1 - F(\hbar\omega_L)]^2 + \pi^2 \left| V(\hbar\omega_L) \right|^4} \tag{26}$$

This ratio can be fairly large at the laser frequency ω_L^m such that

$$\hbar\omega_L^m - \varepsilon_1 = F(\hbar\omega_L^m) \tag{27}$$

as long as

$$F(\hbar\omega_L^m) \gg \left| V(\hbar\omega_L^m) \right|^2 \quad . \tag{28}$$

It should be noticed that the result (26) is only valid near some effective resonant frequency ω_L^m since we have neglected contributions to the scattering cross section from all but the two levels ε_0 and ε_1. We have also neglected the non-resonant part of the matrix element (22). Consequently the frequency dependence of R, far from ω_L^m is not given by the expression (26).

In the limit $2R_z\omega_{sp}/c \ll 1$ and without including the spatial dispersion, the ratio R can be calculated by using Eq. (19). At the resonant frequency ω_L^m, the maximum enhancement ratio is given by

$$R_m \simeq \left(\frac{\hbar\omega_L^m - \hbar\omega_{sp}}{\Gamma} \right)^2 \tag{29}$$

and

$$\hbar\omega_L^m = \varepsilon_1 + \frac{1}{8R_z^3} \left(\mu_z^2 + \frac{1}{2} \mu_\perp^2 \right) \frac{\hbar\omega_{sp}}{\hbar\omega_L^m - \hbar\omega_{sp}} \quad . \tag{30}$$

In Fig. 2, we have plotted log R as a function of laser frequency $\hbar\omega_L$. The parameters have the same values as in Fig. 1.
Without inclusion of spatial dispersion we obtain the solid curve
which is in agreement with the classical image field theory result.[2]
In the presence of the spatial dispersion as given by the dashed
curve, the important contribution to F(E) comes from the states above
the frequency ω_{sp}. Thus the resonant frequency ω_L^m given by Eq. (27)

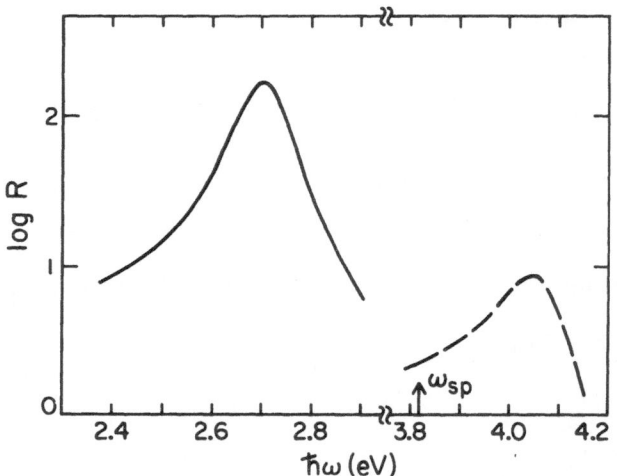

Fig. 2.   Logarithm of the ratio of RS intensity R between adsorbed
           and isolated molecules as a function of frequency. The
           dashed curve is obtained with spatial dispersion, and the
           solid curve neglects spatial dispersion. $R_Z = 1.5$ Å and
           other parameters are the same as in Fig. 1.

has a solution with $\omega_L^m > \omega_{sp}$. For large distance R_z, F(E) becomes
very small (Fig. 1) and $\hbar\omega_L^m$ approaches ε_1. This is a much more
reasonable result than the result of the spatial dispersionless case,
where ω_L^m is always less than ω_{sp}. We believe such an artifact for
the spatially dispersionless case is due to the singular density of
states of surface plasmon at ω_{sp}, which is evidently not physically
reasonable.

V. EFFECT OF "WEAK" SURFACE ROUGHNESS

Due to surface roughness the incident light can excite surface plasmons directly. In the surface enhanced Raman scattering, the excited surface plasmons induce electronic transitions in the adsorbed molecules and thus contributes to the Raman scattering.

Elson and Ritchie[8] have developed a first order perturbation theory to treat the interaction between surface plasmons and light. In this section we shall use their results to qualitatively estimate the contribution of surface roughness to the RS matrix element M of Eq. (23).

Roughness is described by a height function $z = \xi(x,y)$. The average of the Fourier transform

$$\xi_{\vec{q}} = \int e^{-i\vec{q}\cdot\vec{r}} \xi(\vec{r}) d^2\mathbf{r}$$

is assumed to satisfy the conditions:

$$\frac{1}{L^2}\cdot\langle\xi_{\vec{q}}\ \xi_{\vec{q}'}^{*}\rangle_{avg} = \delta_{\vec{q}\vec{q}'},\ \pi\delta^2\sigma^2 \exp(-\frac{\sigma^2 q^2}{4})$$

(31)

and

$$\langle\xi_{\vec{q}}\rangle_{avg} = 0 \quad .$$

(32)

The conditions (31) and (32) greatly simplify Eq. (23) which becomes

$$M = \int \frac{dE}{E-\hbar\omega_L}\left[|\langle 1,g|\hat{e}_i\cdot\vec{p}|0,g\rangle|^2|a(E)|^2\right.$$

$$\left.+ \sum_q |b_{\vec{q}}(E)|^2\left(|\langle 0,g|\hat{e}_i\cdot\vec{p}|0,\vec{q}\rangle|^2\right)_{avg}\right] \quad .$$

(33)

We have neglected the difference between the matrix element $\langle 1,g|\hat{e}_i\cdot\vec{p}|0,g\rangle$ and $\langle 1,g|\hat{e}_s\cdot\vec{p}|0,g\rangle$, since the vibrational frequency is much smaller than the laser frequency.

An important point to notice is that Eq. (33) no longer has interference between the two processes $\langle 1,g|\hat{e}_i\cdot\vec{p}|0,g\rangle$ and $\langle 0,\vec{q}|\hat{e}_i\cdot\vec{p}|0,g\rangle v_q$ which is present in the general expression Eq. (23). This is a direct consequence of the assumption of random weak roughness. If there is some large structure on the surface, then we will expect interference to occur, and higher order terms in the roughness must be included.

In order to estimate the effect of weak roughness on the Raman scattering ratio, we need to obtain the magnitude of the second term in Eq. (33), when roughness is present. This term is absent when roughness is ignored and represents the direct-coupling (photon-surface plasmon) effect. Thus the quantity of interest is given by

$$
M_R = \sum_q \left| \frac{b_{\vec{q}}(E)}{a(E)} \right|^2 \left(\left| \frac{<0,g|\hat{e}_i \cdot \vec{p}|0\vec{q}>}{<0,g|\hat{e}_i \cdot \vec{p}|1g>} \right|^2 \right)_{avg} \quad . \tag{34}
$$

The matrix element $<0g|\hat{e}_i \cdot \vec{p}|0\vec{q}>$ is very complicated in general. For our purpose of estimating the magnitude of M_R, we shall evaluate the quantity $<0g|\hat{e}_i \cdot \vec{p}|0\vec{q}>$ by setting the incident light frequency equal to the surface plasmon frequency ω_q.

For simplicity we consider p-polarized light with polarization in the x direction and incident normal to the plane $z = <\xi(\vec{r})> = 0$. By using the interaction Hamiltonian given by Eq. (20) of Ref. (8), we obtain the ratio of the two matrix elements

$$
\left| \frac{<0,\vec{q}|\hat{e}_i \cdot \vec{p}|0,g>}{<1,g|\hat{e}_i \cdot \vec{p}|0,g>} \right|^2 = \frac{1}{L^2} \frac{|\xi_q|^2}{\mu_x^2} \frac{1}{4\pi p_q} \frac{\hbar\omega_{sp}^2}{c} \cos^2\phi_q
$$

$$
\times (1-\varepsilon)^2 \left(1 + 2\sqrt{\frac{-1-\varepsilon}{-\varepsilon}} \right)^2 \tag{35}
$$

where p_q is defined by Eq. (4) and ϕ_q is the angle between the surface plasmon wave vector q and the x direction. The dielectric constant ε is evaluated at $\omega = \omega_q$. Here we shall only consider the spatially dispersionless case. Thus at E far away from the surface plasmon frequency, we obtain approximately

$$
\frac{|b_q(E)|^2}{|a(E)|^2} \simeq \left(\frac{1}{E-\hbar\omega_{sp}} \right)^2 |V_q|^2 \quad . \tag{36}
$$

By using Eqs. (4), (35) and (36), we obtain

$$
M_R \simeq \left(\frac{\hbar\omega_{sp}}{E-\hbar\omega_{sp}} \right)^2 \sum_q \frac{1}{L^2} |\xi_q|^2 \cos^2\phi_q \left(\frac{\omega_{sp}}{c} \right)^2 \frac{1}{L^2} \frac{1}{p_q^2} (1-\varepsilon)^2
$$

$$
\times \left(1 + 2\sqrt{\frac{-1-\varepsilon}{-\varepsilon}} \right)^2 e^{-2qR_z} \quad . \tag{37}
$$

By using Eqs. (6) and (31), Eq. (37) is approximately reduced to

$$M_R \simeq \left(\frac{\hbar\omega_{sp}}{E - \hbar\omega_{sp}}\right)^2 \frac{\delta^2 \sigma^2}{16} \int_0^\infty e^{-\frac{q^2\sigma^2}{4}} q^3 dq \left(1 + 2\frac{\omega}{cq}\right)^2 e^{-2qR_z}$$

$$\simeq \left(\frac{\hbar\omega_{sp}}{E - \hbar\omega_{sp}}\right)^2 \left(\frac{\omega_{sp}}{c}\delta\right)^2 \tag{38}$$

We have assumed that $\frac{\omega_{sp}}{c}\sigma \gg 1$ and $R_z/\sigma \ll 1$.

Apparently $M_R \ll 1$, since $\frac{\omega_{sp}}{c}\delta \ll 1$. The above perturbative result is derived under the assumption that δ is much less than the photon wavelength λ, $\lambda \sim c/\omega_{sp}$.

Since $M_R \ll 1$, the matrix element M of Eq. (33) is still dominated by the first term. Therefore the enhancement ratio R discussed in the last section is still correct for a weakly rough surface, and the result given in Fig. 2 is expected to apply in presence of this roughness.

VI. SUMMARY

This work (see also refs. 4 and 5 for an earlier version) has two major aspects. First, we have presented a general coupled state quantum formalism for surface enhanced Raman scattering. This formalism can be applied to consideration of coupled surface excitations and discrete molecular excitations. The former may include not only surface plasmons as in the present work but surface states or other excitations.[15] Secondly the model permits clear identification of the important effects and parameters. The wave vector dependence in the dielectric function must be considered in order to obtain physically reasonable results. The three essential parameters are the surface plasmon lifetime Γ, molecular dipole excitation oscillator strength f_{01}, and molecule-surface separation R_z. f_{01} and R_z determine the resonance frequency in RS. Γ is the most crucial parameter to determine the RS enhancement ratio. Using a realistic value of $\Gamma = 0.08$ eV,[14] the image field mechanism gives an enhancement of about 10 at UV laser frequency when molecules are near the metal surface. There is no enhancement at the visible frequency range where $10^4 \sim 10^6$ enhancement is observed in experiments. Surface roughness of order $10 \sim 100$ Å in height has a negligible effect on this result.

We like to emphasize that the model discussed above points out the importance of the short wavelength screening effect of the dielectric function when the molecule is very near the metal surface.

Many other important effects have been neglected, which may be important in understanding the observed RS enhancement ratio.

After the paper was completed we learned of work of Arya and
Zeyher[16] which agrees with our result and conclusion for physiadsorbed molecules on flat surfaces. When chemiadsorbtion and charge
transfer is also taken into consideration, these authors report
possible enhancement ratios in the experimentally observed range of
$10^5 \sim 10^6$ for flat surfaces. Their work supports the view that large
enhancement ratios are possible on flat or weakly roughened surfaces.
Our results support the view that large-scale surface roughness with
scale $>> 10^2$ Å is needed for Raman Scattering enhancement due to
physiadsorbed molecules. At the time of writing, the experimental
results are insufficient to discriminate between the various possible enhancement mechanisms.

ACKNOWLEDGEMENTS

This work was supported in part by ARO Grant No. DDAG29-79-G-
0400, NSF Grant DMR 78-12399, PSC-BHE Faculty Grant 13084 and NSF
Grant PHY77-27084.

REFERENCES

1.   F.W. King, R.P. Van Duyne and G.C. Schatz, Theory of Raman
     scattering by molecules adsorbed on electrode surfaces, J. Chem.
     Phys. 69:4472 (1978).
2.   S. Efrima and H. Metiu, Light scattering by a molecule near a
     solid surface. II Model calculations, J. Chem. Phys. 70:2297
     (1979).
3.   U. Fano, Effects of configuration interaction on intensities and
     phase shifts, Phys. Rev. 124:1866 (1961).
4.   T.K. Lee and J.L. Birman, Molecule adsorbed on plane metal
     surface: Coupled system eigenstates, Phys. Rev. B 22:5953 (1980).
5.   T.K. Lee and J.L. Birman, Quantum theory of enhanced Raman
     scattering by molecules on metals: Surface-plasmon mechanism
     for plane metal surface, Phys. Rev. B 22:5961 (1980).
6.   J.I. Gersten and N. Tzoar, Many-body effects in Auger deexci-
     tation of atoms near solids, Phys. Rev. B 9:4038 (1974).
7.   M.R. Philpott, Effect of surface plasmons on transitions in
     molecules, J. Chem. Phys. 62:1812 (1975).
8.   J.M. Elson and R.H. Ritchie, Photon interactions at a rough
     metal surface, Phys. Rev. B 4:4129 (1971).
9.   J. Nkoma, R. Loudon and D.R. Tilley, Elementary properties of
     surface polaritons, J. Phys. C 7:3547 (1974).
10. G.S. Agarwal, Quantization of the surface polariton field and
     the decay of an atom in the presence of a dielectric, Opt.
     Commun. 13:375 (1975).

11. P. Fuchs and K.L. Kliewer, Surface plasmon in a semi-infinite free-electron gas, Phys. Rev. B 3:2270 (1971).

12. The coupling constant $\lceil V_q \rceil^2$ given by Eq. (15) in Ref. 4 is incorrect by a factor $2\omega_p^2/(\omega_p^2 + 2\omega^2)$. The quantization condition of Eq. (A12) in Ref. 4 should be replaced by

$$\frac{1}{2}\,\hbar\omega = \int \left(\rho\phi + 4\pi\,\frac{\beta^2}{\omega_p^2}\,\rho^2\right)\,dz\,A \quad .$$

   The authors are grateful to Prof. G. Barton for asking an inspiring question which resulted in this correction.

13. R. Loudon, "Quantum Theory of Light," Oxford University, London (1974), p. 188.

14. W. Steinman, Optical plasma resonances in solids, Phys. Status Solidi 28:437 (1968).

15. J.I. Gersten, R.L. Birke and J.R. Lombardi, Theory of enhanced light scattering from molecules adsorbed at the metal-solution interface, Phys. Rev. Lett. 43:147 (1979).

16. K. Arya and R. Zeyher, Enhanced Raman scattering from molecules adsorbed at metal surfaces, preprint. We are very grateful to Dr. Arya and Dr. Zeyher for kindly making a preprint available to us.

THE ROLES OF SURFACE ROUGHNESS

E. Burstein[§+] and S. Lundqvist

Chalmers University of Technology
S-412 96 Gothenburg, Sweden
University of Pennsylvania
Philadelphia, PA 19104

D. L. Mills[++]

University of California
Irvine, CA 92717

I. INTRODUCTORY REMARKS

The effect of surface roughness on the response of metals to electromagnetic fields has been under investigation since the turn of the century when R. W. Wood[1] observed an "anomalous" optical absorption by thin alkali metal films, which he attributed to their granular structure. Until fairly recently, the focus of the investigations has been almost entirely on the effect of surface roughness on the linear optical response of metals, e.g., the reflection, absorption, and elastic scattering of EM radiation,[2] the coupling of EM radiation with surface EM waves,[3] and photoemission by metals.[4] There was some speculation about the possible effect of surface roughness on second-harmonic generation by metals[5] but, until very recently, no experimental study of this effect had been carried out. The observation of a "giant" enhancement ($\sim 10^6$) of the Raman scattering by molecules adsorbed on a Ag electrode[6,7] and the corresponding large enhancement of the inelastic light scattering by rough metal surface[8-10] has provided further impetus to the investigation of the role of surface roughness in non-linear, as well as linear, optical phenomena. The observation of light emission in inelastic electron tunneling between metals is another recently observed optical phenomenon in which surface roughness plays a crucial role.[11-13] It has now been established that, in addition to its role in interconverting EM and surface-EM waves via a grating coupling mechanism, the rough surface exhibits localized collective electron resonances whose excitation by EM fields, and by the electric dipoles of excited molecules, leads to an enhancement of the incident and scattered EM radiation respectively.[14-18] Moreover the rough surface enhances the radiative excitation and recombination of electron-hole pairs and, thereby, leads to an enhanced inelastic scattering of light (e.g., Raman scattering and luminescence) by electron-hole pair excitations in the surface region of the metal.[9]

Various forms of surface roughness (e.g., surface microstructures) have been used to enhance the Raman scattering of molecules adsorbed at Ag, Cu and Au surfaces. These include colloidal suspensions of metal particles,[19] metal island films,[13,20] metal gratings,[21] arrays of axially symmetric bumps on metal surfaces,[22] and lithographically fabricated ordered two-dimensional arrays of metal particles.[23] (The latter has yielded an enhancement of $\sim 10^7$ in the Raman scattering by CN⁻ adsorbed on Ag.) The microstructures enhance both the incident and scattered EM radiation. They are, in effect, passive amplifiers which can be used to enhance a wide variety of optical phenomena not only at metal surfaces, but also at the surface of any low dielectric-loss medium having a dielectric function whose real part is negative. (For a discussion of surface roughness on an atomic scale, the reader is referred to the chaper by Otto, Pockrand, Billmann, and Pettenkofer in this volume.)

In this overview, we will discuss the concepts underlying the several roles played by surface roughness in the inelastic scattering of light by metals and by molecules adsorbed on metal surfaces, and in electromagnetic phenomena at metal surfaces in general. In particular, we will focus on those aspects that come into play because of the non-local dielectric properties of the metal, and specifically because of the radiative-excitation and recombination of electron-hole pairs.

The electromagnetic processes taking place in a metal with a rough surface are not fundamentally different from those occurring in a metal with a smooth surface. We will therefore present in Section II a discussion of the electromagnetic fields at a smooth semi-infinite metal surface, the inelastic light scattering (Raman scattering and luminescence) by the electron-hole pair excitations of the smooth metal, and the RS by molecules adsorbed on a smooth metal surface. We will then discuss the character of rough surfaces in Section III, and in Section IV we will give a discussion of the roles of surface roughness in the inelastic scattering of light by electron-hole pair excitations in metals and in the Raman scattering by molecules adsorbed on metals.

II. SMOOTH SEMI-INFINITE METAL

The excitation of electron-hole pairs at a smooth surface

and their influence on the electromagnetic fields in the surface
region is now reasonably well understood.[24-27] An understanding
of the physics involved should be useful in providing some insight
into the role of non-local effects (e.g., electron-hole pair excitations) in the enhancement of the incident and scattered EM fields
at a rough surface. We therefore undertake in this Section a discussion of the case of a smooth semi-infinite metal. We shall
be particularly concerned with the non-local dielectric response
of the smooth metal, which is associated with the virtual and real
excitation of electron-hole pairs. An understanding of this will
be essential for an understanding of the electromagnetic response
of rough metal surfaces, particularly the modifications of the
EM fields at rough surfaces arising from the excitation of electron-
hole pairs in the surface region. We also give a brief discussion
of the inelastic scattering of light by a smooth metal surface and
the Raman scattering by molecules adsorbed on a smooth metal.

   The theoretical model to be considered in the major part of
our discussions is that of a "free electron" metal. Most of the
theoretical studies of the non-local properties of metals, due to
the excitation of electron-hole pairs, have been restricted to
different versions of the jellium model in which the effects of
the periodic lattice have been ignored. For our purposes, we will
include the major effect of the periodic potential, which is to
introduce energy gaps in the otherwise "free electron" character
of the energy band, so that interband transitions between "free
electron"-like energy bands are included. The particular effects,
due to interband transitions involving d-bands, will also be included where needed.

   Before discussing any effects arising from the excitation of
electron-hole pairs, we briefly summarize some basic surface
electromagnetic properties using classical (e.g., local dielectric
response) theory. Let us consider an EM wave in a dielectric medium
with dielectric constant ε_d, which is incident at an interface
with a metal described by a local dielectric function $\tilde{\varepsilon}_m(\omega)$. The
component of the wave vector parallel to the surface $\underline{k}_\parallel$ is conserved (if we ignore the effects of the periodicity of the metal
lattice) and is real, i.e. $\tilde{k}_{m\parallel} = k_{d\parallel}$. Inside the metal the component of the wave vector perpendicular to the surface, $k_{m\perp}$, is given
by the formula

$$\tilde{k}_{m\perp}^2 = \frac{\omega^2}{c^2} \tilde{\varepsilon}_m(\omega) - k_{d\parallel}^2 \quad . \tag{1}$$

Since the real part of $\tilde{\varepsilon}_m(\omega)$ is negative (as long as ω is smaller
than the plasmon frequency of the metal ω_p), $\tilde{k}_{m\perp}$ is effectively

imaginary (i.e., $\tilde{k}_{m\perp} = i\alpha_m$) and the transmitted EM wave in the metal
is evanescent.

The surface can also support surface-electromagnetic (S-EM)
waves at frequencies where $\varepsilon'_m(\omega) < -\varepsilon_d$. The dispersion relation for
an S-EM is given by[28]

$$\tilde{k}_{s//}^2 = \frac{\omega^2}{c^2} \varepsilon_d \left(\frac{\tilde{\varepsilon}_m(\omega)}{\tilde{\varepsilon}_m(\omega) + \varepsilon_d} \right) \quad . \tag{2}$$

The surface plasmon frequency ω_{sp} is given by $\varepsilon'_m(\omega_{sp}) = -\varepsilon_d$.
At frequencies $\omega < \omega_{sp}$ (where $\varepsilon'_m(\omega) < -\varepsilon_d$, and, therefore $k'_{s//} \leq$
$\omega \varepsilon_d^{1/2}/c$) the perpendicular component of the wave vector for the
S-EM wave inside the metal, and also for the transmitted evane-
scent wave (for which $k'_{m\perp} < \omega \varepsilon_d^{1/2}/c$), is given by

$$\tilde{k}_{m\perp}^2 = \frac{\omega^2}{c^2} \tilde{\varepsilon}_m(\omega) \quad . \tag{3}$$

Thus the penetrating depth into the metal, $\delta = 1/k''_{m\perp}$, is essentially the same for the S-EM wave, as for the transmitted evanescent wave.

When a S-EM wave is excited by a p-polarized EM wave in
a prism-coupling configuration, the field at the surface of a metal,
such as Ag, can be appreciably enhanced relative to that resulting
from an incoming EM wave in the absence of a prism[28]. The spatial
variation of the field inside the metal is, nevertheless, the same
in the two cases.

In the general case of a metal characterized by a non-local
dielectric function, $\tilde{\varepsilon}(\underline{x}, \underline{x}', \omega)$, there will be strong effects due
to electron-hole excitations by the field component normal to
the surface. They give rise to sizeable oscillations in the fields
inside the metal (Fig. 1a and b). Since $E_m/E_d = \varepsilon_d/\tilde{\varepsilon}_m(\omega)$ independent of the nature of the EM waves at the interface, the character
of the oscillations will be the same for both S-EM waves and for
p-polarized evanescent waves.

A characteristic feature of the non-local theory is that one
encounters large differences in the response of the metal to s-
and p-polarized EM radiation. In the case of s-polarized incident
EM radiation, the EM field inside the metal is transverse. Thus,
although the wave vector of the EM field inside the metal does
have Fourier components large enough to excite electron-hole pairs
(since the perpendicular component of the wave vector is not con-

served at the surface), the $\underline{p} \cdot \underline{A}$ momentum matrix element for the
excitation of electron-hole pairs by transverse EM fields is neg-
ligibly small[34].

   In the case of p-polarized EM radiation and surface-EM waves,
the situation is qualitatively different, since the EM fields in-
side the metal have sizeable longitudinal components, e.g. $\nabla \cdot \underline{A} \neq 0$
(where we are using a gauge in which the scalar potential is $\underline{z}$ero).
As a consequnce, there is an appreciable excitation of virtual and
real electron-hole pairs via the relatively strong $\nabla \cdot \underline{A}$ (Coulomb-
interaction) matrix element. The electron-hole pair excitations
produce strong current and charge oscillations that give rise to
characteristic oscillations of the EM field in the surface region
of the metal, which typically extend up to ~100Å from the surface
(Fig. 1).

   The non-local dielectric response has been discussed by several
authors[24-27] using the jellium model. They used different theoretical approaches and made detailed calculations of the fields,
charge oscillations and power absorption for different surface profiles. In the formulations based on the jellium model, the contributions from the wavevector associated with the periodic lattice have
been ignored. However, they are very likely to be small for free-
electron like metals, where interband excitations are weak. The non-
local response theory provides the underlying basis for such phenomena as photoemission[29], reflectivity[30], image dipole effects[31],
etc. The different theoretical results agree in all qualitative
features. However, recent work has shown the importance of using a
self-consistent surface profile and wavefunctions of the forms
proposed by Lang and Kohn[32]. Thus, the use of the Lang-Kohn profile yields more satisfactory agreement with, for example, photoemission data than the step barrier model and other models.[29]

   It is the enhanced excitation and recombination of electron-hole
pairs which is responsible for the observed strong inelastic scattering of light by metals with rough surfaces. We shall therefore give
a brief discussion of the Raman scattering and luminescence by electron-hole pairs in a metal with a smooth surface and also make some
comments about the role of electron-hole pairs in the Raman scattering by molecules adsorbed on a smooth metal surface.

   The (Stokes) Raman scattering by the charge carrier excitations in the metal involves the annihilation of an incident photon,
the creation of a scattered photon and the creation of a real
"intraband" electron-hole pair, i.e., $\omega_{eh} = \omega_i - \omega_s$. The photons
inside the metal are dressed with the polarization associated with
the charge carrier excitations and correspond to polaritons. The
Raman scattering process can be viewed in terms of the incident
polariton and the scattered polariton together constituting a
force that creates the real electron-hole pair. In the case of a

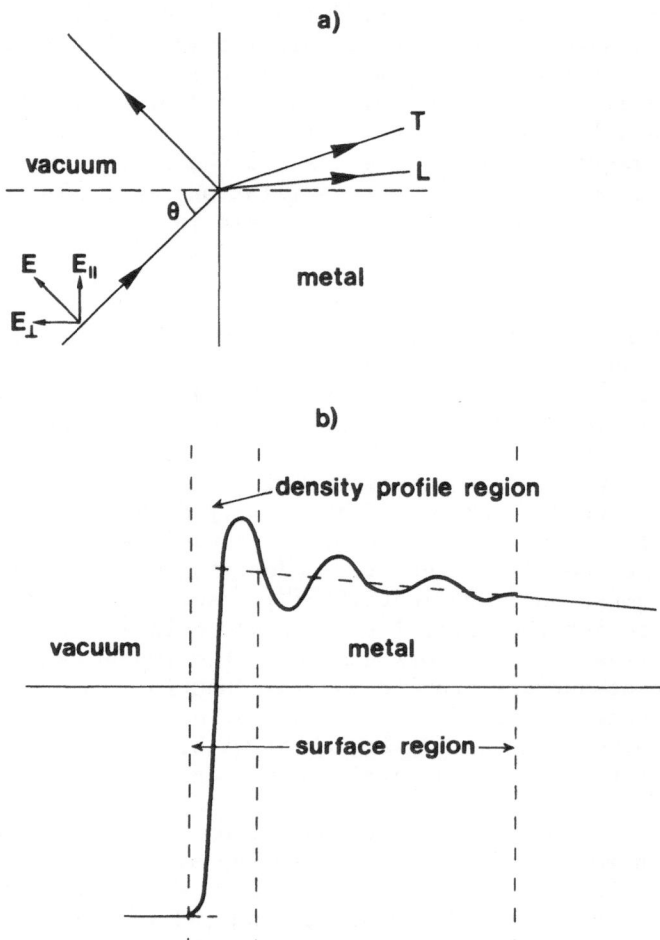

Fig. 1. a) Schematic diagram showing the incident, reflected, and
         transmitted (Transverse and Longitudinal) waves at a metal-
         vacuum interface for the case of p-polarized incident EM
         radiation. b) Schematic diagram showing the variation of
         the real part of the perpendicular component of the elec-
         tric field in the vicinity of the surface metal.

smooth surface $\underline{k}_{/\!/}$ is conserved because of translational invariance, i.e., $\underline{q}_{eh/\!/} = \underline{k}_{i_{/\!/}} - \underline{k}_{s/\!/}$.

There are three basic mechanisms for Raman scattering by electron-hole pair excitations: an A^2 mechanism, a $(\underline{p}\cdot\underline{A})^2$ mechanism and a $(\nabla\cdot\underline{A})^2$ (or Coulomb scattering) mechanism. It should be emphasized that the A^2 scattering is a one-step process, whereas the $(\underline{p}\cdot\underline{A})^2$ and $(\nabla\cdot\underline{A})^2$ scattering are two-step Raman processes in which each step involves the <u>virtual</u> excitation or recombination of an electron-hole pair, which may involve either an intraband or interband transition (Fig. 2a, b and c). The virtual electron-hole pairs remain coupled to photons and the inelastic scattering process therefore retains phase coherence between the incident and scattered photons. The Raman scattering process generates charge density fluctuations which are strongly screened at optical frequencies,

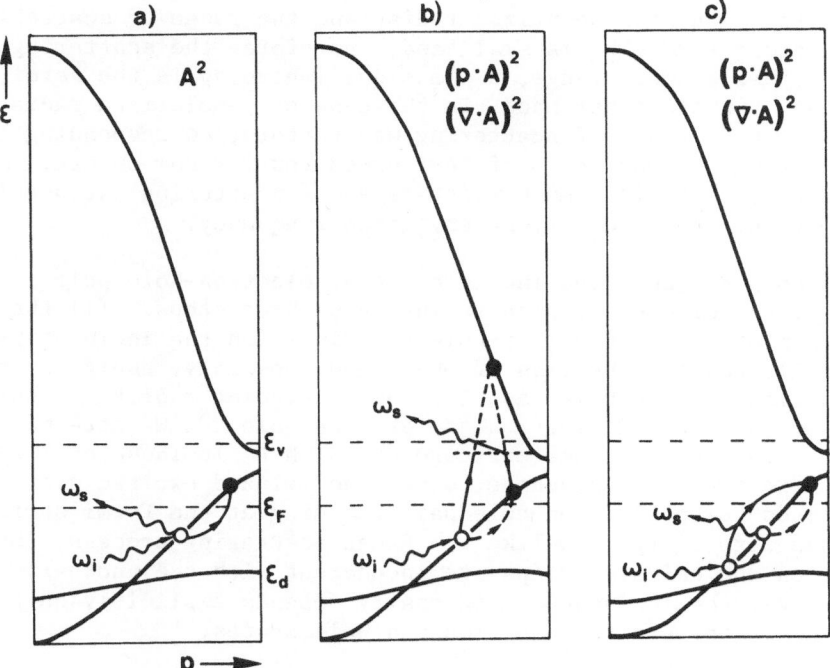

Fig. 2. Schematic diagram of the optical and electronic processes involved in the Raman scattering by electron-hole pair excitations in a "free electron" metal. a) one-step A^2 mechanism; b) two-step $(\underline{p}\cdot\underline{A})^2$ and $(\nabla\cdot\underline{A})^2$ interband mechanisms; c) two-step $(p\cdot A)^2$ and $(\nabla\cdot A)^2$ intraband mechanisms.

i.e., the scattering intensity is proportional to $Im[1/\tilde{\varepsilon}_m(\omega)]$, and the polarization selection rules for all three mechanisms are that $\underline{e}_i \cdot \underline{e}_s = 0$, where $\underline{e}_i$ and $\underline{e}_s$ are the polarization unit vectors of the incident and scattered radiation.[33]

In the case of s-polarized radiation where $\nabla \cdot \underline{A} = 0$, only the $(\underline{p} \cdot \underline{A})^2$ and the A^2 processes will contribute to the scattering. For free electron metals such as Al or Ag at frequencies away from d-band transitions, the $(\underline{p} \cdot \underline{A})^2$ and the A^2 scattering contributions are weak. The intraband and interband $\underline{p} \cdot \underline{A}$ matrix elements are weak in a "free electron"-like metal and the A^2 matrix element is generally weak.

In the case of p-polarized EM (or S-EM) radiation, the $(\underline{p} \cdot \underline{A})^2$ and A^2 mechanisms similarly make only weak contributions to the Raman scattering. The $\nabla \cdot \underline{A}$ matrix element for intraband transitions is appreciably stronger than the corresponding $\underline{p} \cdot \underline{A}$ matrix element[34] and, as has recently been shown, the same is also true for interband transitions in a free electron metal.[29] Thus the $(\nabla \cdot \underline{A})^2$ interband and intraband mechanism makes the dominant contributions.

In the case of s-polarized radiation, the range of scattering wavevectors $\underline{q} = \underline{k}_i - \underline{k}_s$ is small and, therefore, the scattering frequency has a small range, $0 < \omega < qv_F$ (where v_F is the Fermi velocity). On the other hand, in the case of p-polarized radiation, there is a wide range of scattering wavevectors, corresponding to the wide range of surface profile-induced Fourier components. As a consequence, there is a much wider range of scattering frequencies and a much higher intensity at any given frequency.

We consider next the luminescence by electron-hole pair excitations. The luminescence process involves three steps: (1) the excitation of a real electron-hole pair in which the incident photon is annihilated, (2) the loss of phase and, possibly, energy by the excited electron and hole, and (3) the subsequent radiative annihilation of the excited electron and excited hole.[35] We note moreover that the luminescence corresponds to "hot" luminescence, rather than to "relaxed" luminescence, since the relaxed excited state corresponds to an electron-hole pair residing at the Fermi surface, which has zero energy. Unlike the Raman scattering process, the excitation and emission steps are incoherent with one another. Furthermore, the luminescence intensity depends explicitly on the lifetimes of the excited electron and hole states.

Considerations with regard to the role of the polarization of the incident and emitted light are similar to those for Raman scattering. Namely, since electron-hole pairs are not produced by s-polarized incident light, there will only be an excitation of luminescence by p-polarized radiation (or by a S-EM wave). This also applies to the luminescence emission, since the electron-hole pairs can only recombine to create p-polarized radiation (or a S-EM wave which can be converted into EM radiation by a prism).

In frequency regions where d $\rightarrow$ sp interband transitions do not occur, the dominant excitation and recombination transitions, for both Raman scattering and luminescence by electron-hole excitations, will take place via $\nabla \cdot \underline{A}$ matrix elements. This will not be true when d$\rightarrow$sp interband transitions are involved, since for these the $\underline{p} \cdot A$ matrix elements are the dominant ones

We shall now briefly consider the Raman scattering by a molecule at a smooth surface. The electronic structure of the molecule will in general be changed by interaction with the surface. The electron structure of the adsorbed molecule is generally not known. (It can be determined only after detailed experimental and theoretical analysis in each individual case.) We shall therefore be concerned with other aspects and, in particular, with the microscopic electric field at the position of the molecule.

The field at the molecule is made up of several contributions. One is the image dipole field, which occurs because the electric dipole of the excited molecule couples with the collective oscillations of the electrons in the metal. It gives rise to a shift and broadening of the electronic excitations of the adsorbed molecule, which is usually described as a self-energy effect[41,42]. In classical theory, the image dipole field is due to induced charges on the metal surface. In a non-local theory the image dipole field will also include interactions with electron-hole pair excitations. As a result the induced charge distributions will now extend into the metal with a depth corresponding to the non-locality range, which is of the order 50-100 Å[27]. Moreover, the creation of real electron-hole pairs leads to damping of the molecular excitation.

The excitation of electron-hole pairs in the metal by the EM field also modifies the field at the molecule. One aspect may be described as a change in the strength of the EM field arising from the effect of electron-hole pair excitations on the reflectivity of the metal. More interesting is the direct effect of the near field of the charge fluctuations associated with virtual excitations of electron-hole pairs[8]. These localized charge fluctuations will interact strongly with the molecule via their Coulomb field. There are also single-particle aspects of the electron-hole pairs that may be important. An excited electron which tunnels (e.g. penetrates beyond) the surface barrier may be scattered by the field of the vibrating charges of the molecule back into a different excited state and recombine with the hole. Of special interest also is the Raman scattering process where the electron-hole pair excitation involves a charge transfer to the molecule. For weak coupling the charge transfer between the metal and the molecule is a small effect, and the interaction may be described as hopping of electrons

between molecule and metal. For stronger coupling the charge transfer will take place via a strong admixture of molecular orbitals
and electron states in the metal, and this requires a special treatment in each case.

The Raman scattering processes can be divided into four groups:
i) direct interaction of the molecule with the incident and scattered radiation fields, whose strengths are modified by the presence of the surface, ii) indirect interaction via the metal, which
can be described as the interaction between virtual e-h pairs, or
surface plasmons in the metal, with virtual excitations of the molecule, iii) a three-step process in which the excited electron
(hole) tunnels through the surface barrier and is scattered by the
molecular vibration into a new excited state before recombining, and
iv) charge transfer processes in which the virtually excited
electron (hole) is partially localized on the molecule.

Finally, we make some brief remarks about Raman scattering
processes involving d→sp interband transitions which can be excited
by both s- and p-polarized radiation. In the case of s-polarized
light, the range of q values is small, i.e., the interband transitions are esentially vertical. On the other hand, because of the
large q components for p-polarized light, the interband excitations will involve a wide range of momentum transfer, which will
give a higher production of electron-hole pairs having a given
energy. For both s- and p-polarizations the contributions from the
allowed interband transitions to the imaginary part of the dielectric
function is large and therefore the ratio of the imaginary to the
real parts of the dielectric function of the metal will be very much
larger than that encountered in the absence of d→sp interband
transitions.

III.   THE CHARACTER OF ROUGH SURFACES

The random roughness of a flat continuous metal surface can
be characterized geometrically as a random superposition of diffraction gratings having a distribution of spacings (e.g. wave
vectors), amplitudes, orientations and phases[36]. If we let $\xi(\underline{x}_{//})$
be the height of the surface above a point $\underline{x}_{//}$ in a plane parallel
to the average surface we may write

$$\xi(\underline{x}_{//}) = \sum_{\underline{Q}_{//}} \xi(\underline{Q}_{//}) \exp (i\underline{Q}_{//} \cdot x)$$

where $\xi(\underline{Q}_{//})$ is the amplitude of the grating (i.e. spatial Fourier
component) with wave vector $\underline{Q}_{//}$ parallel to the surface.

The gratings serve to couple EM radiation to the surface-
EM modes at the metal-dielectric interface and thereby lead to an

enhancement of the incident and scattered EM fields. The grating
of wave vector $\underline{Q}_{//}$ will "mix" an EM wave with a wave vector $k_{//}$
with the surface-EM mode of wave vector $\underline{k}_{s//} = \underline{k}_{//} + \underline{Q}_{//}$. When the
superposition includes gratings with large wave vectors, the grat-
ings also serve to couple the EM radiation with electron-hole pair
excitations in the metal. Moreover, because of the random orientations of the gratings, the distinction between s- and p-polarized
radiation tends to disappear.

The other aspect of the rough metal surface, that needs to be
taken into account, is the fact that there are geometrical structures (e.g. bumps, pits) that act to "confine" the motion of the
electrons associated with them[37]. The collective oscillations of
these electrons exhibit resonances that correspond to "localized
plasmons". Moreover, the resonance frequencies for oscillations
parallel and perpendicular to the plane of the surface are in
general appreciably different. The localized plasma oscillations
of neighbouring structures are coupled to one another via the
Coulomb fields of their electric dipole moments. They are radiative
modes which interact with the incident and scattered EM radiation
and with surface EM waves. Specifically the "localized plasmons"
which involve electron motion parallel to the surface can couple
with both s- and p-polarized EM radiation, whereas the "localized
plasmons" which involve electron motion perpendicular to the
surface can only couple with p-polarized EM radiation. Both types
of surface plasmons can couple with S-EM waves. However, the interaction of the perpendicular (localized plasma) oscillation with
S-EM waves is appreciably stronger than that of the parallel
oscillations.

It is now well established that, in the Raman scattering of
molecules at rough metal surfaces, the excitation of the localized
plasmons leads to an enhancement of the incident and scattered EM
fields. The theoretical formulation of the field enhancement by
localized plasmons has been carried out by several investigators
using the classical (local) dielectric constant $\varepsilon(\omega)$ for the
metal.[15-18]

We note that there are several characteristic lengths that
determine the properties of surfaces that are rough in a statistically random fashion. One is the rms deviation δ of the surface
from perfect flatness. Another is the transverse correlation length
a, which may be thought of as the mean distance between adjacent
peaks (or "valleys") on the rough surface. Even if the surface is
not rough in a statistical fashion but instead contains regular
features, we may use δ and a to characterize their height and their
transverse scale in an average sense. In this regard, the coupling
of EM waves with surface-EM modes is strongest for frequencies
where the surface-EM modes have vectors $k_{s//} \approx 1/a$. In experimental
studies of the reflectivity of roughened surfaces one sees a pronounced reflectivity dip at these frequencies.

The lateral dimension b of the base of the surface structures
(e.g. bumps), which confine the collective motions of the electrons,
is a third characteristic length. The parallel and perpendicular
resonance frequencies of the "localized plasmas" are determined by
δ/b. When $\delta/b > 1$, the parallel resonance frequency $\omega_{R\parallel}$ will be
greater than the perpendicular resonance frequency $\omega_{R\perp}$ and the
reverse will be true when $\delta/b < 1$.

There is a fourth length that is relevant, namely the wavelength λ of the EM radiation. For many of the surface roughness
geometries of interest the wavelength is very much larger than the
other characteristic lengths, i.e. from the point of view of the
surface λ appears to be infinite. In this limit the mode structure
which controls the EM response of the rough surface depends essentially on the magnitudes of δ and a, b. In the present discussion
we also suppose that δ, a, b are sufficiently large that the substrate may be described as a dielectric continuum characterized by
a continuous dielectric constant $\widetilde{\varepsilon}(\omega, \underline{k})$, i.e. we ignore the
"quantization" of the confined electron motions in the surface
structure.

In the limit $\delta << a,b$, the surface has a gradual modulating
character, and the deviation from perfect flatness may be regarded
as a small perturbation on the properties of an otherwise smooth
interface. The presence of roughness will shift the dispersion
curve of the surface-EM modes towards larger wave vectors, particularly at frequencies near that of the surface plasmon, but
these effects will be relatively small when δ/a and δ/b are small.

When δ, a and/or b are comparable in magnitude, the rough
surface will exhibit a very broad spectrum of localized (i.e.
confined) collective electron resonances whose form is determined
by the range of shapes of the localized surface structures and
by the dipolar interactions of the localized plasma excitations
of neighbouring structures. Associated with the resonances of the
localized structures there will be enhanced electromagnetic fields
in their vicinity and also within the structures themselves. As a
consequence of the surface roughness, there will also be large
shifts in the dispersion curves of the surface-EM modes and also
a large decrease in their lifetimes (or propagation lengths).

It is interesting to note that Ruppin[40] has calculated the
electric fields near a hemispherical bump on the Ag surface. He
finds a series of electromagnetic resonances of the enhanced fields
for both s- and p-polarization. We may understand the origin of
these resonances on the following basis. The hemisphere possesses
no spherical symmetry. As a result of its axial symmetry the electrostatic potential associated with the multipole modes varies as

$\exp(\pm im\phi)$, where ϕ is the azimuthal angle and m is an integer. All modes with m=0 may be excited by an electric field normal to the surface. The modes with $m = \pm 1$ are excited by field components parallel to the surface and are therefore active in both s- and p-polarization. Modes with $|m| > 1$ will not couple to the incident radiation field when $\lambda \gg R$.

We note also that for both s- and p-polarized EM radiation there will be electric field components that are normal to some parts of the surface of the localized structure. This means that for both polarizations there will be sizeable magnitudes of $\nabla \cdot \underline{A}$ in the surface region of the localized structures, and that for both polarizations there will be an appreciable excitation and recombination of electron-hole pairs within the localized structure.

Metal island films represent another form of surface roughness in which the electrons are confined within metal islands (e.g. particles) and whose response to EM fields is determined predominantly by the localized plasma resonances of the metal islands.

The grain boundaries (e.g. internal structure) of the microcrystalline metal film can also confine electrons and thereby exhibit localized plasma resonances. Thus a Ag film, prepared by evaporation onto a substrate held at 240 K has a very fine microcrystalline structure and exhibits a broad absorption band, which is similar to that observed for a Ag island film[38]. The absorption band disappears when the film is annealed (by warming to room temperature) and the grain boundaries become macroscopic in size.

In the typical case of a 50Å (mass thickness) Ag island film, the oblique incidence absorption spectrum exhibits a fairly broad absorption band at $\sim$ 7500Å due to the excitation of the "parallel" resonances of the film, whose frequencies depend sensitively on the magnitude of δ/a[39]. The spectrum also shows a narrow absorption band due to the excitation of the "perpendicular" resonances of the film, which have frequencies close to ω_{sp} that are insensitive to the magnitude of δ/a when δ/a is small. On the other hand, in the case of Cu (and Au) island films, the oblique incidence spectra only exhibit a broad absorption band due to the excitation of the "parallel" resonances of the film, whose frequencies lie below the d → sp interband transitions. The perpendicular resonances of the film, whose frequencies lie above the d → sp interband transition are strongly damped and do not manifest themselves in the absorption spectrum.

IV.   ELECTRON-HOLE PAIRS AT ROUGH SURFACES

It is evident that surface roughness plays <u>several</u> important roles in surface enhanced Raman scattering. It introduces new radiative EM modes at the surface, i.e. the localized plasmons, whose excitations lead to an enhancement of both the incident and scattered EM fields. It provides a mechanism for coupling the incident and scattered radiation to the non-radiative surface-EM modes, which also enhances the EM fields at the surface. It also leads to an enhanced radiative excitation and recombination of electron-hole pairs in the metal by both s- and p-polarized EM radiation.

We now take up the role played by electron-hole pair excitations in the enhancement of the EM fields and also their own role in the Raman scattering and luminescence by the metal. There is, as yet, no theoretical analysis of the non-local electromagnetic properties of localized structures. For simplicity, we consider the case of a spheroidal particle, whose dimensions are much smaller than the wavelength (or penetration depth) of the electromagnetic radiation interacting with it. To begin with, we expect that the field inside the localized surface structure will be modulated and show characteristic oscillations similar to those for a smooth surface.

The localized structure will exhibit a collective resonance with a resonance frequency determined certainly by its shape and possibly also by its size, which will lie somewhere near that calculated using a local $\varepsilon(\omega)$. The field inside the localized structure calculated in a non-local theory will show characteristic modulations along the direction of the polarization, as in the case of a smooth surface. For a sizeable particle (e.g. ≥ 200Å) the oscillations of the field which arise from the excitation and recombination of electron-hole pairs will occur in the vicinity of the parts of the surface on opposite sides of the structure where there are normal components of the field. For much smaller particles (< 100Å), the oscillating behaviour will reach across the particle. The contributions from opposite sides of the particle will superpose and interfere. At resonance the average field inside the particle will be of the same order of magnitude as that using a local dielectric constant $\tilde{\varepsilon}(\omega)$. If anything, we may expect it to be somewhat smaller because of the damping due to excitation of electron-hole pairs.

With this as a basis, we can now discuss the inelastic light scattering by the electron-hole pair excitations in the metal particle and the effect of the electron-hole pair excitations on the Raman scattering by molecules adsorbed on the particle. For the latter, we consider the molecules in the vicinity of those parts of the surface having the maximum magnitudes of the normal components of the field, since they make the major contribution to the scattering.

As in the case of a smooth surface the surface profile introduces longitudinal fields that have high Fourier components and
leads to a production of virtual and real electron-hole pairs. The
fields are the same for the two cases, provided that the dimensions
of the particle are sufficiently larger than the non-locality
length (~ 100Å). There is therefore an enhancement in the production of electron-hole pairs relative to a smooth surface, which is
essentially equal to the enhancement of the average field. This is
one of the factors that contributes to the enhancement of the inelastic scattering by the metal particle.

We now consider the radiation emitted by virtual or real
electron-hole pairs. Inside the particle within ~ 10Å from the
surface there will be an enhancement of the microscopic field
relative to the average by as much as a factor of two. The average
field is enhanced above that for a smooth surface by a factor
approximately given by the local theory. We note moreover that the
ratio of the normal components of the fields inside and outside
the metal particle at the regions of the surface where the polarization is normal to the surface is the same as that for a smooth
surface. Namely, it is given by the continuity of the normal components of the displacement. Again as a consequence of the surface
profile for the normal component of the electric field, as in the
case of the smooth surface, the electron-hole pair radiates into a
polariton near the surface (whose polarization has a component normal to the surface) which radiates out of the particle. The electron-hole pair at the same time excites the dipolar localized plasmon
via the Coulomb interaction, thereby setting up an effective dipole
moment within the particle. The latter is very much larger than that
of the e-h pair and, as a consequence, there is an appreciable enhancement of the scattered electromagnetic radiation. Thus, as in
the case of molecules adsorbed on a spheroidal particle, the enhancement of the inelastic scattering by the metal particle arises from
both an enhancement of the scattered and the incident electromagnetic
fields.

In the Raman scattering of a molecule adsorbed on the particle,
there are similar considerations. For the direct Raman scattering
processes there will be, as in the case of a local dielectric theory,
an enhanced field which arises from the induced electric moment in
the particle and from the near Coulomb field of the electron-hole
pairs that are excited by the incident radiation. (We do not address
the question of the contribution to the self-energy from the image
dipoles.) For those Raman scattering mechanisms which involve the
scattering of electrons or holes or the transfer of electrons (or
holes) from the metal to the molecule, it is the excitation of
electron-hole pairs, whose enhancement determines the Raman scattering intensity. The enhancement of the scattered radiation follows
similar arguments, namely, that the excited molecule induces an
electric moment in the particle by exciting a localized plasmon.

In the charge-transfer mechanism, where electrons (or holes) hop back into the metal, they also act to induce a dipole moment in the particle, which adds to the scattered field.

IV. CONCLUDING REMARKS

The surface microstructures (e.g. gratings, arrays of bumps or particles) can be used to enhance the input and/or output EM fields in a wide range of linear and nonlinear optical phenomena at the surface of metals or, for that matter, at the surface of any medium having a dielectric constant at the frequencies of interest, whose real part is negative and whose imaginary part is relatively small. Thus they can be used at the surface of semiconductors and semimetals (at frequencies below the plasmon frequency of the free carriers) and at the surface of dielectrics, e.g. $BaTiO_3$, (in the frequency region between the LO and TO phonon frequencies). In the latter case, the "localized excitations" correspond to "localized phonons". We note in this connection that the "localized excitations" correspond to longitudinal photons (or polaritons).

Microstructure amplifiers have, in fact, already been used in one form or another to enhance a variety of phenomena. These include: IR absorption by molecules (Ag island film[43]); second harmonic generation by Ag surfaces (Ag electrode,[44] Ag island film[45]); photoemission of a Ag-O-Cs photo-cathode (Ag grating[46]); optical absorption in the visible by dye molecules (Ag island film[47]); light emission by inelastic electron tunneling (Au and Ag island films[11,12] and Ag grating[48]); and luminescence and resonant Raman scattering by dye molecules (Ag island film[49] and axially symmetric bumps on Ag[22]). It has also recently been pointed out that photochemical processes may also be enhanced by surface microstructures.[50]

It should be noted that e-h pair excitations play a vital role in a number of optical phenomena and that, for these, the surface microstructures which exhibit localized plasmons (e.g. arrays of bumps or particles) are particularly advantageous. The phenomena in which e-h pairs play a key role obviously include photoemission where the enhancement of the radiative excitation of e-h pairs greatly enhances the photoelectron yield per unit incident intensity. Luminescence by adsorbed molecules is another phenomenon in which e-h pairs play an important role. Thus, the non-radiative transfer of energy, via Coulomb interaction, from e-h pairs (that are generated in the metal by the incident radiation) to the molecules can serve as an important "channel" for the indirect excitation of the adsorbed molecules.[51] This should be particularly important for phosphorescent molecules, whose optical transitions between the ground and the relevant excited electronic levels are normally forbidden. The corresponding non-radiative transfer of excitation from the excited molecules to e-h pair excitations in the localized structure in turn provides a channel for enhanced emission.

The enhanced light emission by inelastic electron tunneling in Al/Al oxide/metal structures, involving Au (or Ag) island films as the metal overlayer, is another effect in which the excitation and/or recombination of e-h pairs may play a role. The enhancement of the light emission has been attributed to the role played by the Au particle as an antenna for the otherwise non-radiative surface EM-modes in the Al oxide tunnel barrier that are generated by the tunneling electrons.[52] It has also been suggested[20] that the tunneling electrons may actually excite the "localized plasmons" of the Au particles directly. It is possible, on the other hand, that the tunneling electrons (or holes) may either themselves recombine radiatively, or excite e-h pairs within the Au particle and, via the excitation of the localized plasmon of the Au particle, contribute to the wavelength and polarization dependence of the light emission.

ACKNOWLEDGEMENT·

We are indebted to George Ritchie and Dan Whittle of the University of Pennsylvania and Peter Apell and Peter Ahlqvist of Chalmers University of Technology for valuable discussions.

REFERENCES[53]

[§]Guggenheim Fellow (1980-81) and Chalmers University of Technology One Hundred and Fifthieth Anniversary Professor of Physics (Spring 1981).

[+]Research supported in part by ONR and NSF.

[++]Research supported in part by AFOSR.

1. R. W. Wood, A suspected case of the electrical resonance of minute metal particles for light-waves. A new type of absorption, Philos. Mag. & J. Sci. 3:396 (1902). See P. Rouard and A. Meessen, Optical properties of thin metal films, in: "Progress in Optics," Vol. XV, E. Wolf, ed., North-Holland, Amsterdam (1977), p. 79.

2. D. Beaglehole and O. Hunderi, Study of the interaction of light with rough metal surfaces. I. Experiment, Phys. Rev. B 2:309 (1970); O. Hunderi and D. Beaglehole, Study of the interaction of light with rough metal surfaces. II. Theory, Phys. Rev. B 2:321 (1970).

3. E. Kretschmann, Streuung von Licht an rauhen Oberflächen bei der Anregung von Oberflächenplasmonen, Z. Physik 227:412 (1969).

4. J. G. Endriz and W. E. Spicer, Study of aluminum films. II. Photoemission studies of surface-plasmon oscillations on controlled-roughness films, Phys. Rev. B 4:4159 (1971).

5. J. Rudnick and E. A. Stern, Second harmonic radiation from a
    metal surface, in: "Polaritons, Proceedings of the First
    Taormina Research Conference on the Structure of Matter,"
    E. Burstein and F. De Martini, eds., Pergamon Press, New
    York (1974), p. 329.
6. D. L. Jeanmaire and R. P. Van Duyne, Surface Raman spectro-
    electrochemistry. Part I. Heterocyclic, aromatic and
    aliphatic amines adsorbed on the anodized silver electrode,
    J. Electroanal. Chem. 84:1 (1977).
7. M. G. Albrecht and J. A. Creighton, Anomalously intense Raman
    spectra of pyridine at a silver electrode, J. Amer. Chem.
    Soc. 99:5215 (1977).
8. E. Burstein, Y. J. Chen, C. Y. Chen, S. Lundquist, and
    E. Tosatti, "Giant" Raman scattering by adsorbed molecules
    on metal surfaces, Solid State Commun. 29:567 (1979).
9. C. Y. Chen, E. Burstein and S. Lundquist, Giant Raman scat-
    tering by pyridine and CN⁻ adsorbed on silver, Solid State
    Commun. 32:63 (1979).
10. A. Otto, Raman scattering from adsorbates on silver, in:
    "Proceedings of the Conference on Vibrations in Adsorbed
    Layers," Jülich, 1978 (unpublished); Surf. Sci. 92:145
    (1980).
11. J. Lambe and S. L. McCarthy, Light emission from inelastic
    electron tunneling, Phys. Rev. Lett. 37:923 (1976).
12. P. K. Hansma and H. P. Broida, Light emission from gold
    particles excited by electron tunneling, Appl. Phys. Lett.
    32:545 (1978).
13. E. Burstein, C. Y. Chen, and S. Lundquist, Giant Raman scat-
    tering by molecules adsorbed on metals: An overview, in:
    "Proceedings of the USA-USSR Symposium on Light Scattering
    in Solids," J. L. Birman, H. Z. Cummins, and K. K. Rebane,
    eds., Plenum Press, New York (1979), p. 479.
14. J. I. Gersten, The effect of surface roughness on surface en-
    hanced Raman scattering, J. Chem. Phys. 72:5779 (1980);
    Rayleigh, Mie, and Raman scattering by molecules adsorbed
    on rough surfaces, J. Chem. Phys. 72:5780 (1980).
15. S. L. McCall, P. M. Platzman, and P. A. Wolff, Surface en-
    hanced Raman scattering, Phys. Lett. 77A:381 (1980).
16. J. Gersten and A. Nitzan, Electromagnetic theory of enhanced
    Raman scattering by molecules adsorbed on rough surfaces,
    J. Chem. Phys. 73:3023 (1980).
17. D.-S. Wang, H. Chew, and M. Kerker, Enhanced Raman scattering
    at the surface (SERS) of a spherical particle, Appl. Opt.
    19:2256 (1980).
18. C. Y. Chen and E. Burstein, Giant Raman scattering by mole-
    cules at metal-island films, Phys. Rev. Lett. 45:1287 (1980).
19. J. A. Creighton, C. G. Blatchford, and M. G. Albrecht, Plasma
    resonance enhancement of Raman scattering by pyridine ad-
    sorbed on silver and gold solid particles of size com-
    parable to the excitation wavelength, J. Chem. Soc. Faraday

II 75:790 (1979).

20. E. Burstein and C. Y. Chen, Raman scattering by molecules adsorbed at metal surfaces. The role of surface roughness, in: "Proceedings of the VIIth International Conference on Raman Spectroscopy," Ottawa 1980, W. F. Murphy, ed., North-Holland, Amsterdam (1980), p. 346.

21. S. S. Jha, J. R. Kirtley, and J. C. Tsang, Intensity of Raman scattering from molecules adsorbed on a metallic grating, Phys. Rev. B 22:3973 (1980).

22. G. Ritchie, E. Burstein, and R. B. Stephens, Secondary light emission by molecules at metal surfaces with axially symmetric bumps, Bull. Amer. Phys. Soc. 26:359 (1981).

23. P. F. Liao, J. G. Bergman, D. S. Chemla, A. Wokaun, J. Melngailis, A. M. Hawryluk, and N. P. Economou, Surface enhanced Raman scattering from microlithographic silver particle surfaces, Chem. Phys. Lett. (to be published).

24. P. J. Feibelman, Microscopic calculation of electromagnetic fields in refraction at a jellium-vacuum interface, Phys. Rev. B 12:1319 (1975).

25. K. L. Kliewer, Surface photoeffect for metals, Phys. Rev. B 14:1412 (1976).

26. G. Mukhopadhyay and S. Lundqvist, The electromagnetic field near a metal surface, Phys. Scripta 17:69 (1978).

27. P. Apell, Electromagnetic field near a metal surface in the classical infinite barrier model, Phys. Scripta 17:535 (1978).

28. E. Burstein, W. P. Chen, Y. J. Chen, and A. Hartstein, Surface polaritons--propagating electromagnetic modes at interfaces, J. Vac. Sci. Technol. 11:1004 (1974).

29. H. J. Levinson, E. W. Plummer, and P. J. Feibelman, Effects on photoemission of the spatially varying photon field at a metal surface, Phys. Rev. Lett. 43:952 (1979).

30. P. Apell, A simple derivation of the surface contribution to the reflectivity of a metal and its use in the van der Waals interaction, Phys. Scripta (in press).

31. P. J. Feibelman, Local field at an irradiated adatom on jellium--exact microscopic results, Phys. Rev. B 22:3654 (1980).

32. N. D. Lang and W. Kohn, Theory of metal surfaces: Induced surface charge and image potential, Phys. Rev. B 7:3541 (1973).

33. P. M. Platzman and P. A. Wolff, "Waves and Interactions in Solid State Plasmas," Solid State Physics Series, H. Ehrenreich, F. Seitz, and D. Turnbull, eds., Academic Press, New York (1973).

34. E. Burstein and A. Pinczuk, Light scattering by collective excitations in dielectrics and semiconductors, in: "The Physics of Opto-Electronic Materials," W. A. Albers, Jr., ed., Plenum Press, New York (1971) p. 33.

35. K. Rebane and P. Saari, Hot luminescence and relaxation

processes in resonant secondary emission of solid matter, J. Lumin. 16:223 (1978).

36. J. M. Elson and R. H. Ritchie, Photon interactions at a rough metal surface, Phys. Rev. B 4:4129 (1971). Diffuse scattering and surface-plasmon generation by photons at a rough dielectric surface, Phys. Status Solidi B 62:461 (1974).

37. D. W. Berreman, Anomalous reststrahl structure from slight surface roughness, Phys. Rev. 163:855 (1967).

38. O. Hunderi and H. P. Myers, The optical absorption in partially disordered silver films, J. Phys. F.: Metal Phys. 3:683 (1973).

39. T. Yamaguchi, S. Yoshida, and A. Kinbara, Optical effect of the substrate on the anomalous absorption of aggregated silver films, Thin Solid Films 21:173 (1974).

40. R. Ruppin, Electric field enhancement near a surface bump, Solid State Commun. 39:903 (1981).

41. F. W. King, R. P. Van Duyne, and G. C. Schatz, Theory of Raman scattering by molecules adsorbed on electrode surfaces, J. Chem. Phys. 69:4472 (1978).

42. S. Efrima and H. Metiu, Classical theory of light scattering by an adsorbed molecule. I. Theory, J. Chem. Phys. 70:1602 (1979).

43. A. Hartstein, J. R. Kirtley, and J. C. Tsang, Enhancement of the infrared absorption from molecular monolayers with thin metal overlayers, Phys. Rev. Lett. 45:201 (1980).

44. C. K. Chen, A. R. B. de Castro, and Y. R. Shen, Surface-enhanced second-harmonic generation, Phys. Rev. Lett. 46:145 (1981).

45. D. Whittle and J. Simon, private communication.

46. J. G. Endriz, Surface waves and grating-tuned photocathodes, Appl. Phys. Lett. 25:261 (1974).

47. A. M. Glass, P. F. Liao, J. G. Bergman, and D. H. Olson, Interaction of metal particles with adsorbed dye molecules: Absorption and luminescence, Opt. Lett. 5:368 (1980).

48. J. R. Kirtley, T. N. Theis, and J. C. Tsang, Diffraction-grating-enhanced light emission from tunnel junctions, Appl. Phys. Lett. 37:435 (1980).

49. C. Y. Chen, I. Davoli, G. Ritchie, and E. Burstein, Giant Raman scattering and luminescence by molecules adsorbed on Ag and Au metal island films, in: "Proceedings of the 1979 International Conference on Non-Traditional Approaches to the Study of the Solid-Electrolyte Interface." Surf. Sci. 101:363 (1980).

50. A Nitzan and L. E. Brus, Can photochemistry be enhanced on rough surfaces? (to be published).

51. G. Ritchie and E. Burstein, Luminescence of molecules adsorbed at a silver surface, Phys. Rev. B (in press).

52. R. W. Rendell, D. J. Scalapino, and B. Mühlschlegel, Role of local plasmon modes in light emission from small-particle tunnel junctions, Phys. Rev. Lett. 41:1746 (1978).

53. The above list of references is obviously not a complete list
     of all of the work done on the topic. In making reference
     to the literature, we have for the most part generally cited
     the initial investigation(s) of the aspect being referenced.

ELECTROMAGNETIC THEORY: A SPHEROIDAL MODEL

Joel I. Gersten

Physics Department
City College of New York
New York, NY 10031

Abraham Nitzan

Chemistry Department
Tel-Aviv University
Tel-Aviv, Israel

The phenomenon of surface enhanced Raman scattering (SERS) provides a dramatic example for the modification of the optical properties of molecules near solid state surfaces. There is some experimental evidence that much, if not all, of the enhancement is associated with surface roughness in the range of 10 to 10,000 Å.[1-13]
Since the electrodynamic characteristics of a rough surface may
differ considerably from those of a smooth surface, it is suggestive
to regard SERS as being primarily of electrodynamic origin. In this
chapter, we study a simple model[14] for a rough surface, analyze the
electrodynamic properties of the model, and explore some ramifica-
tions for the optical properties of nearby molecules.

A rough surface may be characterized as a set of protrusions
which stick out of an underlying planar substrate. The protrusions
will likely have a wide assortment of sizes and shapes, making the
exact analysis of the surface rather difficult. Let us simplify
matters by assuming the protrusions to be hemispheroids (half of an
ellipsoid of revolution) whose symmetry axis is oriented normal to
the substrate plane. These spheroids are characterized by a circular
cross section of radius b and a semi-major axis of length a. For
prolate spheroids, the aspect ratio a/b is larger than one, while for
oblate spheroids it is less than one. In the special case a/b = 1,
the spheroid degenerates to a sphere. The geometry is illustrated
in Fig. 1. The model is further simplified by assuming that neigh-

boring hemispheroids are sufficiently well separated (i.e., their
distance apart is large relative to their characteristic size) so
that their mutual interaction may be disregarded, to zeroth order.
This assumption limits the quantitative predictions of this theory
to colloidal solutions of dielectric particles or to low coverage
surface island films. We expect that the qualitative aspects of
the theory will not change also in the general case. What remains
is an isolated protrusion on a base plane. For now, the molecule
will be allowed to be at an arbitrary position outside the solid
and the incident laser field may have an arbitrary direction of
incidence and polarization.

In order to understand the electrodynamic properties of the
above model, let us consider two limiting cases. If all we had was
a smooth planar surface, the basic electronic excitation of interest
would be a surface plasmon. This plasmon would have a frequency, ω,
which depends on the two-dimensional wave vector parallel to the
surface $k_{//}$. Its amplitude would be delocalized over the surface
and the allowed values of ω would span a continuous range.

In the other limit, we may imagine an isolated spheroid. Such
an object would possess a set of discrete plasma frequencies representing localized excitations of the spheroid. The primary excita-

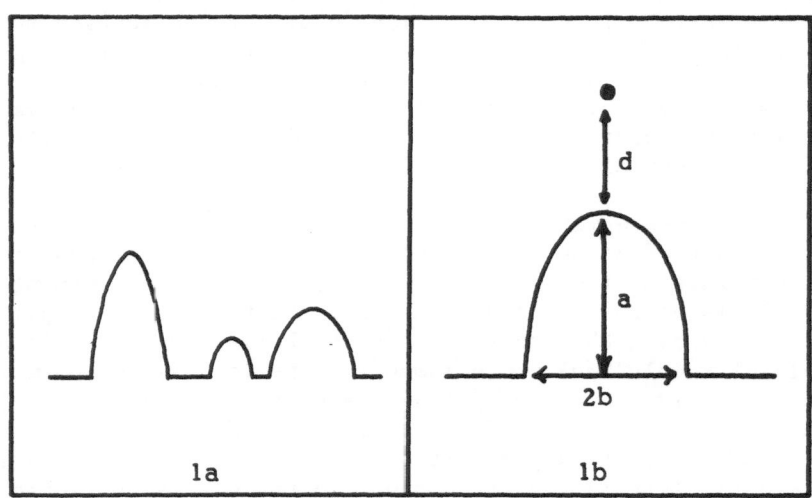

Fig. 1.   Replacement of a "realistic" surface (Fig. 1a) by an
         isolated hemispheroidal protrusion (Fig. 1b). A molecule
         is assumed to reside at a distance d above the apex.

tion of interest to us here is the dipolar plasma oscillation, since it may directly couple to the optical fields.

In the general case of a rough surface, we would expect our model to possess attributes of both limiting cases. However, it is well known that, when a discrete excitation (the dipolar plasma oscillation) is embedded in a continuum of excitations (the substrate surface plasmons), what results is a resonance. (In such a situation, retardation effects would have to be included.) This resonance is characterized by having a high field strength on the spheroidal bump and a low field strength delocalized over the remainder of the surface. It is this bump resonance that plays a dominant role in the electrodynamic theory of SERS.

In the neighborhood of the bump, the bump resonance and the localized plasma oscillation of an isolated spheroid are expected to be quite similar. We therefore begin by studying an isolated prolate spheroid. The isolated spheroid model is also relevant for the cases of colloidal solutions of dielectric particles and for island films deposited on an inert substrate. (An inert substrate is characterized by a dielectric constant approximately equal to that of the medium above the surface.) It will be henceforth assumed that the physical dimensions of the spheroid are small compared with the wavelength of light. Under such conditions, retardation effects may be neglected and the electrodynamic properties are obtained by solving an electrostatic problem.

The free oscillations of spherical and spheroidal particles have been extensively studied in relation to light scattering by such particles.[15,16] It will be useful to review here some properties of these resonances. Consider the free oscillations of a prolate spheroid. The electrostatic potential is given in prolate coordinates by

$$\Phi = \begin{cases} \Phi_0 P_1(\xi) P_1(\eta) & \xi < \xi_0, \\ \Phi_0 Q_1(\xi) P_1(\eta) P_1(\xi_0)/Q_1(\xi_0), & \xi > \xi_0, \end{cases} \tag{1}$$

where ξ and η are the spheroidal coordinates, ξ_0 denotes the surface of the spheroid, and P_1 and Q_1 are Legendre functions of the first and second kind, respectively. The parameter ξ_0 is given by $\xi_0 = a/f$, where f is a scale parameter defined by $f = (a^2 - b^2)^{1/2}$. Equation 1 corresponds to the axially symmetric dipolar plasma oscillation. Another set of degenerate modes exists, corresponding to an azimuthal excitation

$$\Phi = \begin{cases} \Phi_0 P_1^{(1)}(\xi) P_1^{(1)}(\eta) \exp(\pm i\phi), \\ \Phi_0 Q_1^{(1)}(\xi) P_1^{(1)}(\eta) P_1^{(1)}(\xi_0) \exp(\pm i\phi)/Q_1^{(1)}(\xi_0), \end{cases} \tag{2}$$

where $P_1^{(1)}$ and $Q_1^{(1)}$ are associated Legendre functions of the first and second kind, respectively. The potentials in Eqs. (1) and (2) are chosen to be continuous across the boundary. Imposition of the boundary condition that the normal component of the electric displacement vector be continuous leads to the following condition for the axially symmetric mode:

$$\epsilon(\omega) = \xi_0 Q'_1(\xi_0)/Q_1(\xi_0) \equiv -\bar{\epsilon}_1, \tag{3}$$

and for the azimuthal modes:

$$\epsilon(\omega) = P_1^{(1)}(\xi_0) Q'^{(1)}_1(\xi_0)/[Q_1^{(1)}(\xi_0) P'^{(1)}_1(\xi_0)] \equiv -\bar{\epsilon}_1^{(1)}. \tag{4}$$

Here the dielectric constant of the spheroid is denoted by the complex number $\varepsilon(\omega)$. The above formulas are also applicable to the case of oblate spheroids if an analytic continuation is made.

A plot of the right hand sides of Eqs. (3) and (4) as a function of the aspect ratio is given in Fig. 2. By using the experimental dielectric constant as a function of frequency, Eqs. (3) and (4) may be used to translate Fig. 2 into a plot of the resonance frequency as a function of aspect ratio. Such a plot is given in Fig. 3 for silver.[17] The plasma frequency is a sensitive function of the aspect ratio of the spheroid. The axially symmetric mode is

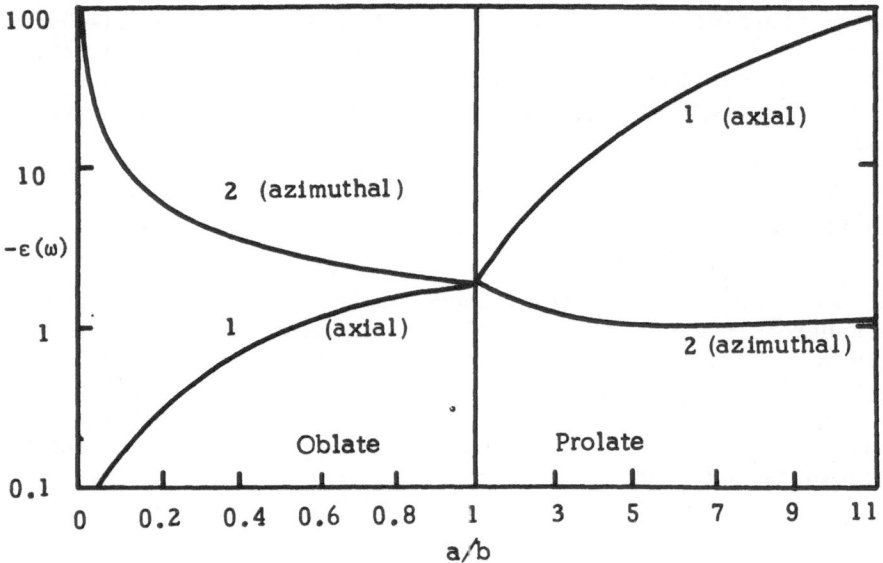

Fig. 2. The real part of the dielectric constant for axially symmetric and azimuthal modes.

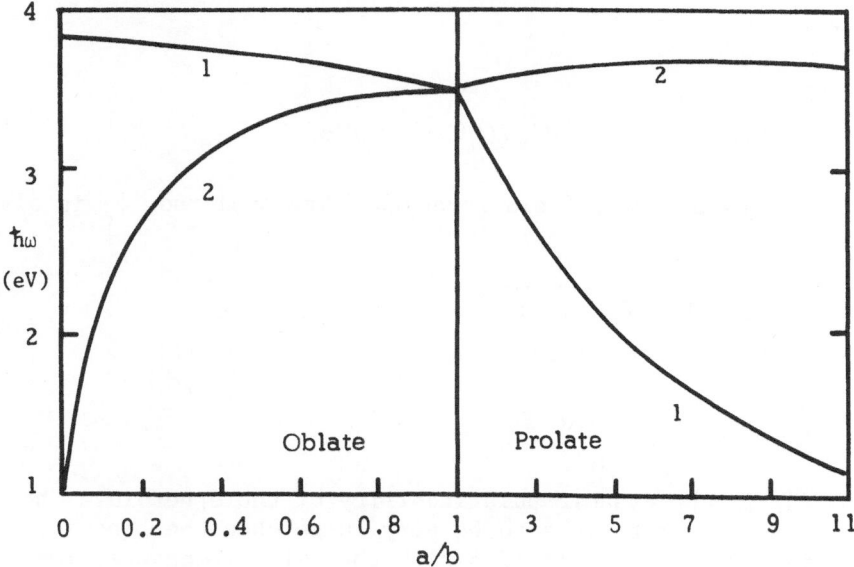

Fig. 3. Plasma energies for silver spheroids, ℏω, as a function
of the aspect ratio a/b. The axially symmetric mode is
denoted by 1 and the azimuthally excited (doubly
degenerate) modes are denoted by 2.

associated with charge separation along the symmetry axis. The
aximuthally excited mode corresponds to charge separation perpendicular to the symmetry axis. In the case of highly prolate spheroids, the axial mode frequency is seen to fall sharply with increasing aspect ratio. This contrasts with the case of highly oblate
spheroids, where it is the azimuthally excited mode which drops in
frequency. This behavior may be understood qualitatively in terms
of the lines of electric flux joining charges at the opposite ends
of the spheroid. When the tip of a prolate spheroid (or the edge
of an oblate spheroid) is sharp, few flux lines pass through the
solid to link charges at the opposite ends, and the restoring force
needed to sustain the plasma oscillation is weakened. Since the
oscillation frequency is determined by the square root of the ratio
of restoring forces to inertial terms, this accounts for the drop
off in frequency.

For the sake of brevity, let us limit our attention to the
case of prolate spheroids. Consider what happens if a uniform time
varying electric field, such as is produced by a laser, is applied.
If the polarization of the field is along the symmetry axis, we
expect a strong coupling to the axial mode. The potential will now

be given by:

$$\Phi = \begin{cases} \Phi_0 P_1(\xi) P_1(\eta) \\ (\Phi_0 + E_0 f) Q_1(\xi) P_1(\eta) P_1(\xi_0)/Q_1(\xi_0) - E_0 f \xi \eta, \end{cases} \tag{5}$$

where E_0 is the external field strength. The amplitude Φ_0 is given by:

$$\Phi_0 = -E_0 \frac{f}{(\xi_0^2 - 1) Q_1(\xi_0) [\epsilon(\omega) + \bar{\epsilon}_1]}. \tag{6}$$

The dipole moment of the spheroid is

$$D = \frac{\xi_0 f^3 (\epsilon - 1) E_0}{3 Q_1(\xi_0) [\epsilon + \bar{\epsilon}_1]} = \alpha_s(\omega) E_0, \tag{7}$$

where $\alpha_s(\omega)$ is the dynamic polarizability of the spheroid. The dynamic polarizability gets to be very large when the frequency of the incident radiation coincides with the axial resonance frequency of the spheroid. At that frequency, we may obtain substantial polarization with even weak incident fields. The polarized spheroid itself acts as a source of electric field. The local field in the neighborhood of the spheroid thus has two contributions to it: the incident field and the field due to the spheroid. At resonance, the local field may be substantially larger than the incident field. An expression for this local field at a point on the major axis (coordinates $\xi = \xi_1$, $\eta = 1$) is obtained from Eqs. (5) and (6):

$$E = \left[1 - \frac{3 \alpha_s(\omega) Q'_1(\xi_1)}{f^3} \right] E_0 \equiv A(\omega) E_0. \tag{8}$$

Here $A(\omega)$ may be regarded as an amplification factor arising from the presence of the spheroid. A plot of $A(\omega)$ for a silver spheroid is given in Fig. 4 for the case where $\xi_1 = \xi_0$ (the surface field). Results are given for several aspect ratios. In this case,

$$A(\omega) \rightarrow \epsilon(\omega) [1 + \bar{\epsilon}_1] [\epsilon(\omega) + \bar{\epsilon}_1]^{-1}.$$

The results for oblate spheroids are obtained by analytic continuation. We see from Fig. 4 that the amplification factor grows dramatically near resonance and is small off resonance. As the aspect ratio increases, the peak amplification factor also increases. This is due to the "lightning rod effect" which will be discussed shortly.

Finally, it should be noted that the amplification factor depends on the distance from the spheroid surface [expressed by ξ_1 in the expression for $A(\omega)$] (see Fig. 5). Also the terms contributing to $A(\omega)$ may add constructively or destructively, so that it

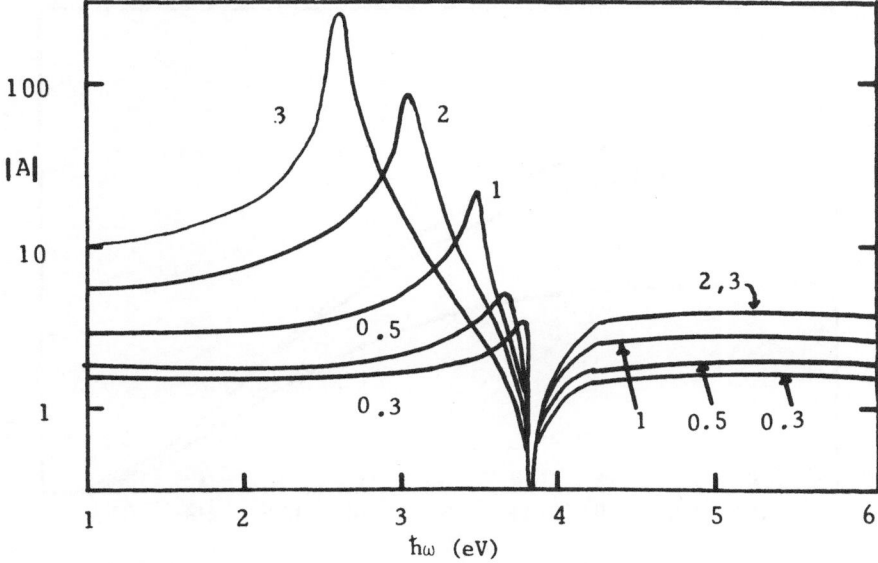

Fig. 4. The modulus of the field amplification factor on the tip
         of a silver spheroid as a function of photon energy for
         several aspect ratios: 3, 2, 1, 0.5, and 0.3.

is possible, in principle, that at some intermediate distance from
the surface $A(\omega)$ will be smaller than unity as a result of destructive interference.

     The local field outside the spheroid is by no means uniform.
Near the tips of a prolate spheroid or the edges of an oblate spheroid, the local field is much stronger than at other positions. The
larger the departure from sphericity, the stronger the field will
be. This is the same principle underlying the operation of the
lightning rod, where the corona discharge takes place near the
sharpest feature of the rod. The reason for the enhanced field is
simple. Consider the case of a perfect conductor. In order to
maintain an equipotential surface, an image charge (or charges) would
have to lie fairly close to a point of high curvature. The field due
to this charge would be intense outside the surface. The same
phenomenon occurs with a finite dielectric replacing a perfect conductor. A description may still be given in terms of appropriately
arrayed image charges. An illustration of the lightning rod effect
is given in Fig. 5, where the magnitude of the electric field is
given as a function of position along the major axis in the vicinity
of a prolate spheroid. The field is seen to be intense in a region
near the tip whose radius is $\sim b^2/a$. This is just the radius of curvature of the tip. Likewise we would find the field to be strong

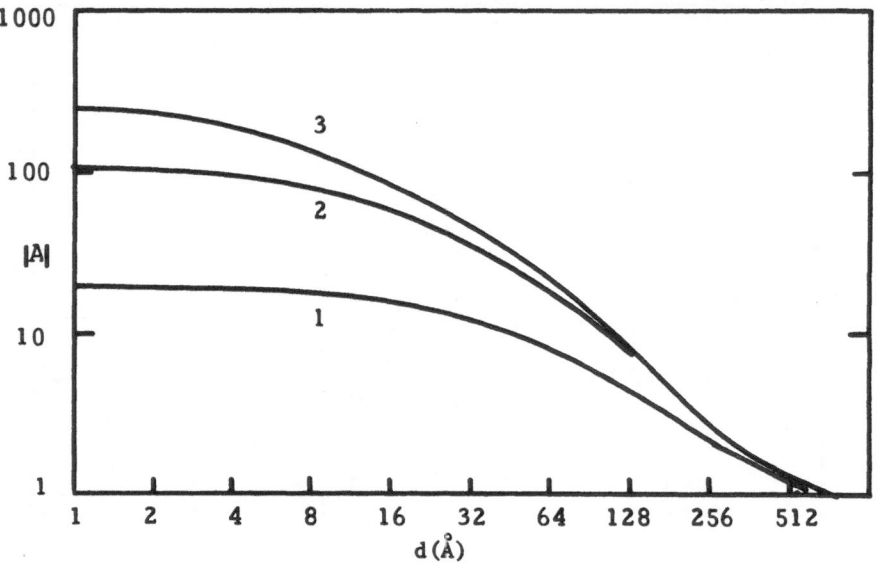

Fig. 5.   Amplification factor as a function of distance from the
          tip of a silver spheroid. Here a = 200 Å and curves for
          three values of a/b are given: a/b = 3, 2, and 1.

near the edge of an oblate spheroid. In Fig. 6, the amplification
factor on the surface is given as a function of polar angle along
the surface for several aspect ratios. The factor is seen to be
strongest near the tip of the spheroid and to fall off appreciably
as the equator is approached. Nevertheless, substantial enhancement
remains near the equator, particularly for higher aspect ratios.

     Before proceeding further, it is perhaps worthwhile to state
what limitations apply to the present description.

     (a) <u>Finite size limitations</u>: We have used the dielectric pro-
perties of the bulk metal in describing the spheroid. This is valid
provided the mean free path of the electrons is smaller than the
size of the spheroid (the major axis, when the field is parallel to
it). If this condition is not obeyed, Im ε may be augmented by
phenomenological terms involving the effect of wall collisions. Such
effects become significant for a $\lesssim$ 100 Å.

     (b) <u>Effect of the substrate plane</u>: In those cases where the
substrate dielectric function is significantly different from that
of the medium above it, the substrate plays two major roles. First,
it provides a decay channel for the dipolar plasmon of the spheroid,
transforming it from a bound excitation into a resonance. The

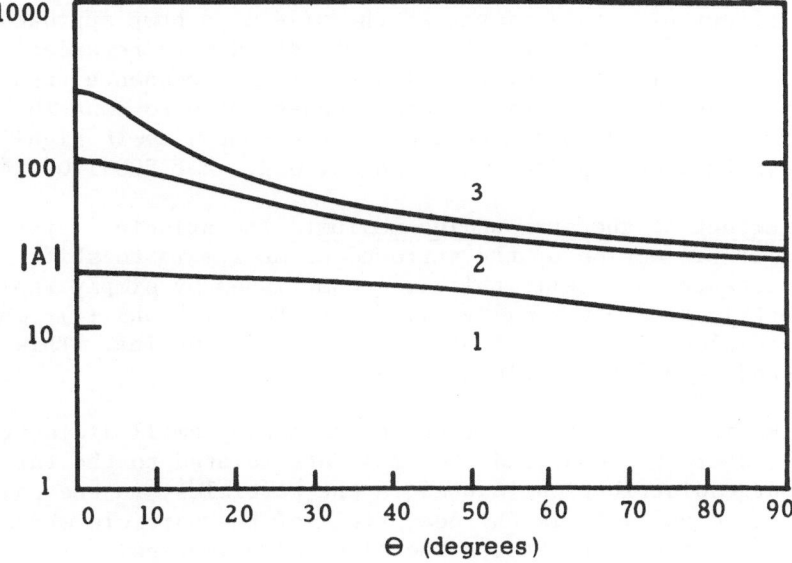

Fig. 6.   Amplification factor as a function of the polar angle
          (measured with respect to the symmetry axis) along the
          surface of a silver spheroid. Here a = 200 Å and curves
          for the three values of a/b are given: a/b = 3, 2, and 1.

quality factor (Q) of the resonance is determined by Im ε as well as
by the coupling of the localized bump plasma oscillation to the delocalized plasmons of the surface. In addition, this coupling
causes a shift in the resonance frequency. A second effect of the
substrate is that it allows an electromagnetic wave to couple to the
localized plasma resonance of the spheroid through the intermediate
step of exciting a delocalized surface plasmon. Thus radiation may
excite a surface plasmon, and the surface plasmon may then excite the
bump dipolar plasmon. Since the surface fields associated with the
surface plasmon may be larger than the incident field, this effect
may give an added enhancement. This effect is not included in the
present analysis.

(c) Effect of neighboring bumps: The near fields of neighboring
bumps will influence both the plasma resonance frequency and amplification factors of a given bump. One expects this effect to be most
severe when (1) neighboring bumps are in close proximity (i.e., a
distance of the order of or smaller than the characteristic size;
(2) when the shape of neighboring bumps coincides with that of a
given bump (so they are resonantly coupled); and (3) the bumps are
oblate and the field is parallel to the surface (a configuration
which maximizes electric flux linkage to the neighboring bumps).
Multi-bump effects are not treated in this chapter.

(d) Effect of retardation: As the size of a bump approaches the wavelength of light, interference effects due to retardation become important. The role of the dipolar plasma resonance then becomes less pronounced and higher order modes begin to contribute. Our theory is limited to the case where the bump size is significantly less than the optical wave length, e.g., a $\lesssim$ 500–1,000 Å.

(e) Effect of the surrounding medium: The principal effect of the index of refraction of the surrounding medium is to shift the plasma resonance. Formally this is accomplished by simply replacing the dielectric constant $\varepsilon(\omega)$ by $\varepsilon(\omega)/\varepsilon_s$ in Eqs. (3) and (4), where ε_s is the dielectric constant of the surrounding medium. This effect is neglected here for simplicity's sake.

Unlike in the theory of light scattering by small dielectric particles, where the observed phenomena are related to the far field of the charge distribution induced in the particles, in the calculation described below it is the near field of the particle which plays a major role because it affects the adsorbed molecules.

Consider a molecule located a distance d above the apex of the spheroid as shown in Fig. 1b. Let the molecule have a dipole moment μ which will be assumed to point perpendicular to the surface. This collinear geometry is chosen both for the sake of mathematical simplicity and because it is the most important configuration in SERS. Again let us ignore the base plane and treat the bump as a full spheroid.

The potential given by Eq. (5) now has an added term due to the presence of the dipole:

$$\Phi = \begin{cases} \sum_n A_n P_n(\xi) P_n(\eta), \\[2ex] \sum_n B_n Q_n(\xi) P_n(\eta) - E_0 f \xi \eta + \dfrac{\vec{\mu} \cdot [\vec{r} - (a+d)\hat{k}]}{|\vec{r} - (a+d)\hat{k}|^3} \end{cases} . \tag{9}$$

This formula is complicated by the fact that all multipoles are induced on the spheroid by the presence of the molecular dipole, and all of them are important in the near field of the spheroid. Using the Green's function,

$$\frac{\vec{\mu} \cdot [\vec{r} - (a+d)\hat{k}]}{|\vec{r} - (a+d)\hat{k}|^3} = \mu f^{-2} \sum_n P_n(\xi) P_n(\eta) Q_n(\xi_1), \tag{10}$$

where $\xi_1 = (a+d)/f$ and $\xi_0 < \xi < \xi_1$, it is possible to again solve the boundary value problem and obtain expressions for the elementary physical quantities. Skipping the details, the dipole moment of the system (spheroid plus molecule) turns out to be given by

$$\vec{D} = A(\omega')\vec{\mu},$$ (11)

where $A(\omega')$ is the same factor that appeared in Eq. (8). We write ω' rather than ω, anticipating the fact that in Raman scattering the emission frequency ω' is shifted from the incident frequency ω. Equation 11 demonstrates that the molecular dipole produces a system dipole which is amplified by the factor A. In order to satisfy the boundary conditions on the spheroid, a polarization must develop on the particle which may, at times, be rather large (see Fig. 4).

The molecular dipole is related to the local field through the dynamic molecular polarizability $\alpha(\omega)$:

$$\vec{\mu} = \alpha(\omega)\vec{E},$$ (12)

where $\alpha(\omega)$ is assumed to be a scalar for the sake of simplicity. The local field is obtained from Eq. (9) and is

$$\vec{E} = A\vec{E}_0 + \frac{\Delta(\omega)\vec{\mu}}{\alpha(\omega)},$$ (13)

where $\Delta(\omega)$ is a complex parameter defined by

$$\Delta(\omega) = \alpha(\omega)\frac{\epsilon - 1}{f^3}\sum_n \frac{(2n+1)P_n(\xi_0)P'_n(\xi_0)[Q'_n(\xi_1)]^2}{\epsilon(\omega)Q_n(\xi_0)P'_n(\xi_0) - P_n(\xi_0)Q'_n(\xi_0)}.$$ (14)

In Eq. (13) the local field is expressed as the sum of the amplified incident field, AE_0, plus a contribution induced by the molecular dipole. The emission dipole of the system (molecule plus particle) is thus

$$D = \frac{A(\omega)A(\omega')\alpha(\omega)E_0}{1 - \Delta(\omega)}.$$ (15)

In the semiclassical description of Raman scattering, the polarizability is taken to fluctuate at the frequncy of a normal mode coodinate q. Since the polarizability of the molecule is taken to be a classical parameter, the semiclassical description may only be applied for nonresonant situations. For resonant Raman scattering, a more careful quantum mechanical approach should be followed. We henceforth restrict our attention to the nonresonant case. Thus the Raman dipole is [using Eqs. (14) and (15)]:

$$\Delta D = (\partial D/\partial q)\Delta q$$

$$= \frac{A(\omega)A(\omega')E_0}{[1 - \Delta(\omega)]^2}(\partial\alpha/\partial q)\Delta q.$$ (16)

The Raman cross section is:

$$\sigma = \frac{8\pi\omega^3}{3\hbar c^4}|\Delta D/E_0|^2$$

$$= \frac{8\pi\omega^3}{3\hbar c^4}[\Delta q \partial\alpha/\partial q]^2 \frac{|A(\omega)A(\omega')|^2}{|1 - \Delta(\omega)|^4}. \tag{17}$$

Let us define the enhancement ratio as the above cross section divided by that of a free molecule. For a free molecule, $A \to 1$ and $\Delta \to 0$ so the enhancement ratio is

$$R = |A(\omega)A(\omega')|^2/|1 - \Delta(\omega)|^4. \tag{18}$$

The physical meaning of the parameter $\Delta(\omega)$ is obtained from the following argument. Suppose that, instead of Eq. (12), we had introduced a classical oscillator model for the dipole:

$$(\omega_0^2 - i\omega\gamma - \omega^2)\vec{\mu} = \alpha_0\omega_0^2\vec{E}, \tag{19}$$

where α_0 is the D.C. polarizability. Insertion of Eq. (12) into Eq. (19) would give:

$$\mu = \frac{\alpha\omega_0^2 A(\omega)E_0}{\tilde{\omega}_0^2 - i\omega\tilde{\gamma} - \omega^2}, \tag{20}$$

where the modified molecular resonance frequency and damping are given by

$$\tilde{\omega}_0^2 = \omega_0^2[1 - \mathrm{Re}\{\alpha_0\Delta(\omega)/\alpha(\omega)\}] = \omega_0^2 - \Delta\omega^2, \tag{21a}$$

$$\tilde{\gamma} = \gamma + \mathrm{Im}\frac{\alpha_0\omega_0^2\Delta(\omega)}{\omega\alpha(\omega)}. \tag{21b}$$

The molecule thus behaves near the surface as a polarized dipole with polarizability

$$\tilde{\alpha}(\omega) = \frac{\alpha_0\omega_0^2}{\tilde{\omega}_0^2 - \omega^2 - i\omega\tilde{\gamma}}, \tag{22a}$$

while the polarizability of the free molecule is [from Eq. (19)]:

$$\alpha(\omega) = \frac{\alpha_0\omega_0^2}{\omega_0^2 - \omega^2 - i\omega\gamma}. \tag{22b}$$

We note in passing that the correction to the molecular resonance frequency and width given by Eq. (21) are the surface induced shift and damping:[18,19]

$$\Delta\omega^2 = \alpha_0\omega_0^2 \text{Re}\Delta(\omega)/\alpha(\omega),$$ (23a)

$$\Gamma = \text{Im}\frac{\alpha_0\omega_0^2\Delta(\omega)}{\omega\alpha(\omega)}.$$ (23b)

A plot of these quantities as a function of d for a resonant spheroid is presented in Figs. 7 and 8. It should be noted that the present classical theory which is based on a point dipole model for the molecule predicts very large surface induced damping and shifts at distances of the order of 1-2 Å from the surface. However, the theory is not valid at such distances and more rigorous calculations[20-22] show $\Delta\omega^2$ and Γ to saturate at relatively small values (of the order of the classical result at d ∿ 3-4 Å).

We now insert the above relations into Eq. (15), noting that

$$1 - \Delta(\omega) = \alpha(\omega)/\tilde{\alpha}(\omega) = \frac{\tilde{\omega}_0^2 - \omega^2 - i\omega\tilde{\gamma}}{\omega_0^2 - \omega^2 - i\omega\gamma}.$$ (24)

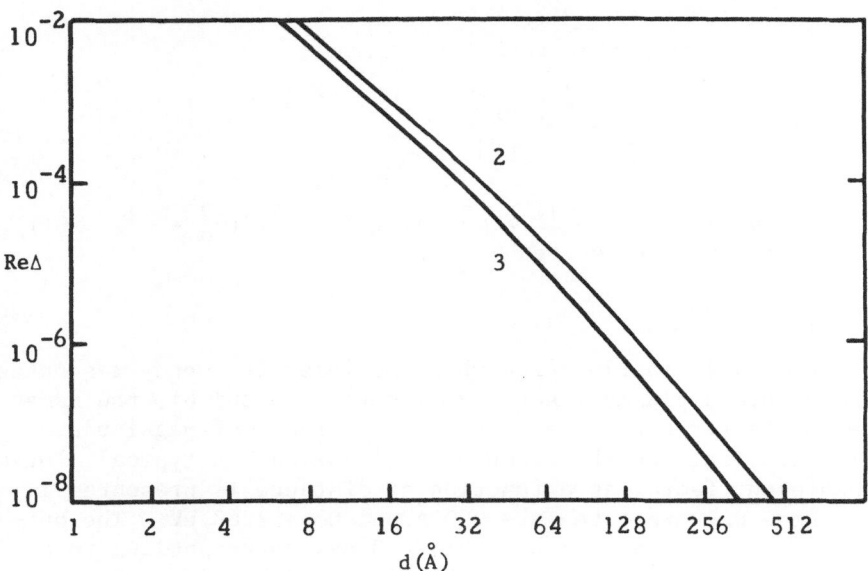

Fig. 7. Plot of the level shift correction, ReΔ vs molecule-spheroid separation, d. Two graphs are given - for a/b = 2, respectively. Here a = 200 Å and the static molecular polarizability was taken to be α_0 = 10 Å³.

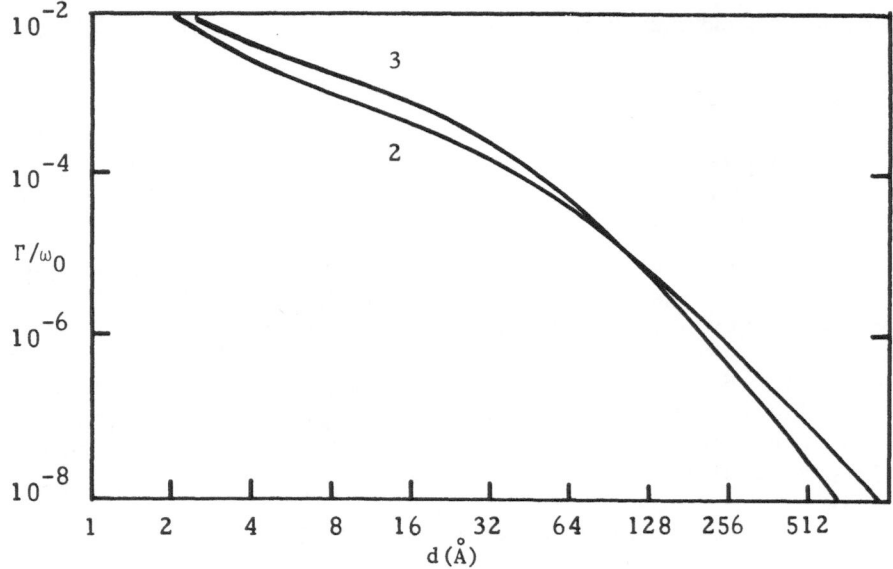

Fig. 8.   Plot of the decay width in units of the molecular fre-
          quency: Γ/ω_0, vs molecule-spheroid separation, d.   The
          parameters are the same as in Fig. 7.

The enhancement factor then takes the form

$$R = |A(\omega)A(\omega')|^2 \left| \frac{\omega_0^2 - \omega^2 - i\omega\gamma}{\tilde{\omega}_0^2 - \omega^2 - i\omega\tilde{\gamma}} \right|^4 . \tag{25}$$

Far from resonance $|\omega_0^2 - \omega^2| \gg \omega\tilde{\gamma}$ and $(\omega_0^2 - \omega^2)/(\tilde{\omega}_0^2 - \omega^2) \sim O(1)$.
The enhancement factor is then

$$R \rightarrow |A(\omega)A(\omega')|^2 . \tag{26}$$

Since $A(\omega)$ is fairly large when the laser frequency resonates
with the spheroid plasma frequency (see Figs. 5 and 6), the Raman
enhancement factor can be as large as 4–9 orders of magnitude,
depending precisely on the frequencies involved. A typical plot of
the enhancement factor as a function of distance is presented in
Fig. 9.   Here we have taken a = 200 Å and $\hbar\omega$ = 2.63 eV.   The outgoing photon energy was taken to be 2.50 eV, corresponding to a
Stokes shift of $\sim$ 1000 cm^{-1}.   For the silver spheroid, a resonant
configuration (with the incident photon) corresponds to an aspect
ratio of a/b = 3.   The enhancement is seen to be a fairly long
range effect, particularly for less eccentric spheroids.   Note that
for a/b = 4 there is a manifestation of destructive interference

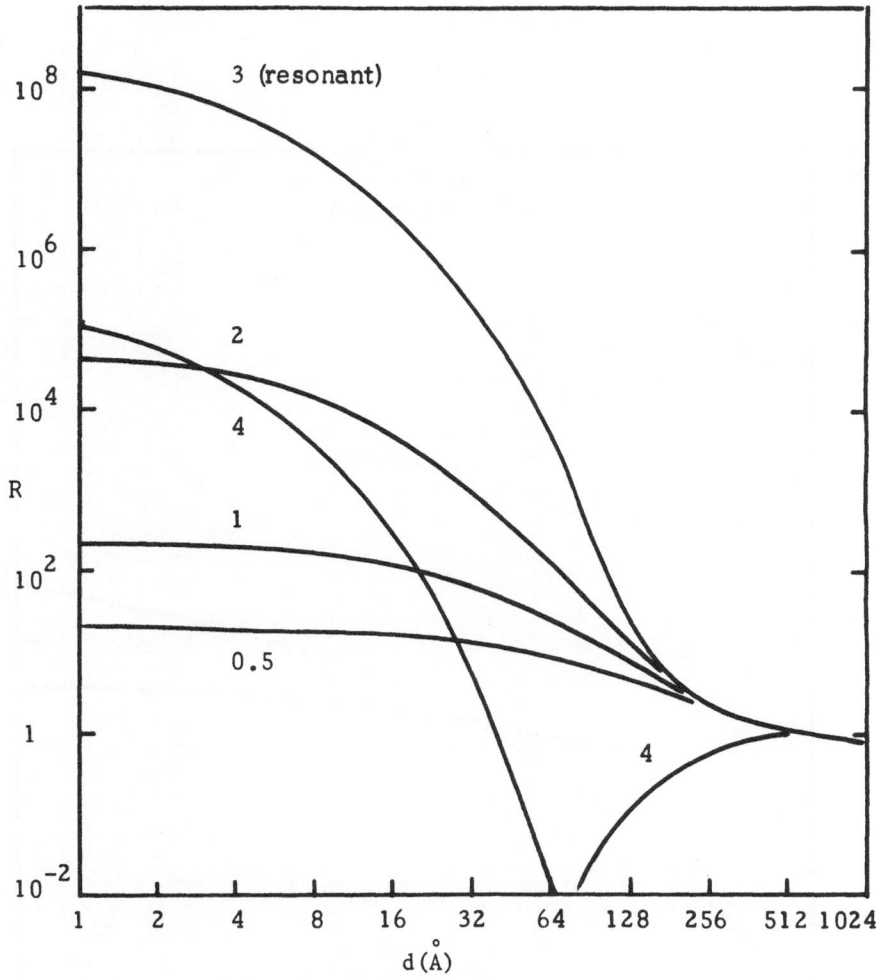

Fig. 9. Plot of the SERS enhancement factor as a function of the
molecule-spheroid distance, d.

between the two terms in the enhancement factor [see Eq. (8)].

In Fig. 10, a plot is made of the SERS enhancement factor, R,
as a function of the semi-major axis, a, for several aspect ratios.
Here the molecule-spheroid distance is held fixed at $d = 2$ Å. The
incident photon energy is again chosen so as to be in resonance with
a spheroid whose aspect ratio is $a/b = 3$. The basic trend is that
the enhancement ratio grows with increasing aspect ratio. The rate

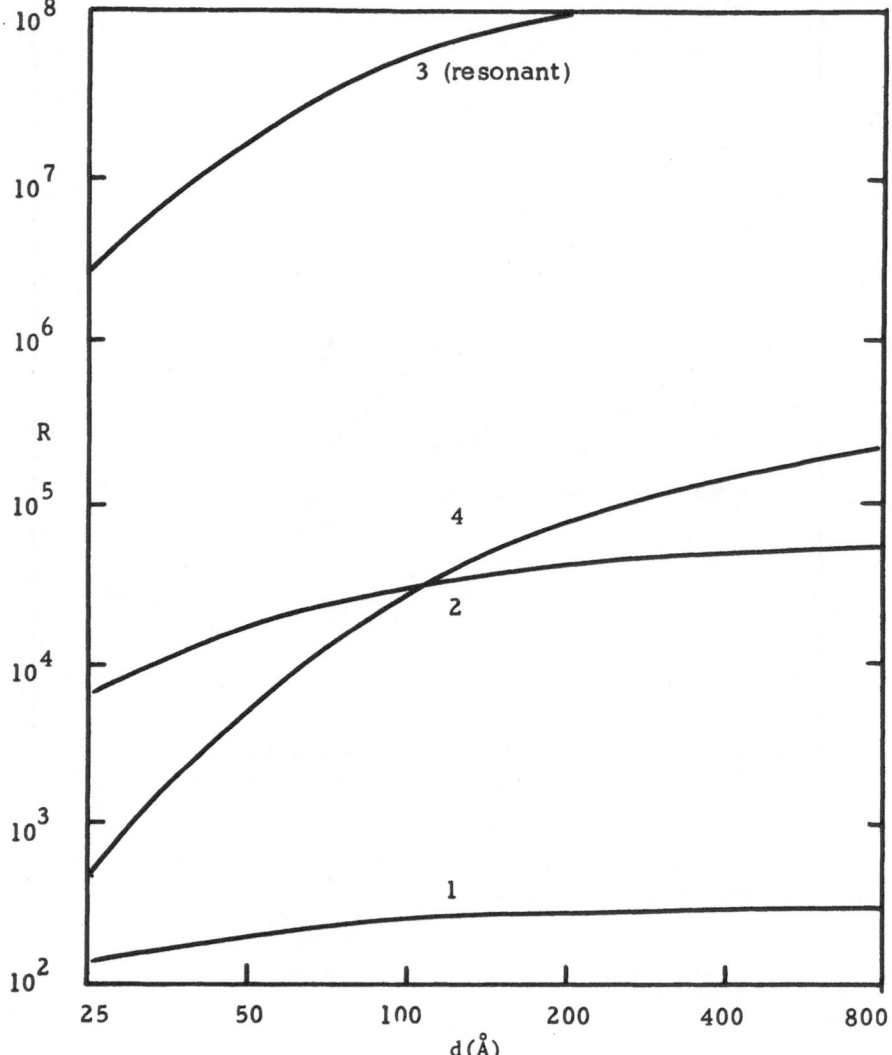

Fig. 10.   SERS enhancement factor, R, as a function of the semi-
           major axis a.   Here $\hbar\omega$ = 2.63 eV, $\hbar\omega'$ = 2.50 eV and
           d = 2 Å.   The aspect ratios corresponding to these curves
           are a/b = 1, 2, 3, and 4.

of growth is largest for the sharpest structures, i.e., those with the largest aspect ratios. The role played by the bump plasmon resonance is again evident in that R for a/b = 3 is much larger for other aspect ratios. Omitted in Fig. 10 are the effects of wall collisions and the effects of retardation. The former effect would tend to decrease R at small values of a, while the latter effect would decrease R at large values of a.

To conclude our discussion, we point out the following observations:

(a) We have identified two main factors involved in the electromagnetic enhancement of the Raman scattering cross section of a molecule adsorbed on a suitable dielectric surface. These are the surface plasmon resonance, where the plasmon is excited in protrusions on the rough surface, and the lightning rod effect, in which the local electric field is enhanced by the sharpness of the spheroid. Examination of Figs. 4, 5, and 6 shows that these two effects are of comparable magnitude when the aspect ratio is approximately equal to three (for a molecule situated along the long ellipsoid axis and parallel to it).

(b) The level shift caused by the interaction between the dipole and its image charge distribution in the dielectric affects the magnitude of the enhancement and, when the experiment is done with constant incident energy and varying molecule-surface distance, may increase or decrease the enhancement as described by Eq. (25). However, this is a secondary effect; the main electromagnetic effect of the surface is in inducing the plasmon and the lightning rod enhancements.

(c) Even though this work has focused on SERS, the theory presented implies immediately that any phenomenon associated with electromagnetic interactions will be markedly affected on rough surfaces. Indeed recent experimental results indicate that absorption and fluorescence by adsorbed molecules are affected, and a theoretical investigation of these effects has been presented.

REFERENCES

1. M. Moskovits, Surface roughness and the enhanced intensity of Raman scattering by molecules adsorbed on metals, J. Chem. Phys. 69:4159 (1978). Enhanced Raman scattering by molecules adsorbed on electrodes. A theoretical model, Solid State Commun. 32:59 (1979).
2. J. A. Creighton, C. G. Blatchford, and M. G. Albrecht, Plasma resonance enhancement of Raman scattering by pyridine adsorbed on silver or gold sol particles of size comparable to the excitation wavelength, J. Chem. Soc. Faraday II 75:790 (1979).

3. D. A. Zwemer, C. V. Shank, and J. E. Rowe, Surface enhanced
    Raman scattering as a function of molecule-surface separa-
    tion, Chem. Phys. Lett. 73:201 (1980).
4. D. A. Weitz, T. J. Gramila, A. Z. Genack, and J. I. Gersten,
    Anomalous low-frequency Raman scattering from rough metal
    surfaces and the origin of surface-enhanced Raman scatter-
    ing, Phys. Rev. Lett. 45:355 (1980). Inelastic Mie scatter-
    ing from rough metal surfaces: Theory and experiment, Phys.
    Rev. B 22:4562 (1980).
5. A. Girlando, M. R. Philpott, D. Heitmann, J. D. Swallen, and
    R. Santo, Raman spectra of thin organic films enhanced by
    plasmon surface polaritons on holographic metal gratings,
    J. Chem. Phys. 72:5187 (1980).
6. P. N. Sanda, J. M. Warlaumont, J. E. Demuth, J. C. Tsang,
    K. Christmann, and J. A. Bradley, Surface-enhanced Raman
    scattering from pyridine on Ag(111), Phys. Rev. Lett.
    45:1519 (1980).
7. J. C. Tsang, J. R. Kirtley, and J. A. Bradley, Surface-enhanced
    Raman spectroscopy and surface plasmons, Phys. Rev. Lett.
    43:772 (1980). J. C. Tsang, J. R. Kirtley, and T. N. Theis,
    Surface plasmon polariton contributions to Stokes emission
    from molecular monolayers on periodic Ag surfaces, Solid
    State Commun. 35:667 (1980).
8. C. Y. Chen and E. Burstein, Giant Raman scattering by molecules
    at metal-island films, Phys. Rev. Lett. 45:1287 (1980).
9. D.-S. Wang, H. Chew, and M. Kerker, Enhanced Raman scattering
    at the surface (SERS) of a spherical particle, Appl. Opt.
    19:2256 (1980). M. Kerker, D.-S. Wang, and H. Chew, Surface-
    enhanced Raman scattering (SERS) by molecules adsorbed at
    spherical particles, Appl. Opt. 19:4159 (1980).
10. S. L. McCall, P. M. Platzman, and P. A. Wolff, Surface enhanced
    Raman scattering, Phys. Lett. A 77:381 (1980).
11. S. S. Jha, J. R. Kirtley, and J. C. Tsang, Intensity of Raman
    scattering from molecules adsorbed on a metal grating, Phys.
    Rev. B 22:3973 (1980).
12. J. I. Gersten, The effect of surface roughness on surface en-
    hanced Raman scattering, J. Chem. Phys. 72:5779 (1980).
    Rayleigh, Mie, and Raman scattering by molecules adsorbed on
    rough surfaces, J. Chem. Phys. 72:5780 (1980).
13. C. A. Murray, D. L. Allara, and M. Rhinewine, Silver-molecule
    separation dependence of surface-enhanced Raman scattering,
    Phys. Rev. Lett. 46:57 (1981).
14. J. I. Gersten and A. Nitzan, Electromagnetic theory of enhanced
    Raman scattering by molecules adsorbed on rough surfaces,
    J. Chem. Phys. 73:3023 (1980).
15. M. Kerker, "The Scattering of Light and Other Electromagnetic
    Radiation," Academic Press, New York (1969).
16. H. C. van de Hulst, "Light Scattering by Small Particles,"
    Wiley, New York (1957).

17. P. B. Johnson and R. W. Christy, Optical constants of the noble metals, Phys. Rev. B 6:4370 (1972).
18. R. R. Chance, A. Prock, and R. Silbey, Molecular fluorescence and energy transfer near interfaces, Adv. Chem. Phys. 37:1 (1978).
19. J. I. Gersten and A. Nitzan, Spectroscopic properties of molecules interacting with small dielectric particles, J. Chem. Phys., to be published (1981).
20. P. R. Hilton and D. W. Oxtoby, Surface enhanced Raman spectra: A critical review of the image dipole description, J. Chem. Phys. 72:6346 (1980).
21. G. Korzeniewski, T. Maniv, and H. Metiu, The interaction between an oscillating dipole and a metal surface described by a jellium model and the random phase approximation, Chem. Phys. Lett. 73:212 (1980).
22. P. J. Feibelman, Local field at an irradiated adatom on jellium-exact microscopic results, Phys. Rev. B 22:3654 (1980).

ENHANCED RAMAN SCATTERING BY MOLECULES ADSORBED AT THE SURFACE
OF COLLOIDAL PARTICLES

M. Kerker, D.-S. Wang, H. Chew, O. Siiman, and L.A. Bumm

Clarkson College of Technology

Potsdam, N.Y. 13676

1. INTRODUCTION

The observation of SERS by molecules adsorbed at the surface of colloidal particles has led to analysis of classical electromagnetic field effects which contribute to enhancement of the Raman signals. Such effects alone can give rise to large enhancements which are sensitive to particle size, shape and optical constants. The theory is formulated in such a way that specific changes in Raman polarizability which occur upon adsorption can be incorporated, when these are known.

The suggestion by Moskovits[1] that colloidal metal spheres covered with adsorbate and isolated in a dielectric medium might display Raman signals similar to SERS at roughened electrode surfaces was followed by the measurements of Creighton et al.[2] which displayed a 250-fold enhancement for pyridine adsorbed on colloidal silver. Creighton suggested that the effect was in some way related to the absorption and scattering of light by the silver particles and our contribution to this dialogue has been to articulate such an electrodynamic mechanism.

Since we had described a model for the effect of embedment of Raman and fluorescent molecules within colloidal particles[3-6] which was consistent with experimental observations,[7-11] it seemed reasonable to extend that analysis to the case where the inelastically scattering molecules are located outside of the particle.

The molecules are treated as classical electric dipoles. The electromagnetic fields are solutions of Maxwell's equations subject to
the appropriate boundary conditions at the surface of the colloidal
spheres. In calculating the enhancement, the Raman polarizability
at the surface is taken to be the same as for the isolated molecule
in the fluid medium. However, this presumption can be amended by
inserting the polarizability appropriate to the adsorbed molecules
whenever that is known.

The model does indeed predict the main features of SERS.[12-13]
Furthermore, by utilizing a colloidal silver hydrosol for which the
theory predicts enhancements of the order of 10^5 to 10^6, we have
measured Raman signals[14] which are actually enhanced to the same
extent. Although theory and experiment show quite different dependence of the enhancement upon excitation wavelength, the discrepancy
can be resolved by considering that the colloidal particles may
deviate from spherical symmetry.[15]

The study of dilute colloidal hydrosols has an advantage not
only that these provide a more tractable morphology for theoretical
analysis, but that the critical sensitivity of the enhancement to
the scale of "roughness" of a macroscopic surface becomes trans-
formed in the case of colloids to a dependence of the enhancement
upon particle size and shape, quantities which can be observed and
which can be described deterministically.

In section 2 we review the theoretical model for colloidal
spheres. The limiting expressions for small spheres are given in
section 3. Section 4 displays some numerical results and section 5
describes some preliminary experiments which are then compared with
theory. The extension of the theory to spheroids which are small
compared to the wavelength and numerical results are presented in
section 6. The conclusions are stated in section 7.

2. RAMAN SCATTERING BY MOLECULES LOCATED OUTSIDE OF A SPHERE INCLUDING LOCATIONS ON THE SURFACE

In order to emphasize the effects of the optical and geometrical properties of the particle upon the electromagnetic fields, a
simple model is assumed in which the active molecule located at $\vec{r}'$
emits photons through electric-dipole transitions. The dipole
moment at the shifted frequency ω, $\vec{p}(\vec{r}',\omega)$ is given by

$$\vec{p}(\vec{r}',\omega) = \overset{=}{\alpha} \cdot \vec{E}_p(\vec{r}',\omega_o) \tag{1}$$

where $\overset{=}{\alpha}$ is the polarizability tensor and $\vec{E}_p(\vec{r}',\omega_o)$ is the field at
the exciting frequency ω_o at the location of the molecule. It can
be seen at the outset that the effect of the colloidal particle
upon the molecular dynamics is deferred by incorporating this into

the phenomenological parameter $\bar{\bar{\alpha}}$. Also, the model can be extended to magnetic-dipole and higher-order transitions should these be indicated.

The exciting field in turn is comprised of the incident field $\vec{E}_i(\vec{r}',\omega_o)$, which may be taken to be that of a plane wave, plus the elastically scattered field $\vec{E}_{LM}(\vec{r}',\omega_o)$ as calculated by Lorenz-Mie theory.[16]

$$\vec{E}_p(\vec{r}',\omega_o) = \vec{E}_i(\vec{r}',\omega_o) + \vec{E}_{LM}(\vec{r}',\omega_o) \ . \qquad (2)$$

It should be noted that in the neighborhood of the particle, as in this case, the molecule is situated in the near field rather than in the radiation zone. There will also be a magnetic field corresponding to each electric field, which in MKS units is given by

$$\vec{B} = \frac{1}{i\omega}(\nabla \times \vec{E}) \ . \qquad (3)$$

The Raman radiation $\vec{E}_R(\vec{r},\omega)$ at the observer coordinate $\vec{r}$ is given by

$$\vec{E}_R(\vec{r},\omega) = \vec{E}_{DIP}(\vec{r},\omega) + \vec{E}_{SC}(\vec{r},\omega) \ . \qquad (4)$$

$\vec{E}_{DIP}(\vec{r},\omega)$ is the field of the oscillating dipole $\vec{p}(\vec{r}',\omega)$ in the absence of the particle and $\vec{E}_{SC}(\vec{r},\omega)$ is a secondary or scattered field that must be computed by solving the appropriate boundary value problem at the Raman frequency. Evaluation of $\vec{E}_{SC}(\vec{r},\omega)$ corresponds to determining the elastic scattering of radiation from a dipole source located near a sphere. This can be viewed as the general problem for which the Lorenz-Mie solution is a limiting case when the dipole source is located very distant from the sphere. An earlier solution of this problem[17] in which the dipole was located along the axis of the incident ray that passes through the center of the sphere, has been generalized for any location in the outer region. The geometry and the various fields are depicted in Fig. 1. The reader is referred to Ref. 13 for the formulation of the various electric fields specified in the model.

For a distribution of molecules, it is necessary to sum the time-average power or the electric fields associated with each molecule dependent upon whether the emission is incoherent or coherent, respectively. Unless the molecules are isotropically polarizable, their orientation with respect to the particle surface must be specified. Also, if the orientation of the emission dipole at ω differs from that of the polarizable dipole at ω_o, this must also be considered. When $\vec{r}$ is sufficiently distant, the far-field approximation can be utilized. The enhancement is taken to be the

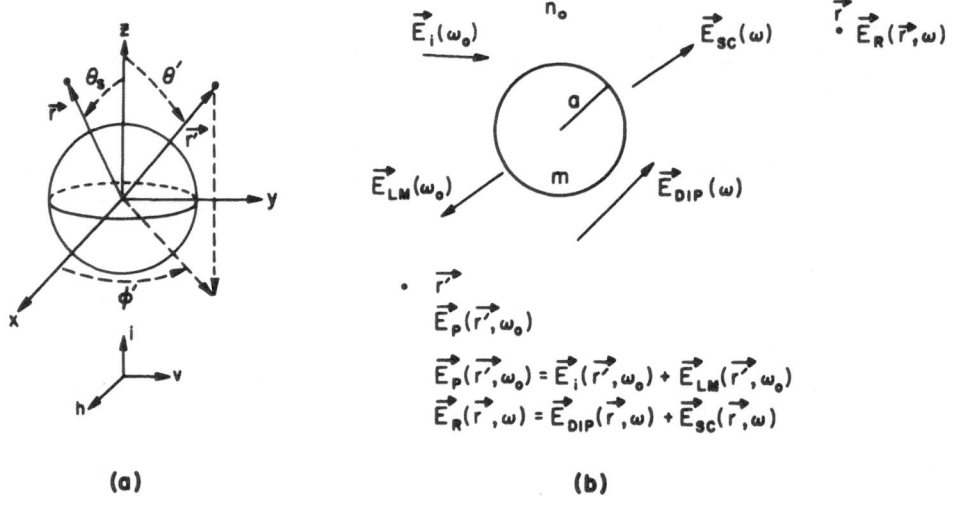

$$\vec{E}_P(\vec{r}',\omega_o) = \vec{E}_i(\vec{r}',\omega_o) + \vec{E}_{LM}(\vec{r}',\omega_o)$$
$$\vec{E}_R(\vec{r},\omega) = \vec{E}_{DIP}(\vec{r},\omega) + \vec{E}_{SC}(\vec{r},\omega)$$

(a)   (b)

Fig. 1. (a) Coordinate system showing incident direction i along
positive z axis. (b) Schematic view of electric fields near sphere
of radius a and refractive index m. At the incident frequency ω_o
there are the incident, Lorenz-Mie, and primary fields, $E_i(\omega_o)$,
$E_{LM}(\omega_o)$, and $E_p(r',\omega_o)$. At the shifted frequency ω there are the
dipole, scattered, and Raman fields, $E_{DIP}(\omega)$, $E_{SC}(\omega)$, and $E_R(r,\omega)$,
where r' and r are the positions of the dipole and observer,
respectively.

ratio of the Raman scattering intensity at $\vec{r}$ per molecule to the
intensity of a molecule with the same polarizability in the absence
of the sphere. Since such a molecule is randomly oriented in the
medium, the intensity scattered by the free molecule is averaged
over all orientations.

Because some radiation at the Raman frequency is scattered
back to the molecule there will be feedback effects leading to
higher harmonics.[18] At least with smaller particles, for which the
enhancement is greatest, the contribution of such feedback is ex-
ceedingly small and so it has been neglected. We have been unable
to estimate it in the general case.[13] Also, the mutual electromagnetic interaction between neighboring dipoles has been neglected,
an aspect that will be investigated in future work. Keeping these
limitations in mind, the solution contains complete information
about the amplitude, phase and polarisation of the inelastically
scattered fields at any location for molecules at any position
outside or at the surface of a spherical particle of any size or
optical constants. Earlier work[3-6] provides similar results for
molecules embedded within particles.

Also, the model may be applied to fluorescent scattering.

Accordingly, one may anticipate corresponding enhancements for adsorbed fluorescent molecules.

3. SMALL SPHERE LIMITS

Before proceeding to review some numerical results, we consider the limiting case for spheres which are small compared to the wavelength.[12] In this so-called Rayleigh limit, the electric field at ω_o and $\vec{r}'$ may be considered equivalent to the field of an electric dipole at the center of the sphere with dipole moment

$$\vec{p}_o = g_o a^3 \vec{E}_i (\vec{r}', \omega_o) \tag{5}$$

where $\vec{r}'$ is a vector pointing from the center of the sphere to a field point and $g_o = (m_o^2 - 1)/(m_o^2 + 2)$ with m_o the refractive index of the sphere relative to that of the medium. If there is a molecule located at $\vec{r}'$ it can be described by an electric dipole with dipole moment

$$\vec{p}_1 = \bar{\bar{\alpha}} \cdot \vec{E}_p (\vec{r}', \omega_o) \tag{6}$$

where $\vec{E}_p(\vec{r}', \omega_o)$ is the incident field plus the Lorenz-Mie field, which is now due to the dipole $\vec{p}_o$. This depends on $(a/r')^3$ in the near zone. Since we normally choose the observer coordinate so that $\vec{r} \gg \vec{r}'$, $\vec{r}$ can be considered as a vector pointing from the center of the sphere. It can then be shown that $\vec{E}_{sc}(\vec{r}, \omega)$ is the field of an electric dipole also located at the center of the particle with dipole moment also depending on $(a/r')^3$

$$\vec{p}_2 = ga^3 \vec{E}_{DIP}(-\vec{r}', \omega) . \tag{7}$$

The Raman signal is obtained from $\vec{p}_1 + \vec{p}_2$ giving rise to an enhancement for a monolayer on the surface of

$$G = \left| 1 + 2g_o + 2g + 4gg_o \right|^2 . \tag{8}$$

Accordingly, the enhancement in this limiting case does not depend explicitly on the size of the sphere.

Equation (8) can be shown to be the small particle (or low frequency) limit of the general treatment in the previous section by expanding in power series the Hankel and Bessel functions which are contained in the expansion coefficients for the fields. For a sufficiently small sphere only the leading expansion coefficients are significant and for these only the leading term in the power series usually need be used for sufficiently small particles.

However, there will be resonance conditions which will be referred to shortly for which g or g_o may increase without limit. Then it is necessary to take second-order terms in the expansion coefficients into consideration leading to the following expression for g_o

$$g_o = \frac{[(m_o^2-1)-((k_o a)^2/10)(m_o^4-1)]}{[(m_o^2+2)-((k_o a)^2/10)(10-9m_o^2+m_o^4)]} \tag{9}$$

with a corresponding expression for g and m. Since g_o and g are now dependent on the particle size, the enhancement also varies with size, even for this limiting case. Also, with this expression the enhancement will not increase without limit.

4. NUMERICAL RESULTS

     The numerical examples presented in this section are hardly exhaustive of the various phenomena that are predicted to occur when Raman scattering molecules are located near small particles and these results are merely intended to illustrate some of the principal features of SERS. Thus in this case the molecules will be described by an electric dipole with dipole moment p oriented normal to the surface. All observations are taken to occur in the x-z plane (Fig. 1). For the figures in this section a single dipole is located on the surface along the positive y-axis. The incident radiation is polarized with its electric vector parallel to the y-axis. For this case, the Raman radiation will be polarized in the same direction and the Raman signal will be independent of scattering angle. The reader is referred to Ref. 13 for illustrations of the effect of location of the dipole at various positions on the particle surface or, when the surface is covered by a monolayer, how the polarization and intensity of the Raman radiation vary with scattering angle.

     The Raman shift has been selected at 1010 cm^{-1} corresponding to the well-known pyridine line. The upper value of particle radius a = 500 nm has been dictated by computational considerations rather than any limitations of the theoretical analysis. Two sets of refractive indices have been selected. The value m = 1.50 at both the exciting and shifted wavelengths is illustrative of a typical dielectric solid in air. Although inclusion of dispersion of the refractive index with wavelength changes the quantitative results somewhat, such dispersion does not give rise to any interesting new qualitative features as long as one is dealing with dielectrics.

The refractive indices reported by Johnson and Christy[19] for silver have been used to illustrate the effects for metal surfaces. Silver, which has been utilized in most SERS studies, is particularly interesting because in the appropriate wavelength region and at the appropriate particle sizes it exhibits striking enhancements of the inelastic scattering signals. Since most experiments with silver have been carried out in aqueous solution, including ours[14] and those of Creighton et al.,[2] the refractive index relative to that of water has been used. For particles with radii less than $\sim$10 nm the phenomenological optical constants may differ from the bulk values due to surface collisions by the conduction electrons. However, any anticipated differences[20] did not affect the main qualitative features of our results. As noted earlier, enhancement will be defined by comparison with an isolated randomly oriented dipole located in the ambient medium.

Figure 2 depicts the enhancement over the 350-650 nm excitation wavelength range for silver spheres with particle radii a = 5, 50, 500 nm. The smallest particle exhibits a sharp peak somewhat greater than 10^6 at 382 nm which drops off rapidly at both higher and lower wavelengths. Yet even these lesser values are greater than 10^3. For the 50 nm particle the enhancement peak of nearly 10^4 is much broader and it is shifted to a longer wavelength. For the still larger 500 nm particle the very much smaller enhancement oscillates in the range of 10^1 to 10^2 throughout the visible and then decreases to values of less than unity in the ultraviolet.

The effect of silver particle size is shown in more detail in Fig. 3 where the enhancement is plotted for two wavelengths, λ_o = 382 and 514.5 nm. The former is the one at which sufficiently small silver particles show the peak enhancement. The latter corresponds approximately to the peak enhancement for the 50 nm particle. The enhancement oscillates mainly in the 10^1 to 10^4 range except for radii <30 nm at λ = 382 nm for which the enhancement increases sharply to about 10^6 at $\sim$ 5 nm. For still smaller sizes, the enhancement remains constant at this value.

Although these results have been calculated using the full expressions for the electromagnetic fields, insight into the origin of the large enhancement for the smallest particle can be obtained by reference to eq. (8). The condition for the enhancement to become very large is that either g_o or g becomes very large and this will occur whenever either refractive index m or m_o is nearly a purely imaginary number with value $\sqrt{2}i$ (or the dielectric constant is -2) in which case the denominator (m^2+2) becomes zero. This is the condition for excitation of a dipolar surface plasmon. For silver at 382 nm m_o = 0.04-1.41 i. Of course as noted above the enhancement will never blow up to infinity as suggested by eq. (8) since for sufficiently small particles and values of m or m_o very

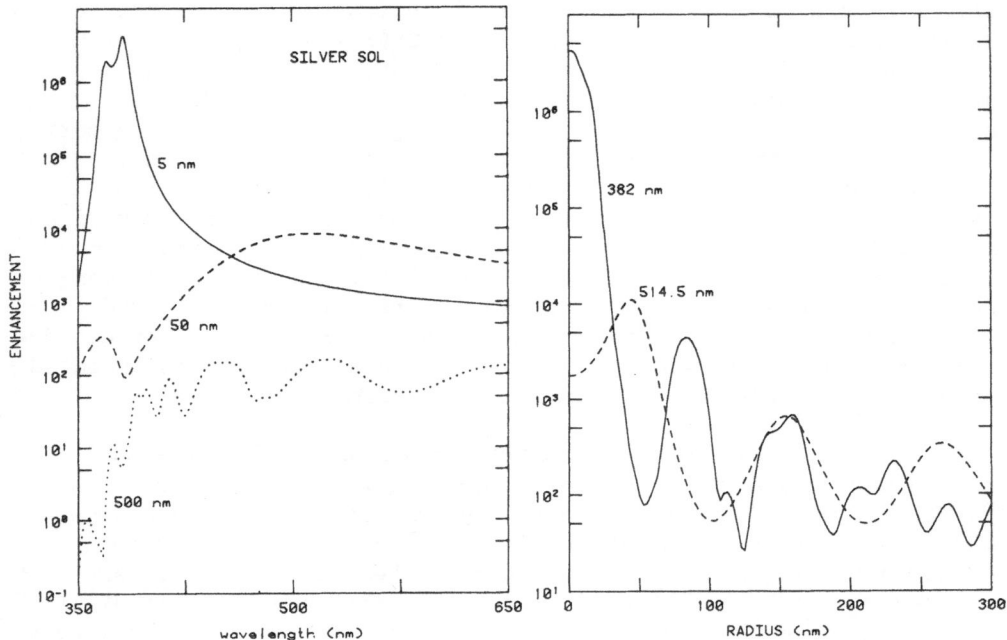

Fig. 2. Enhancement vs excitation wavelength for silver
particles of radii 5, 50, and
500 nm. Dipole is located on
the surface along positive y
axis, observer is in the far
field along positive x axis,
Raman band is 1010 cm^{-1}.

Fig. 3. Enhancement vs radius
of silver particle at λ = 382
and 514.5 nm. Other conditions
same as Fig. 2.

close to $\sqrt{2}i$ a higher approximation such as eq. (9) or the full
formalism of eq. (4) must be utilized.

The manner in which the enhancement falls off with distance
of the molecular dipole from the surface is shown in Fig. 4.

In Fig. 5 the enhancement of a dielectric sphere (m = 1.50)
is shown as a function of excitation wavelength for particle
radii 5, 50, and 500 nm. Significant albeit not gigantic enhancements can be obtained even for such a dielectric, indicating the
effect of the local electromagnetic fields upon this process.

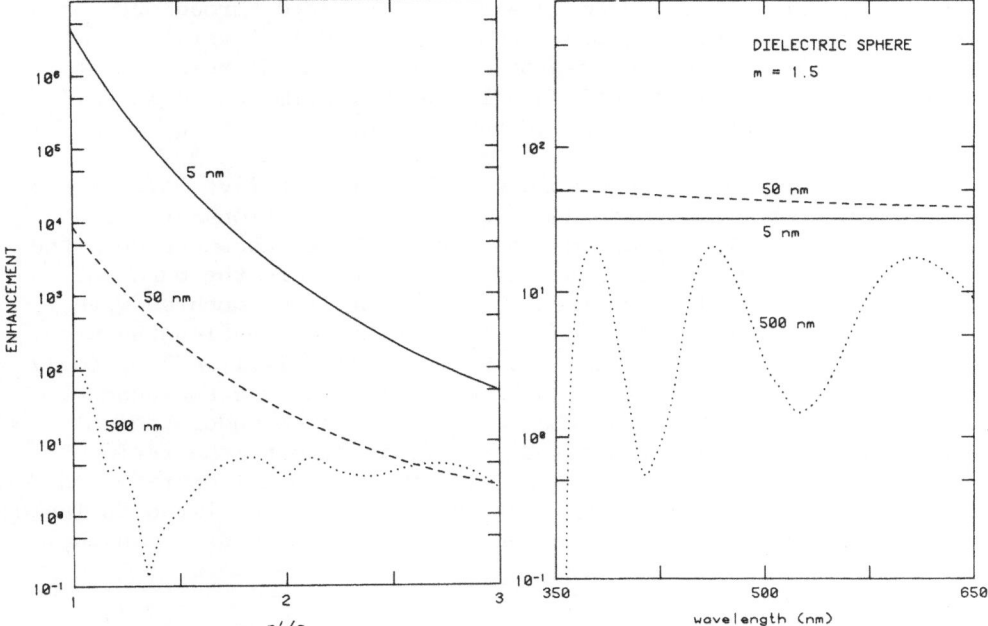

Fig. 4. Enhancement vs distance of dipole from origin (in
units of particle radius) for
silver particles with a = 5 nm
(λ_o = 382 nm), a = 50 nm (λ_o =
511 nm), and a = 500 nm (λ_o =
528 nm).

Fig. 5. Same as Fig. 2 for
dielectric particle, m = 1.50.

5. EXPERIMENTAL RESULTS

We now report initial measurements of Raman signals enhanced
greater than 10^5-fold from citrate ion adsorbed on colloidal
silver.[14] Although this confirms the theoretical model just discussed in that SERS comparable to that observed on roughened
macroscopic surfaces also occurs in colloidal dispersions, the
measured wavelength dependence of the Raman intensity differs
substantially from that predicted by the theoretical model.

A dense red silver hydrosol, prepared by Carey Lea's method[21]
was comprised of 21 nm average radius particles as estimated by
comparison of the measured optical extinction with calculated
values. The particles have been shown to be covered by a monolayer
of citrate, each ion occupying an area of 38Å^2 so that by assuming

all of the silver introduced as $AgNO_3$ is retained throughout the
preparative procedure and converted to colloidal silver, we were
able to calculate a concentration of $6.4x10^{-6}$ M adsorbed citrate
in our samples. This is undoubtedly an overestimate and it will
result in underestimation of the enhancement.

Raman spectra of these sols were measured at five wavelengths.
Measurements were also made under identical conditions of 1.37 M
sodium citrate. The strong dependence of the enhancement upon the
incident wavelength is illustrated in Fig. 6 where the bands at
1400 cm^{-1} from the 1.37 M citrate on the left are compared with
those from the sol on the right. The excitation wavelengths proceeding down the figure from A to E are 647.1, 514.1, 457.9, 406.7
and 350.7 nm, respectively. Numerical values of the enhancement
are obtained by multiplying the ratio of the area under the sol
Raman band to that under the sodium citrate band by the ratio of
the sodium citrate concentration (1.37 M) to the sol citrate concentration ($6.4x10^{-6}$M). The enhancements which are listed in Table
1 vary from less than $3x10^3$ in the ultraviolet to $6.0x10^5$ in the
red.

Also listed in the table are values of the enhancements for
the 1400 cm^{-1} band calculated for 20 and 50 nm radii. Both theory
and experiment give maximum enhancements in the range 10^5 to 10^6;
however, the wavelength dependences differ. Whereas the theoretical
values for 20 nm particles show a sharp peak of greater than 10^5
at 400 nm, the measured enhancements do not display this peak but
appear to increase monotonically with increasing wavelength.

Table 1.   Measured and Calculated Raman Enhancements

Wavelength (nm)	Integrated band intensity relative to 1.37 M Na citrate	Measured enhancement	Calculated enhancement for various radii (nm)	
			20	50
350.7	< 0.014	< $3.1x10^3$	$3.5x10^2$	$5.4x10^1$
406.7	0.14	$3.1x10^4$	$5.2x10^4$	$5.6x10^2$
457.9	1.5	$3.2x10^5$	$1.7x10^3$	$1.2x10^3$
514.5	1.7	$3.6x10^5$	$5.6x10^2$	$1.7x10^3$
647.1	2.8	$6.0x10^5$	$2.1x10^2$	$6.0x10^2$

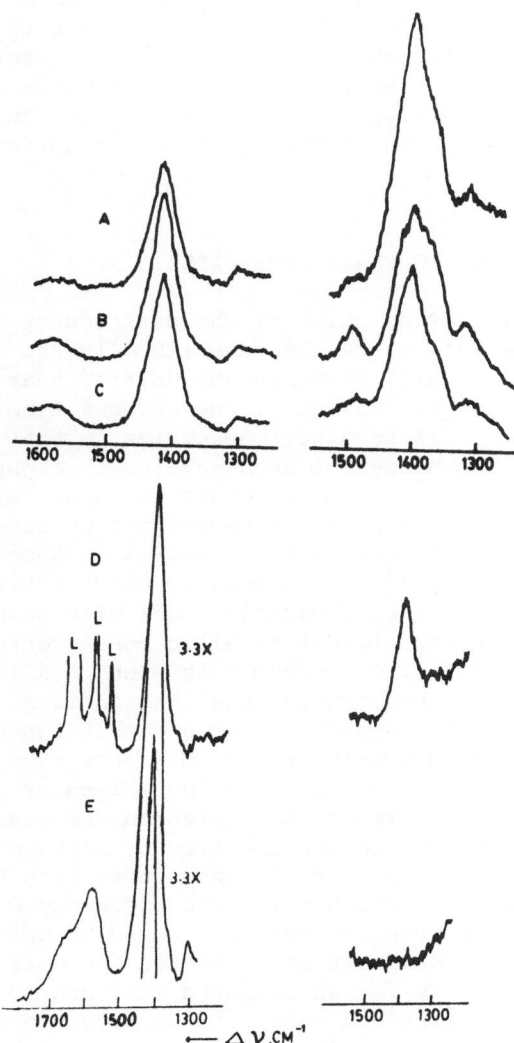

Fig. 6. Raman spectra of 1.37 M sodium citrate, LHS, and of silver sol with adsorbed citrate, RHS, in 1400-cm^{-1} region. Experimental conditions: laser power, 50 mW; spectral slit width, 10 cm^{-1}; scan speed, 1 cm^{-1}/sec.

	Excitation wavelength, nm	Sensitivity	Time constant, sec
A	647.1 Kr$^+$	4000 cps	2
B	514.5 Ar$^+$	4000 cps	2
C	457.9 Ar$^+$	4000 cps	2
D	406.7 Kr$^+$	LHS, 1 µA; RHS, 3 nA	4.4
E	350.7 Kr$^+$	LHS, 1 µA; RHS, 3nA	4.4

Subsequent analysis of the extinction spectrum of the sol as well as electron microscopic examination[22] indicate that the particles in this sol were probably coagulated into clusters. A sol prepared subsequently by a somewhat different procedure appears to be comprised of isolated particles. This yellow and considerably more stable preparation is the subject of a current investigation.

6. EXTENSION OF THEORY TO SMALL SPHEROIDS

A clue to possible resolution of the discrepancy between the experimental and theoretical results described in the above section is the suggestion by McCall, Platzman and Wolff[23] that the surface plasmon resonance condition varies if one changes from an isolated sphere (for which $m = \sqrt{2}i$) to other geometries so that the dependence of the net Raman enhancement upon wavelength might be obtained by an appropriate distribution of resonant frequencies, i.e., an appropriate distribution of particle geometries or arrangements. This notion was pursued in much greater detail at about the same time by Gersten and Nitzan[24] who solved the electrostatic model (i.e. low frequency limit; sufficiently small dimensions of the boss) for a prolate hemispheroidal metallic boss protruding perpendicularly from an infinitely conducting plane. A further simplification was achieved by assuming that the incident light beam propagates parallel to the symmetry axis and that the Raman active molecule is located on this axis at some distance from the spheroidal boss on the illuminated side with the molecular dipole oriented parallel to the axis. The surface plasmon resonance frequencies, and hence the excitation spectrum of SERS, depend on the shape of the boss, a finding which is consistent with the apparently conflicting results obtained for the frequency dependence of SERS, since differently prepared surfaces would be characterized by different surface irregularities. Gersten and Nitzan also present analytical results for an isolated full spheroid subject to the same simplifications and Adrian[25] has provided a calculation for this model showing how the SERS factor varies with eccentricity for a particular excitation and Raman frequency.

We have relaxed the above restrictions for an isolated spheroid in the electrostatic limit so that the spheroid may be oriented in any manner with respect to the incident beam and the molecule may be located at any position outside of or on the surface of the spheroid.[15] Both prolate and oblate spheroids are treated. The analysis and detailed numerical results are provided in ref. 15 for a spheroid which is randomly oriented with respect to the incident beam and is covered with a monolayer of dipoles having their axes oriented perpendicular to the spheroidal surface. As before, enhancements are expressed with regard to randomly oriented iso-

lated molecules having Raman polarizabilities identical to those
of adsorbed molecules.

Figure 7 depicts the enhancement of the 1010 cm^{-1} band over
the excitation wavelength range 350 to 650 nm for randomly oriented,
monolayer-covered silver prolate spheroids in water with the axial
ratios equal to 1.0, 1.5, 2.0, 2.5 and 3.0. (The curve for the
sphere, axial ratio 1, differs from the 5 nm curve in Fig. 2 by a
factor of five because the former curve is for a monolayer rather
than for a single molecule located on the positive y-axis). The
maximum enhancement increases in magnitude and shifts toward longer
excitation wavelengths as the axial ratio increases. Accordingly,
one might synthesize almost any excitation profile by an appropriate distribution of shapes. Similar results are obtained with
oblate spheroids and also for other Raman shifts.

An interesting aspect of each excitation curve is the two
closely spaced peaks which are separated precisely by the Raman
shift, in this case 1010 cm^{-1}. Also, whereas the separation of
the two peaks varies with the Raman shift, the second of each pair
of peaks (that at the longer wavelength) always occurs at the excitation wavelength.

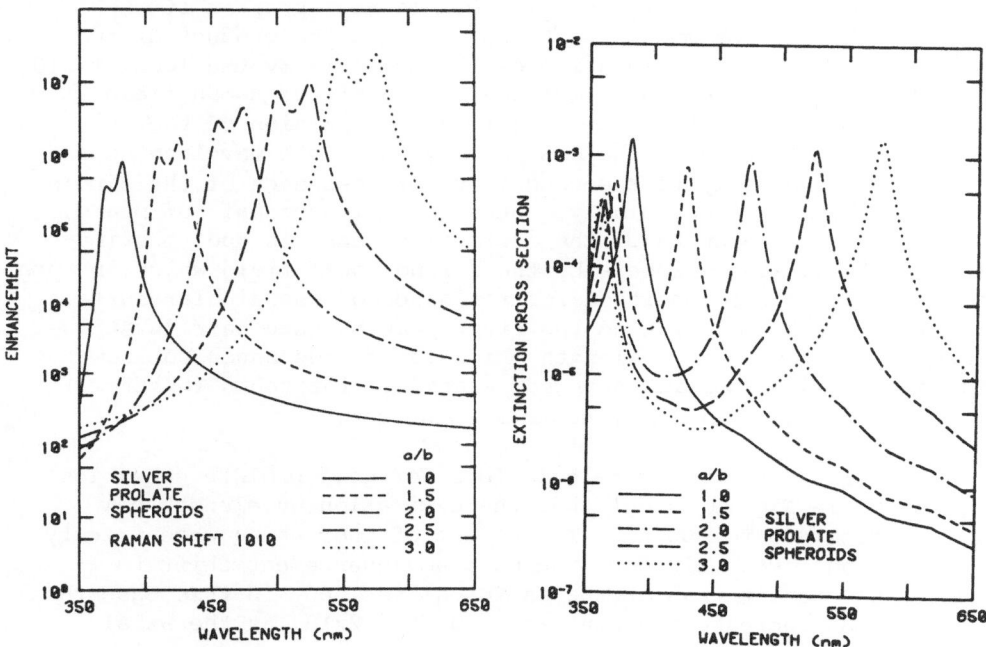

Fig. 7. Enhancement for silver
spheroids of various axial ratios
vs excitation wavelength.

Fig. 8. Extinction for silver
spheroids of various axial
ratios vs excitation wavelength.

In order to explore these effects more closely, the extinction cross sections for these randomly oriented silver spheroids are shown in Fig. 8 over the same range of wavelengths and axial ratios as in Fig. 7. Appropriate expressions for this calculation were given by Gans.[26]

The central features are the two sharp extinction peaks for the spheroids and the single peak for the sphere. These peaks are due to excitation of the two dipolar plasmon resonances corresponding to components of the electric field parallel to the axes of the spheroid; that at the higher wavelength corresponds to the axis of symmetry, at the lower wavelength to the two-fold axes. It should be noted that for small particles absorption makes the predominant contribution to extinction so that high extinction implies a strong electric field within the particle. Since, in this case, the field is uniform, the fields at the surface, both within and outside of the particle are correspondingly high. Therefore high extinction corresponds to high local fields for stimulation of the Raman process.

Comparison of Figs. 7 and 8 shows that the second peak in each of the SERS curves is located at the same excitation wavelength as the longer of the two wavelengths for maximum extinction. The physical origin of the double peak is now apparent if one recalls that in our model the Raman field is the product of two processes, viz. (1) stimulation of the molecule by the local field at the incident wavelength, (2) scattering of the Raman field at the shifted wavelength by the particle. Dispersion of the refractive index obviates the possibility that both wavelengths can correspond precisely to the condition for resonance of the dipolar surface plasmon. Accordingly, the second of each pair of peaks, being at the wavelength of the extinction peak, is due to stimulation of the Raman process by the strong local field which in turn corresponds to the condition for resonance of the dipolar surface plasmon. On the other hand the first peak of each pair is stimulated by an incident wavelength which is off resonance but now the Raman shifted signal is enhanced because it interacts (scatters) with the particle at the resonance condition.

Similar enhancement calculations for gold prolate spheroids are shown in Fig. 9 except that the excitation wavelength now ranges from 400 to 800 nm. The effect of increasing eccentricity is striking. Not only does the maximum enhancement shift to longer excitation wavelengths as for silver, but in this case the enhancement increase sharply from 10^3 to 2×10^6 as the axial ratio changes from 1.0 to 3.0. The extinction, shown in Fig. 10, exhibits a corresponding pattern. The extinction maxima occur at the same wavelengths as the enhancement maxima with the peak extinction cross sections rising nearly an order of magnitude as the particle is extended from a sphere to a prolate spheroid with

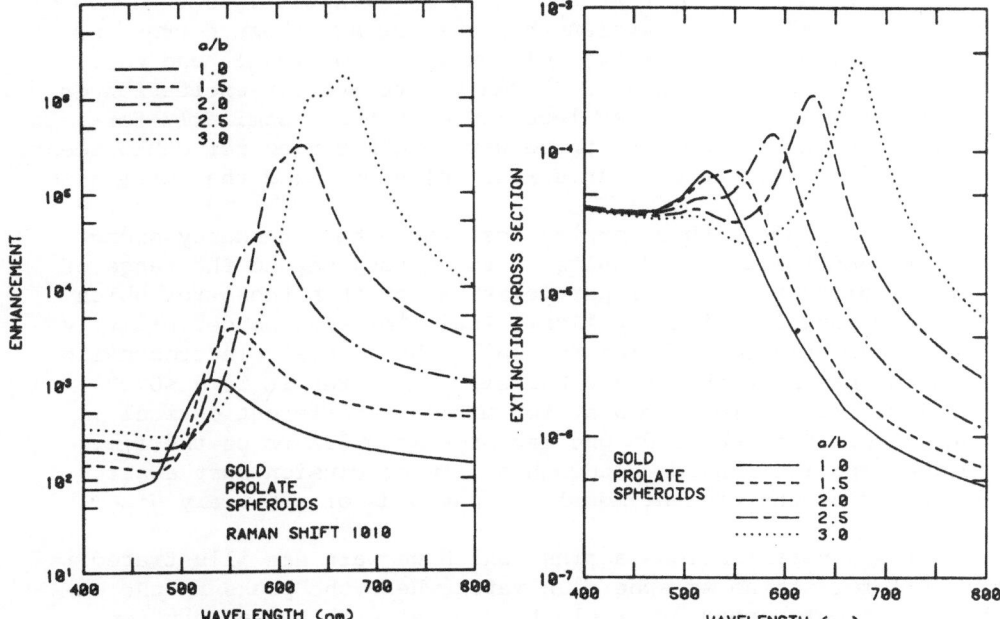

Fig. 9. Enhancement for gold spheroids of various axial ratios vs excitation wavelength.

Fig. 10. Extinction for gold spheroids of various axial ratios vs excitation wavelength.

axial ratio 3.0. Another aspect is that only the most eccentric spheroid shows a double peak. This is because of the broader extinction curves so that either excitation into the extinction peak or Raman scattering into this peak does not result in a significantly greater Raman enhancement than when either of these processes occurs at neighboring wavelengths. The emergence of the double peak for axial ratio 3.0 results from the narrowing of the extinction curve with increasing axial ratio. Calculations for copper display qualitative features similar to those for gold except that in this case the enhancement and extinction maxima are not displaced to longer wavelengths as sharply as gold. The double peak which Au exhibits at axial ratio 3.0 fails to emerge with Cu, a result which is consistent with the somewhat broader extinction curve for Cu. The corresponding peak enhancements for Cu are less than those for gold which is also consistent with the lower corresponding values of the extinction cross section.

The longer wavelength extinction maxima for Ag in Fig. 8 as well as those displayed for Au in Fig. 10, correspond to resonant

excitation along the axis of symmetry. There are additional extinction maxima for both Au and Cu at wavelengths less than 400 nm
and just as the lower wavelength peaks for Ag, these correspond
to excitations along the two-fold axes. Accordingly, one would
expect corresponding enhanced Raman scattering for excitation or
for Raman emission at these lower wavelengths. Similarly for
ellipsoids there should be three wavelength ranges for enhancement,
corresponding to the dipolar surface plasmon along the three axes.

We conclude with a warning lest there be a tendency among
some to generalize the results of this study beyond the range of
its validity, namely, for particles larger than those for which
the electrostatic limit applies. That limit can be definitively
prescribed only for spheres for which the general electrodynamic
solution has been obtained. For silver spheres it is a $<0.02\lambda$;
the limit may differ somewhat for markedly different optical
constants. For spheroids one may presume 0.02λ to be the limit
on the longer dimension although a firm conclusion must await
extension of the present model to spheroids of arbitrary size.

The generalizations against which we warn are illustrated in
Fig. 11 for a 5 nm Ag sphere in water where the peaks in the enhancement correspond precisely to the peaks in the absorption,
scattering and extinction peaks, i.e., excitation is into the dipolar surface plasmon resonance or the Raman emission is into·
this resonance.

With larger particles one is dealing with a superposition
of multipolar fields[16] rather than with a dipolar field. The
extinction and Raman enhancement are not linked in a simple linear
fashion. The absorption cross section is related to the distribution of lossy sinks throughout the particle which in turn is
determined by what may be a very complicated distribution of the
electric field.[27] The scattering cross section is the spatial
average of the squared modulus of the field in the radiation zone.
The Raman signal on the other hand arises in two steps. The molecule is stimulated by a local field comprised of the near field
elastically scattered by the particle coherently added to the
incident field; then the observer views the superposition of the
dipolar field of the Raman molecule also coherently added to a
field elastically scattered by the particle at the Raman wavelength.

Figure 12 illustrates the various cross sections for a 50 nm
radius Ag sphere. The SERS excitation is the same as in Fig. 2.
The absorption, scattering and extinction spectra were calculated
with the Lorenz-Mie equations.[16] Although there is a sharp extinction peak at 380 nm, this does not translate into a correspondingly large enhancement. Indeed the Raman enhancement shows a
minimum at this wavelength. The peaks in the SERS excitation curve

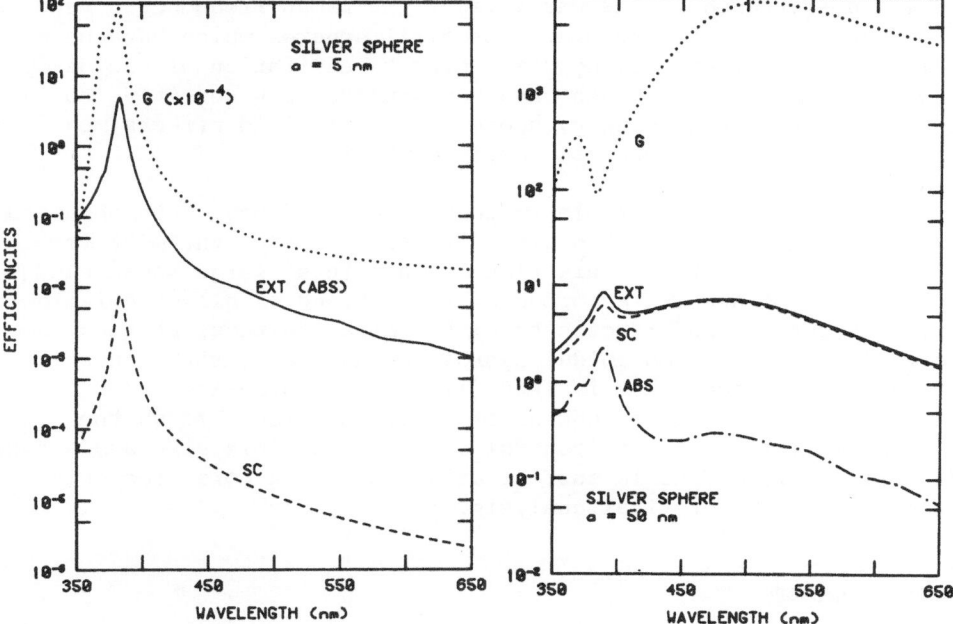

Fig. 11. Comparison of wavelength dependence of extinction,
scattering and absorption cross
sections, as well as enhancement
of the 1010 cm^{-1} Raman shift, for
an adsorbed monolayer on 5 nm
Ag spheres in water.

Fig. 12. Same as for Fig. 11
for 50 nm Ag spheres.

and those in the Lorenz-Mie cross sections appear to be quite un-
related for still larger particles. This is because the electromagnetic fields at the incident and the Raman wavelengths, each of
which depend in a complicated way upon ratio of size to wavelength
and the refractive index, are no longer connected in a simple
fashion.

7. CONCLUSION

     Many of the principal features of SERS are exhibited by a
classical electromagnetic model in which an isolated colloidal
sphere of arbitrary size and optical constants, coated with a monolayer of classical electric dipoles, is stimulated by a local field
comprised of the incident plus the Lorenz-Mie near field. The
Raman signal is comprised of the direct dipole reradiation plus

radiation scattered by the particle at the Raman frequency. Very large enhancements are obtained for small spheres which have been stimulated at a frequency corresponding to excitation of the dipolar surface plasmon. Such enhancements are very sensitive to particle shape. Accordingly, these classical field effects must be considered in any complete theory of SERS.

Enhancements comparable to those predicted have been observed with colloidal silver coated with citrate. Perhaps the most important point to note at this time is that these large Raman band enhancements of species adsorbed at the surface of silver colloids permit structural and kinetic studies similar to those at roughened electrode surfaces. Colloidal hydrosols, as a substrate, have a number of advantages that include (1) ease of formation and manipulation, (2) continuous renewal of sample by flow through the light beam, (3) ability to control and vary particle size and shape, (4) more easily definable surface area, and (5) a more tractable morphology for theoretical analysis.

ACKNOWLEDGMENTS

This work has been supported in part by Army Research Office Grant DAAG-29-79-C-0059, National Science Foundation Grant CHE-8011444, and DOE Contract DE-ACO2-77EV04361.

REFERENCES

1. M. Moskovits, Surface roughness and the enhanced intensity of Raman scattering by molecules adsorbed on metals, J. Chem. Phys. 69:4159 (1978).
2. J. A. Creighton, C. G. Blatchford, and M. G. Albrecht, Plasma resonance enhancement of Raman scattering by pyridine adsorbed on silver and gold particles of size comparable to the excitation wavelength, J. Chem. Soc., Faraday Trans. II. 75:790 (1979).
3. H. Chew, P. J. McNulty, and M. Kerker, Model for Raman and fluorescent scattering by molecules embedded in small particles, Phys. Rev. A 13:396 (1976).
4. H. Chew, M. Kerker, and P. J. McNulty, Raman and fluorescent scattering by molecules embedded in concentric spheres, J. Opt. Soc. Am. 66:440 (1976).
5. H. Chew, D. D. Cooke, and M. Kerker, Raman and fluorescent scattering by molecules embedded in dielectric cylinders, Appl. Opt. 19:44 (1980).
6. D.-S. Wang, M. Kerker, and H. Chew, Raman and fluorescent scattering by molecules embedded in dielectric spheroids, Appl. Opt. 19:1573 (1980).

7. J.P. Kratohvil, M.-P. Lee, and M. Kerker, Angular distribution of fluorescence from small particles, Appl. Opt. 17:1978 (1978); also P.J. McNulty, S.D. Druger, M. Kerker, and H. Chew, Fluorescent scattering by anisotropic molecules embedded in small particles, Appl. Opt. 18:1484 (1979); also M. Kerker, P. Hsu, J. P. Kratohvil, M. van Dilla, J. Grey and A. Brunsting, paper in preparation.

8. E. H. Lee, R. E. Benner, J. B. Fenn, and R. K. Chang, Angular distribution of fluorescence from monodispersed particles, Appl. Opt. 17:1980 (1978).

9. E. H. Lee, R. E. Benner, J. B. Fenn, and R.K. Chang, Angular distribution of fluorescence from liquids and monodisperse spheres by evanescent wave excitation, Appl. Opt. 18:862 (1979).

10. R. E. Benner, J. F. Owen, and R. K. Chang, Radiation patterns of inelastic reemission from microparticles in homogeneous surroundings and near dielectric or metal surfaces, J. Phys. Chem. 84:1602 (1980).

11. R. E. Benner, P. W. Barber, J. F. Owen, and R. K. Chang, Observation of structure resonances in the fluorescence spectra from microspheres, Phys. Rev. Lett. 44:475 (1980).

12. D.-S. Wang, M. Kerker and H. Chew, Enhanced Raman scattering at the surface (SERS) of a spherical particle, Appl. Opt. 19:2256 (1980).

13. M. Kerker, D.-S. Wang, and H. Chew, Surface enhanced Raman scattering (SERS) by molecules adsorbed at spherical particles, Appl. Opt. 19:4159 (1980).

14. M. Kerker, O. Siiman, L. A. Bumm and D.-S. Wang, Surface enhanced Raman scattering (SERS) of citrate ion adsorbed on colloidal silver, Appl. Opt. 19:3253 (1980).

15. D.-S. Wang and M. Kerker, Enhanced Raman scattering (SERS) by molecules adsorbed at the surface of colloidal spheroids, Phys. Rev. In press.

16. M. Kerker, "The Scattering of Light and other Electromagnetic Radiation," Academic Press, New York (1969).

17. H. Chew, M. Kerker, and D. D. Cooke, Electromagnetic scattering by a dielectric sphere in a diverging radiation field, Phys. Rev. A 16:320 (1977).

18. S. Efrima and H. Metiu, Classical theory of light scattering by an adsorbed molecule. 1. Theory, J. Chem. Phys. 70:1602 (1979).

19. P. B. Johnson and R. W. Christy, Optical constants of the noble metals, Phys. Rev. B 6:4370 (1972).

20. U. Kreibig and P. Zacharias, Surface plasma resonances in small spherical silver and gold particles, Z. Physik. 231: 128 (1970).

21. G. Frens and J.Th.G. Overbeek, Carey Lea's colloidal silver, Kolloide Zs. u. Zs. f. Polymere. 233:922 (1969).

22. Electron micrographs were kindly supplied by Professor R. K. Chang.

23. S. L. McCall, P. M. Platzman, and P. A. Wolff, Surface
     enhanced Raman scattering, Phys. Lett. 77A:381 (1980).
24. J. Gersten and A. Nitzan, Electromagnetic theory of enhanced
     Raman scattering by molecules adsorbed on rough surfaces,
     J. Chem. Phys. 73:3023 (1980).
25. F. J. Adrian, SERS by surface plasmon enhancement of EM
     fields near spheroidal particles on a roughened metal
     surface, Chem. Phys. Lett. 78:45 (1981).
26. R. Gans, On the shape of ultramicroscopic gold particles,
     Ann. d. Physik. 37:881 (1912).
27. P. W. Dusel, M. Kerker, and D. D. Cooke, Distribution of
     thermal sources within irradiated spheres, J. Opt. Soc. Am.
     69:55 (1979).

A MODEL INCLUDING EXCITATION OF SURFACE PLASMON POLARITONS

Sudhanshu S. Jha

Tata Institute of Fundamental Research
Homi Bhabha Road, Bombay 400005
India

I. INTRODUCTION

The surface-enhanced Raman effect is the observation that Raman cross sections for excitations of many vibrational modes of certain molecules like pyridine are enhanced by a factor $10^2 - 10^6$ or more when adsorbed or deposited on a metal like Ag, as compared to their values in the free state. The problem is to understand the physical processes responsible for such selective and specific enhancements. For several years now, a general theoretical formulation for this effect,[1-5] observed with a variety of substrate shapes and structures in different types of systems, has been hindered by somewhat conflicting experimental results, mostly due to difficulties in characterizing these samples and differences in molecular environment obtained in such experiments. However, during the last year or so,[6-11] a better theoretical picture is emerging which is capable of analyzing experimental data on substrates with well characterized surface profiles. We will present here salient features of such a theoretical model for the surface-enhanced Raman scattering (SERS).

There is enough experimental evidence[12-14] to show that the submacroscopic surface roughness of the metal and the geometrical shape of the substrate play a significant role in SERS. This arises because of resonant excitation of the surface plasmon polaritons (SPP) of the substrate by the light wave and consequent enhancement of the induced dipole-moment of the molecule, as well as the modification in the propagation of the Stokes scattered field. Since the SPP mode frequencies and the efficiency of their excitation by a transverse light wave depend considerably on the shape and nature of the metal surface, the net enhancement via SPP resonances is a strong function of the surface morphology of the substrate. However, the classical enhancement via SPP resonances does not tell the whole story for the first one or two molecular monolayers on the surface. When the distance d of the molecule from the metal surface is of the order of atomic distances (a few Å), there can be an additional short range enhancement due to (1) local changes in the ground state electronic charge density distribution near the surface arising from bonding or tunneling of metal electrons to the molecular site, and (2) the new excitation spectrum of the metal-molecule complex, including the tunneling and charge-transfer states.

The classical enhancement due to SPP resonances is relatively long range, essentially governed by the distance scale over which the SPP field amplitude decays appreciably outside the metal. A significant classical enhancement can thus occur as long as the molecular distance d is small compared to the wavelength $\lambda = 2\pi c/\omega$ of light. Note that, in contrast to the additional short range enhancement, the long range part of the enhancement is independent of the nature of the specific metal-molecule interaction and the adsorption properties of the surface. No changes in molecular polarizability have to be considered because of the metal-molecule interaction, if d is large compared to atomic distances. The long range SPP enhancement is essentially a consequence of the interaction of the incident and scattered photons with the metal, in the presence of the polarizable molecule. In terms of the phenomenological optical dielectric function of the metal and the polarizability of an isolated molecule, this enhancement may be calculated using classical electrodynamics. A more detailed quantum calculation which takes into account the metal-molecule interaction is, however, necessary to obtain the very interesting short range enhancement.

In what follows, we will first describe the theory of long range resonant SPP enhancement calculation for a molecule deposited on a metal with an arbitrary shape and surface profile. Explicit results will be presented for a plane surface with small amplitude surface roughness and also for a sphere. A phenomenological model will then be presented in Section III, for calculating the additional short range quantum enhancement for the first one or two monolayers of adsorbed molecules. Note that the resonant SPP enhancement will also be present for the molecules at atomic distances from the metal substrate.

II.   LONG RANGE CLASSICAL SPP ENHANCEMENT FACTOR

A.   Raman Intensity for an Isolated Molecule

Before we consider the enhancement factor arising from the resonant SPP excitation of the metal substrate, it is logical to go through the familiar classical electrodynamic calculation of Raman intensity from an isolated molecule in free space. Let

$$\vec{E}_{in}(\vec{r},t) = \vec{E}_{inc}(\vec{r},\omega)e^{-i\omega t} = \vec{E}_i e^{i\vec{k}\cdot\vec{r}}e^{-i\omega t}; \ k = \frac{\omega}{c} \tag{1}$$

represent the incident light wave on the molecule located at $\vec{r} = \vec{r}_o$. For calculating the radiation from the induced dipole at the optical frequency ω, the molecule can be represented by a point dipole, with the polarization

$$\vec{P}(\vec{r},\omega) = \vec{p}_o(\omega)\delta(\vec{r} - \vec{r}_o) \quad , \tag{2}$$

where $\vec{p}_o(\omega)$ is the induced dipole moment in the molecule. In free space, the field pattern at the incident frequency ω is then obtained by the solution of Maxwell's equation

$$- \nabla^2\vec{E}_o(\vec{r},\omega) + \vec{\nabla}\vec{\nabla}\cdot\vec{E}_o(\vec{r},\omega) - \frac{\omega^2}{c^2}\vec{E}_o(\vec{r},\omega) = \frac{4\pi\omega^2}{c^2}\vec{p}_o(\omega)\delta(\vec{r} - \vec{r}_o) \ , \tag{3}$$

where, in the absence of the dipole, $\vec{E}_o$ is given by $\vec{E}_{inc}(\vec{r},\omega)$.

Equation (3) can be solved in general by using the free space Green function

$$\vec{\vec{G}}_o(\vec{r},\vec{r}_o,\omega) = (\vec{\vec{I}} + \frac{c^2}{\omega^2}\vec{\nabla}\vec{\nabla}) \frac{e^{i\omega|\vec{r} - \vec{r}_o|/c}}{|\vec{r} - \vec{r}_o|} \quad , \tag{4}$$

which satisfies the equation

$$[- \nabla^2 + \vec{\nabla}\vec{\nabla}\cdot - \frac{\omega^2}{c^2}] \ \vec{\vec{G}}_o(\vec{r},\vec{r}_o,\omega) = 4\pi\vec{\vec{I}}\delta(\vec{r} - \vec{r}_o) \quad , \tag{5}$$

where $\vec{\vec{I}}$ is the unit tensor. The solution is

$$\vec{E}_o(\vec{r},\omega) = \vec{E}_{inc}(\vec{r},\omega) + \frac{\omega^2}{c^2}\vec{\vec{G}}_o(\vec{r},\vec{r}_o,\omega)\cdot\vec{p}_o(\omega) \quad . \tag{6}$$

Since, in terms of bare molecular polarizability $\vec{\vec{\alpha}}_o$,

$$\vec{p}_o(\omega) = \vec{\vec{\alpha}}_o(\omega)\cdot\vec{E}_o(\vec{r}_o,\omega) \quad , \tag{7}$$

the solution (6) implies that the form

$$\vec{p}_o(\omega) = [\vec{\vec{I}} - \frac{\omega^2}{c^2}\vec{\vec{\alpha}}_o(\omega)\cdot\vec{\vec{G}}_o(\vec{r}_o,\vec{r}_o,\omega)]^{-1}\cdot\vec{\vec{\alpha}}_o(\omega)\cdot\vec{E}_{inc}(\vec{r}_o,\omega) \tag{8}$$

is self-consistent. Here, for the sake of taking the inverse, a
second-rank tensor is assumed to be represented by the corresponding
3 x 3 matrix. Since the experimental molecular polarizability tensor
$\overset{\leftrightarrow}{\alpha}(\omega)$ for an isolated molecule at $\vec{r} = \vec{r}_o$ is defined by

$$\vec{p}_o(\omega) = \overset{\leftrightarrow}{\alpha}(\omega) \cdot \vec{E}_{inc}(\vec{r}_o,\omega) \quad , \tag{9}$$

one gets the renormalization relation

$$\overset{\leftrightarrow}{\alpha}(\omega) = [\overset{\leftrightarrow}{I} - \frac{\omega^2}{c^2} \overset{\leftrightarrow}{\alpha}_o(\omega) \cdot \overset{\leftrightarrow}{G}_o(\vec{r}_o,\vec{r}_o,\omega)]^{-1} \cdot \overset{\leftrightarrow}{\alpha}_o(\omega) \quad . \tag{10}$$

Note that even in free-space, the experimental polarizability is
different than the bare polarizability. One has to assume that
$\overset{\leftrightarrow}{\alpha}(\omega)$ is well defined, even though both $\overset{\leftrightarrow}{G}_o(\vec{r}_o,\vec{r}_o,\omega)$ and the bare
polarizability $\overset{\leftrightarrow}{\alpha}_o$ may not be very well defined. The trick is to
work always with the experimental polarizability $\overset{\leftrightarrow}{\alpha}(\omega)$.

Because of molecular vibrations, $\overset{\leftrightarrow}{\alpha}(\omega)$ is a function of the
normal coordinates $\vec{Q} e^{i\omega_\nu t}$, $\omega_\nu \ll \omega$, where ω_ν is the particular mode
frequency. This implies that one now has an induced dipole moment
at Stokes frequency $\omega - \omega_\nu$, having the value

$$\vec{p}_{Ram}(\omega_s = \omega - \omega_\nu) = \vec{Q} \cdot \left[\frac{\partial}{\partial \vec{Q}} \overset{\leftrightarrow}{\alpha}(\omega,\vec{Q}) \right]_o \cdot \vec{E}_{inc}(\vec{r}_o,\omega) \quad . \tag{11}$$

The corresponding Maxwell's equation at the frequency ω_s then leads
to the Stokes field

$$\vec{E}_{os}(\vec{r},\omega_s) = \frac{\omega_s^2}{c^2} \overset{\leftrightarrow}{G}_o(\vec{r},\vec{r}_o,\omega_s) \cdot \vec{p}_{Ram}(\omega_s) \quad , \tag{12}$$

there being no Stokes field in the absence of $\vec{p}_{Ram}$. Since in the
radiation zone, (4) leads to

$$\overset{\leftrightarrow}{G}_o(\vec{r} \to \infty,\vec{r}_o,\omega_s) \cdot \vec{p}_{Ram}(\omega_s) \to \frac{e^{i\frac{\omega_s}{c}r}}{r} \hat{n}_s \times \hat{n}_s \times \vec{p}_{Ram}(\omega_s) \tag{13}$$

where $\hat{n}_s$ is a unit vector along the scattering direction $\vec{r}$, the
Stokes power radiated per unit solid angle, with a given unit
polarization vector $\hat{e}_s$, becomes

$$\frac{dP_s}{d\Omega}(\text{isolated}) = \frac{c}{8\pi} \left| r^2 \vec{E}_{os}(\vec{r}, \omega_s) \right|^2_{r\to\infty}$$

$$= \frac{\omega_s^4}{8\pi c^3} \left| [\vec{Q} \cdot \frac{\partial}{\partial \vec{Q}}] \, \hat{e}_s \cdot \vec{\alpha} \cdot \vec{E}_{inc}(\vec{r}_o, \omega) \right|^2 \quad . \tag{14}$$

The scattering cross section per molecule, per unit solid angle is then given by

$$\frac{d\sigma_s}{d\Omega}(\text{isolated}) = [\frac{dP_s}{d\Omega}] \bigg/ [\frac{c}{8\pi} |\vec{E}_{inc}|^2]$$

$$= \frac{\omega_s^4}{c^4} \left| [\vec{Q} \cdot \frac{\partial}{\partial \vec{Q}}] \, \hat{e}_s \cdot \vec{\alpha} \cdot \hat{e}_i \right|^2 \tag{15}$$

where $\hat{e}_i$ is a unit vector along the incident field.

For spontaneous Raman scattering, $\langle\vec{Q}\rangle$ corresponds to the zero-point vibration amplitude $(\hbar/M\omega_v)^{1/2}$, M being the ionic mass, and the nonresonant molecular polarizability α is of the order of atomic volume $(1 - 10 \text{ Å}^3)$. Thus typical values of $\langle\vec{Q}\rangle$ are 10^{-10} to 5×10^{-10} cm and for allowed Raman transitions $\partial\alpha/\partial\vec{Q} \sim 10^{-16}$ cm^2 $- 10^{-15}$ cm^2. This implies that, for an isolated molecule, Raman cross sections (per unit solid angle) range from 10^{-29} to 10^{-32} cm^2 for visible incident light. The molecular polarizability, and hence the Raman cross section, can of course be enhanced considerably if there is a resonance between the incident light frequency and an electronic excitation level of the molecule, i.e., in the case of resonant Raman scattering (RRS). In this article we will, however, assume that we are not dealing with such an intrinsic resonance process.

B. Raman Intensity for an Adsorbed Molecule

In the presence of a metallic medium of dielectric function $\varepsilon(\omega)$ and the molecular point dipole, the electric field anywhere is now determined by

$$- \nabla^2 \vec{E}(\vec{r}, \omega) + \vec{\nabla}\vec{\nabla} \cdot \vec{E}(\vec{r}, \omega) - \frac{\omega^2}{c^2} \varepsilon(\vec{r}, \omega)\vec{E}(\vec{r}, \omega) = \frac{4\pi\omega^2}{c^2} \vec{p}(\omega)\delta(\vec{r} - \vec{r}_o)$$

$$\tag{16}$$

where $\varepsilon(\vec{r}, \omega)$ is equal to 1 outside the metal. In the absence of the dipole [right hand side of (16)], the field pattern now is given by $\vec{E}_L(\vec{r}, \omega)$, which is the solution of the equation

$$- \nabla^2 \vec{E}_L(\vec{r},\omega) + \vec{\nabla}\vec{\nabla}\cdot\vec{E}_L(\vec{r},\omega) - \frac{\omega^2}{c^2}\varepsilon(\vec{r},\omega)\vec{E}_L(\vec{r},\omega) = 0 \qquad (17)$$

with the given incident wave (1). In other words $\vec{E}_L(\vec{r}_0,\omega)$ is the usual local field at the molecular site, in the absence of the dipole.

In terms of the Green function $\vec{\vec{G}}(\vec{r},\vec{r}_0,\omega)$, defined by

$$\vec{\vec{G}}(\vec{r},\vec{r}_0,\omega) = \vec{\vec{G}}_0(\vec{r},\vec{r}_0,\omega) + \vec{\vec{G}}_s(\vec{r},\vec{r}_0,\omega) \qquad , \qquad (18)$$

$$[- \nabla^2 + \vec{\nabla}\vec{\nabla}\cdot -\varepsilon(\vec{r},\omega)\frac{\omega^2}{c^2}]\vec{\vec{G}}(\vec{r},\vec{r}_0,\omega) = 4\pi\delta(\vec{r} - \vec{r}_0)\vec{\vec{I}} \qquad , \qquad (19)$$

the complete solution of (16) is now given by

$$\vec{E}(\vec{r},\omega) = \vec{E}_L(\vec{r},\omega) + \frac{\omega^2}{c^2}\vec{\vec{G}}(\vec{r},\vec{r}_0,\omega)\cdot\vec{p}(\omega) \qquad . \qquad (20)$$

With the definition

$$\vec{p}(\omega) = \vec{\vec{\alpha}}_0(\omega)\cdot\vec{E}(\vec{r}_0,\omega) \qquad (21)$$

and the relation (10), the self-consistent solution of (20) and (21) leads to

$$\vec{p}(\omega) = \vec{\vec{\alpha}}_{eff}(\omega)\cdot\vec{E}_L(\vec{r}_0,\omega) \qquad , \qquad (22)$$

where the effective renormalized polarizability of the molecule in the presence of the metallic medium is given by

$$\vec{\vec{\alpha}}_{eff}(\omega) = [\vec{\vec{I}} - \frac{\omega^2}{c^2}\vec{\vec{\alpha}}(\omega)\cdot\vec{\vec{G}}_s(\vec{r}_0,\vec{r}_0,\omega)]^{-1}\cdot\vec{\vec{\alpha}}(\omega) \qquad . \qquad (23)$$

Note that in (23) only $\vec{\vec{G}}_s = \vec{\vec{G}} - \vec{\vec{G}}_0$ and the polarizability $\vec{\vec{\alpha}}$ of an isolated molecule appear. The pre-factor in (23) leads to the so-called "image-enhancement," at certain $\vec{r}_0$.

It is now straightforward to find the Raman part of the polarizability and hence the Stokes scattered field from the ad-sorbed molecule. One gets

$$\vec{E}_{Stokes}(\vec{r},\omega_s = \omega - \omega_\nu) = \frac{\omega_s^2}{c^2} \vec{\vec{G}}(\vec{r},\vec{r}_o,\omega_s) \cdot [\hat{Q} \cdot \frac{\partial}{\partial \vec{Q}}] \vec{\vec{\alpha}}_{eff}(\omega)$$

$$\cdot \vec{E}_L(\vec{r}_o,\omega) \quad . \tag{24}$$

As compared to the Raman intensity for an isolated molecule [see Eq. (10)] for a given vibrational mode, we thus find a very general expression for the classical enhancement factor in SERS in the form[15]

F(Classical)

$$= \frac{\left| \vec{\vec{G}}(\vec{r} \to \infty, \vec{r}_o, \omega_s) \cdot [\hat{Q} \cdot \frac{\partial}{\partial \vec{Q}}] [\vec{\vec{I}} - \frac{\omega^2}{c^2} \vec{\vec{\alpha}} \cdot \vec{\vec{G}}_s(\vec{r}_o,\vec{r}_o,\omega)]^{-1} \vec{\vec{\alpha}} \cdot \vec{E}_L(\vec{r}_o,\omega) \right|^2}{\left| \vec{\vec{G}}_o(\vec{r} \to \infty, \vec{r}_o, \omega_s) \cdot [\hat{Q} \cdot \frac{\partial}{\partial \vec{Q}}] \vec{\vec{\alpha}} \cdot \vec{E}_{inc}(\vec{r}_o,\omega) \right|^2} \quad .$$

$$\tag{25}$$

Note that the classical enhancement ratio F is made of three distinct factors: (1) The incident field at the molecular site is changed from $\vec{E}_{inc}$ to the local field $\vec{E}_L(\vec{r}_o,\omega)$ obtained by solving Maxwell's equation in the presence of the metallic medium but in the absence of the molecule, (2) the experimental polarizability $\vec{\vec{\alpha}}(\omega)$ of an isolated molecule is renormalized to $\vec{\vec{\alpha}}_{eff}(\omega)$ of (23) due to the presence of the metal, (3) the propagator (Green function) for the scattered field is modified from the free space Green function $\vec{\vec{G}}_o$ to $\vec{\vec{G}} = \vec{\vec{G}}_o + \vec{\vec{G}}_s$, where $\vec{\vec{G}}_s$ is the additional contribution to the Green function due to the presence of the metal. Because of the possibility of resonant excitation of surface-plasmon-polariton (SPP) of the metallic substrate, each of these factors may contain SPP resonance denominators. We next discuss these possibilities for specific geometries of the metal substrate.

Plane Surfaces. For a metal occupying the half space z < 0, with its surface defined by the plane z = 0, it is well known that an incident light wave, with a maximum possible wave vector parallel to the surface less than or equal to ω/c, cannot excite the corresponding SPP. This is because, in such a geometry, the wave vector parallel to the surface is still a conserved quantity, and since the SPP dispersion relation in this case is given by

$$K_{sp}^2 = \frac{\omega^2}{c^2} \left[\frac{\varepsilon_1(\omega)}{\varepsilon_1(\omega) + 1}\right] > \frac{\omega^2}{c^2} \quad ; \quad \varepsilon_1(\omega) < -1 \quad , \tag{26}$$

where K_{sp} is the parallel wave vector of the SPP and $\varepsilon(\omega) = \varepsilon_1(\omega) + i\varepsilon_2(\omega)$ is the dielectric function of the metal. For a given frequency ω, K_{sp} is always greater than ω/c. Thus, in such a geometry the local field $\vec{E}_L(\vec{r}_0,\omega)$, related to $\vec{E}_{inc}$ by the usual Fresnel expression, has no SPP resonance. The same thing is true for the scattering Green function $\vec{G} = \vec{G}_o + \vec{G}_s$, so that there is no resonance in even the out-coupling (scattering) channel. The only significant difference in this case can come from the renormalization denominator for $\vec{\alpha}_{eff}(\omega)$. Indeed, apart from a numerical factor A of order unity, which is determined by the detailed symmetry of the polarizability tensor $\vec{\alpha}$ under consideration, in this case

$$\alpha_{eff} \simeq \left[1 - \left(\frac{\varepsilon - 1}{\varepsilon + 1} \right) \frac{\alpha}{4d^3} A \right]^{-1} \alpha \quad , \tag{27}$$

where d is the distance of the molecule from the metal surface. Since α is of the order of atomic volume $(1 - 10 \times 10^{-24} \text{ cm}^{-3})$, this implies that except for d of the order of 10^{-8} cm, we may also ignore the renormalization effect on α (this is called the image enhancement in earlier literature[1,2]) for a plane surface. In any case, for such small d where the image enhancement is important, our classical calculation has no validity. A quantum calculation (to be described in Section III) is then an essential aspect of the problem.

Surface with Small Amplitude Roughness. Next, let us consider the case of a metal whose surface departs from the plane z = 0 by a small amount.[7] The roughness in this case may be represented by defining the actual surface as a superposition of sinusoidal gratings of the form

$$z = \xi(x,y) = \sum_{\vec{g}} \xi_g e^{i\vec{g}\cdot\vec{r}_{||}}; \quad \vec{r}_{||} = x\hat{i} + y\hat{j} \tag{28}$$

in which each ξ_g is small compared to the wavelength of light λ. One can then use Rayleigh's first approximation to obtain the field distribution $\vec{E}_L$ and the additional Green function $\vec{G}_s$ in this geometry up to the first order in ξ_g/λ. For a single component (sinusoidal grating of periodicity $\vec{g}$), this problem has been tackled by Jha, Kirtley, and Tsang. See Ref. 7 for details. In this case, one has a secondary SPP field with parallel wave vector $\vec{K}_{sp} = \vec{k}_{||} \pm \vec{g}$, which can be excited resonantly for the incident angle so that (26) is satisfied. Note that for an incident angle θ, $|\vec{k}_{||}| = (\omega/c)\sin\theta$, so that for a given ω in the range in which $\varepsilon_1(\omega) < -1$, (26) can be satisfied for a fixed angle θ. For a grating with $\vec{g}$ parallel to incident $\vec{k}_{||}$, only p-polarized incident light can resonantly excite the SPP field, but for an arbitrary orientation of $\vec{g}$ with respect to $\vec{k}_{||}$, both p- and s-polarized light can excite SPP resonantly.

The general expression for the SPP field amplitude[7] is quite complicated. However, for $\varepsilon_1(\omega) \ll -1$, near the resonance, the SPP part of the field can be written approximately as[7]

$$\frac{|\vec{E}_L(\vec{r}_o,\omega)|^2_{SPP}}{|\vec{E}_{inc}(\vec{r}_o,\omega)|^2} \simeq 4\xi_g^2 K_{sp}^2 |\varepsilon_1(\omega)| |g_{Res}(\omega,K_{sp})|^2 e^{-2(K_{sp}^2 - \frac{\omega^2}{c^2})^{1/2} d}$$

$$(29a)$$

$$g_{Res}(\omega,K_{sp}) \equiv \frac{(K_{sp}^2 - \frac{\omega^2}{c^2})[\varepsilon_1(\omega) - 1]}{\{[K_{sp}^2(\varepsilon_1(\omega)+1) - \frac{\omega^2}{c^2}\varepsilon_1(\omega)] + i(K_{sp}^2 - \frac{\omega^2}{c^2})\varepsilon_2(\omega)\}}$$

$$(29b)$$

$$K_{sp}^2 - \frac{\omega^2}{c^2} = -(\varepsilon_1(\omega) + 1)^{-1}\frac{\omega^2}{c^2} \quad \text{(at resonance)} \quad , \qquad (29c)$$

where d is the distance of the molecule from the plane $z = 0$. For silver in the visible, the peak resonant enhancement can be as large as 10^4 for $\xi_g K_{sp} \simeq 0.2$ and d small compared to the wavelength of light. Note that this SPP field enhancement decays exponentially on the scale of the wavelength of light so that, compared to the atomic scale, it is relatively long ranged; the molecule need not be in direct contact with the metal.

At an arbitrary scattering angle, for a given sinusoidal grating there is no resonance in $G(r\to\infty,\vec{r}_o,\omega_s)$, so that the net classical enhancement factor is close to the expression (29), except for the renormalization of the polarizability at atomic distances d. The renormalization is, of course, not very important for $d \gg 10 - 20$ Å, even when there is resonance in α_{eff}. However, for particular scattering angles for which the near field at frequency ω_s can couple out resonantly, additional enhancement in the scattering propagator must be considered.

Spherical Surface. For a molecule on a metallic sphere of radius $a \ll \lambda$, the calculation of $\vec{G}$ and $\vec{E}_L$ is known from the well known Mie theory. For the molecular coordinate

$$\vec{r}_o = r_o\hat{z} = (a + d)\hat{z}, \quad d \ll \lambda \quad , \qquad (30)$$

in this case[8,9,15]

$$\vec{E}_L(\vec{r}_o,\omega) = \vec{E}_{inc}(\vec{r}_o,\omega) + \frac{a^3}{r_o^3} g_1(\omega,a)[3\hat{r}\hat{r}\cdot\vec{E}_{inc} - \vec{E}_{inc}] \quad , \qquad (31)$$

where

$$g_1(\omega,a) \simeq \frac{\varepsilon(\omega) - 1}{[\varepsilon(\omega) + 2 + \frac{12}{5}\frac{\omega^2}{c^2}a^2]} \qquad (32)$$

gives the resonant enhancement of the field when the dipolar ($\ell = 1$) SPP resonance condition

$$\varepsilon_1(\omega) + 2 + \frac{12}{5}\frac{\omega^2}{c^2}a^2 = 0 \qquad (33)$$

for the sphere is satisfied. For the incident wave polarized along $\vec{r}_o$, and scalar α, the full classical enhancement factor in this case is given by[15]

F(small sphere, classical)

$$= \frac{\left|1 + 2(\frac{a}{a+d})^3 g_1(\omega_s,a)\right|^2 \left|1 + 2(\frac{a}{a+d})^3 g_1(\omega,a)\right|^2}{\left|1 - 2a^3\alpha\left[\frac{\varepsilon-1}{\varepsilon+1}\right]\left[\frac{1}{(2a+d)^3 d^3} + \frac{\varepsilon(\omega)}{\varepsilon(\omega)+2}\frac{1}{(d+a)^6}\right]\right|^4} . \quad (34)$$

In the above expression, the denominator represents the renormalization factor (squared) for the polarizability and the two factors in the numerator are due to modifications in the scattering propagator (out coupling) and the local field enhancement, respectively. Because of the smallness of α, the denominator in (34) does not become important even when $g_1(\omega,a)$ is resonant, unless $a + d$ < 20 Å. Note also that except for the incident and Stokes frequencies very close to each other, F has two distinct resonance peaks as a function of ω. Only for very small Stokes shift, they overlap substantially to lead to a double resonance in F.

Similar calculations of the enhancement factor due to SPP resonance can be performed for other interesting geometries, some of which are discussed by different authors in this book.

III. SHORT RANGE ENHANCEMENT INCLUDING SPP RESONANCE

As already discussed in the Introduction, when the molecule

is in contact with the metal surface, one has to consider the
quantum tunneling and bonding of the metal electrons with the
electrons of the molecule which can give rise to a new distribution
of the ground state electric charge density and a completely modified
excitation spectrum for the molecule-metal complex. For an isolated
molecule, the polarizability $\vec{\alpha}(\omega)$ involved in the Raman cross section
(11) is given by the well known quantum expression

$$\vec{\alpha}(\omega) = \sum_i f_i \sum_n \left(\frac{<i|e\vec{r}|n><n|e\vec{r}|i>}{E_n - E_i - \hbar\omega - i\Gamma_{ni}} + \frac{<i|e\vec{r}|n><n|e\vec{r}|i>}{E_n - E_i + \hbar\omega + i\Gamma_{ni}} \right)$$

$$(35)$$

where f_i is the ground state occupation probability of the single-
particle electronic state $|i>$, with energy E_i of the molecule.
Note that there is an intrinsic resonance in $\vec{\alpha}(\omega)$ when one of the
excited state energy E_n differs from E_i by exactly $\hbar\omega$, giving rise
to the so-called resonance Raman scattering (RRS). However, we
are not considering here such intrinsic resonances for the isolated
molecule.

In the presence of the metal, we must now consider three
important differences for molecules at atomic distances from the
surface, as compared to the classical long-range calculation.
These are:

(1) The first difference involves the calculation of the
changed single-particle-like electronic excitation spectrum of the
metal-molecule complex. The new states will be a hybrid of the
conduction-electron states of the metal (tunneling states in the
new self-consistent barrier potential between the metal and molecule
or the corresponding free states) and the unperturbed states of the
isolated molecule. Apart from varying shifts in the excited states
of the molecule, these would get considerably broadened due to the
interaction with the metal electrons. Although not very accurate
at small distances, the classical renormalization factor for the
polarizability discussed in the last section can be shown to lead
to such shifts and broadening of electronic oscillators in the
molecule. In between these broadened molecular states, one would
have mostly metal-like tunneling states, with the possibility of an
additional broad bound charge-transfer state. Because of the new
set of almost continuous excited states n in this case, the summation over n in (35) can lead to a substantial increase in $\alpha(\omega)$.
This would, however, have no pronounced structure as a function of
ω, unless we are close to the broad charge-transfer state in the
excited spectrum or close to a shifted molecular electronic state.

(2) The second difference comes from the fact that the ground
state charge distribution and the number of electronic oscillators
interacting with the light field are now different. Because of
tunneling of metal electrons, one has many more oscillators now as
compared to the case for an isolated molecule.

(3) The third difference arises from the fact that even in the
absence of the molecule the induced local electromagnetic field
distribution at atomic distances from the metal surface (in particular, the component of the field normal to the surface) differs
considerably from the classical calculation of Section II. In
addition, it varies rapidly on the atomic scale, so that the expres-
sion (35) for $\overset{\leftrightarrow}{\alpha}(\omega)$ based on the electric dipole-approximation for
the field cannot be used, although we may plug in the new set of
ground and excited state energies and wave functions in the expression (35) (as detailed in differences 1 and 2 above). The self-
consistent calculation of the local induced electromagnetic field
at such distances is quite complicated, even in the absence of the
molecule, for a realistic quantum mechanical model for the motion
of the conduction electrons at the metal surface. The dielectric
function is no longer a local function, so that the "image-
enhancement" factor (renormalization factor for the polarizability)
discussed in the last section in terms of local dielectric function
$\varepsilon(\omega) = \varepsilon_1(\omega) + i\varepsilon_2(\omega)$ for the metal cannot be taken too seriously
for d $\sim$ 1 or 2 Å.

In view of the difficulties in determining the local electromagnetic field at short distances from the metal surface (in the
absence of the molecule), there is a lot of confusion and sometimes
quite unphysical models in the literature to describe the short
range part of SERS. We propose that actual induced effective
dipole moment for the metal-molecule system should again be defined
operationally by the relation

$$\vec{p}(\omega) = \overset{\leftrightarrow}{\alpha}_s(\omega) \cdot \vec{E}_L(\vec{r}_o, \omega) \quad , \tag{36}$$

where $\vec{E}_L(r_o, \omega)$ is still the well defined classical local field of
the last section, calculated for a sharp discontinuity in $\varepsilon(\omega)$ at
the metal surface. Here $\vec{r}_o$ is the coordinate of the molecular site.
The short range $\overset{\leftrightarrow}{\alpha}_s$, of course, differs from α and the classical
$\alpha_{eff}(\omega)$ of the last section due to reasons (1) – (3), discussed
above. In general, we can further split the short range $\overset{\leftrightarrow}{\alpha}_s(\omega)$ as

$$\overset{\leftrightarrow}{\alpha}_s(\omega) \equiv \overset{\leftrightarrow}{\alpha}(\omega) + \overset{\leftrightarrow}{\beta}_s(\omega) \quad , \tag{37}$$

where $\overset{\leftrightarrow}{\beta}_s(\omega)$ is the additional contribution to $\overset{\leftrightarrow}{\alpha}_s(\omega)$ at short distances (first few monolayers), which has to be determined both experimentally and theoretically.

Because of severe theoretical complications in carrying out the calculation of effects (1) – (3) to determine $\vec{\beta}_s$, not a single complete microscopic calculation for any specific metal-molecule system is available in the literature. Therefore, at this stage, it may be advisable to try to model the nature of the induced effective dipole moment $\vec{p}(\omega)$ of (36) in terms of certain physical quantities. In fact, one has the general relation

$$\vec{p}(\omega) = \int \delta\rho(\vec{r},\omega)\vec{r}d^3r \quad , \tag{38}$$

where the induced charge density

$$\delta\rho(\vec{r},\omega) = \frac{1}{4\pi} \vec{\nabla} \cdot \vec{E}(\vec{r},\omega) \quad . \tag{39}$$

Now classically, there is no overlap of the charge densities of the metal and the molecule, and the induced charge density in the molecule is related to the calculation of this quantity in an isolated molecule. However, even classically, there is an induced charge density at the metal surface of the form

$$\delta\rho_{metal}(\vec{r},\omega) = \frac{1}{4\pi} [E_n(z = \xi^+) - E_n(z = \xi^-)]\delta(z - \xi) \quad , \tag{40}$$

where $z = \xi$ is assumed to define the surface of the metal and where $E_n(z = \xi^+) - E_n(z = \xi^-)$ is the classical discontinuity in the normal component of the field across this metal surface. Because of the delta-function in (40), the corresponding classical $\vec{p}_{metal}(\omega)$ is of course zero, via (38), with the origin at the metal surface. However, the delta-function in (40) is a consequence of the classical step-function model of $\varepsilon_1(\omega)$ at the metal surface. In reality, $\varepsilon_1(\omega)$ varies smoothly from $\varepsilon(\omega)$ in the bulk metal to 1 in the vacuum, on the atomic scale. Thus, more correctly the delta-function in (40) should be replaced by a more general function $F(\vec{r})$:

$$\delta(z - \xi) \to F(\vec{r}) \quad . \tag{41}$$

One still has, because of charge conservation, the constraint

$$4\pi \int dz \; \delta\rho_{metal}(\vec{r},\omega) = [E_n(z = \xi^+) - E_n(z = \xi^-)] \tag{42}$$

on $F(\vec{r})$, where the right-hand side of the above equation is the usual classical discontinuity in the electric field.

Note that any phenomenological model for $F(\vec{r})$ and the consequent calculation of $\vec{\alpha}_s$, and hence $\vec{\beta}_s$, only tries to take into account the differences (2) and (3), discussed earlier. It ignores the additional contribution to $\vec{\alpha}(\omega)$ due to the change in the electronic excitation spectrum, including an additional possible charge-transfer state, of the molecule. In a complete calculation, this also has to be taken into account. Otto in his "adatom model" (see Ref. 6a for details) has considered this effect separately but seems to have ignored the differences (2) and (3).

Jha et al.[7] have assumed that the excitation spectrum of the molecule and the ground state electronic charge distribution inside the molecule have not changed considerably in the systems considered by them. In particular, if the vibrational frequencies of the adsorbed molecules are more or less the same as for the isolated molecule, this may not be a bad approximation after all. In such a situation, one can just model $F(\vec{r})$ of (40) and (41) in terms of parameters of physical importance and calculate the additional short range polarizability $\vec{\beta}_s(\omega)$ arising from it. The short range enhancement in such a case is given by

$$
F(\text{short range}) = \left| 1 + \frac{[\hat{Q}\cdot\frac{\partial}{\partial\vec{Q}}]\vec{\beta}_s(\omega)}{[\hat{Q}\cdot\frac{\partial}{\partial\vec{Q}}]\vec{\alpha}(\omega)} \right|^2 \frac{|\vec{E}_L(\vec{r}_o,\omega)|^2}{|\vec{E}_{inc}(\vec{r}_o,\omega)|^2} \quad . \tag{43}
$$

Note that even the short range enhancement contains the usual classical local field enhancement due to the SPP resonance. In addition, one now has a new polarizability contribution, possibly with a different symmetry than that in an isolated molecule.

In the model of Jha, Kirtley, and Tsang (JKT model)[7] they replace $F(\vec{r})$ of (40) and (41) by two decaying exponentials, inside and outside the metal, respectively. Using the WKB approximation, inside the metal the decay constant is parameterized by the quantity

$$
K_{in}(r) = 2\left(\frac{\omega}{v_f} + \frac{q_{FT}}{2}\right) \quad , \tag{44}
$$

where v_f is the local Fermi velocity of the conduction electrons and q_{FT} is the Thomas-Fermi wave vector. Outside the metal, they use the decay constant

$$
K_{out}(\vec{r}) = 2\left(\frac{2m}{\hbar^2}\right)^{1/2} [\Phi_B(\vec{r},\vec{R})]^{1/2} \quad , \tag{45}
$$

where $\Phi_B(\vec{r},\vec{R})$ is the electronic surface barrier potential for tunneling of the conduction electrons to the molecular site. $\Phi_B(\vec{r},\vec{R})$ depends on the coordinates of the effective ionic charges Z_j at sites $\vec{R}_j$ in the molecule:

$$\Phi_B(\vec{r},\vec{R}) = \Phi_M - \sum_j \left\langle Z_j e^2 \left[\frac{1}{|\vec{r} - \vec{R}_j|} - \frac{1}{|\vec{r} - \vec{R}_j + \hat{z}(R_{jz} - \xi)|} \right] \right\rangle .$$

(46)

Here Φ_M is the barrier potential in the absence of the molecule, and the additional term is due to the Coulomb interaction between the electron and the ionic charges, including their images in the metal. The modulation of β_s by molecular vibration arises because of the modulation of Φ_B and hence of $\delta\rho_{metal}$ and $\vec{p}(\omega)$ by the molecular coordinates. In terms of phenomenological parameters defined by an effective step-function equilibrium barrier potential between the electron and adsorbed molecule and the effective ionic charges in the molecule, one can then obtain the complete expression for $\beta_s(\omega)$ due to this tunneling mechanism. This is given in Ref. 7.

In general, in the JKT model one finds that only the component $\partial\beta_{nn}/\partial Q_n$, where n is the local normal to the metal surface, is significant. In other words, only the molecular vibration normal to the local surface has a significant additional short range enhancement. For a flat surface where the SPP resonance is not possible, this is of course the only enhancement. In the JKT model, this additional short range enhancement of polarizability falls off rapidly with the distance d of the molecule from the local metal surface, with no appreciable enhancement beyond about 10 Å. The first layer close to the surface gets the maximum additional enhancement. This additional enhancement also depends critically on the effective barrier potential for electronic tunneling from the metal to the molecular site. The lower this barrier potential, the higher is the enhancement. For the first monolayer, one can get an additional short range enhancement of almost 10^3 or more for the barrier potential less than about 0.5 eV. In such a case, if there is also the usual classical local field enhancement due to SPP resonance, one can obtain a net enhancement of 10^5 – 10^7 for the first monolayer of adsorbed molecules.

IV. CONCLUSIONS

The theory of SERS presented in the last two sections shows that there is a relatively long range classical enhancement of Raman intensity for molecules adsorbed or deposited at a metal surface of the dielectric constant $\varepsilon(\omega) = \varepsilon_1(\omega) + i\varepsilon_2(\omega)$, due to the possibility of resonant excitation of surface plasmon polaritons of the metal by the incident light. In the visible frequency region, for a single sinusoidal grating surface, this enhancement can be as high as 10^4 for Ag for molecular distance d which is small compared to the wavelength of light. It is less for Au and Cu. The enhancement can be even higher for molecules on single small metallic spheres or ellipsoids. Experimental results[12-14,16] on controlled surfaces are in very good agreement with this prediction of the theory.

An additional short range polarizability enhancement is possible for the first few monolayers only for certain vibrational modes and specific metal-molecule systems. This greatly depends on the new electronic excitation spectrum of the metal-molecule system, e.g., whether there is a well defined additional charge-transfer state or whether the new continuous excited tunneling states have strong matrix elements with the ground electronic state or not. One additional possible mechanism for enhancement of short range $\partial \vec{\beta}_s(\omega)/\partial \vec{Q}$ is the modulation of the induced surface-dipole moment for the tunneled conduction electrons by the molecular vibration. This is large for small metal-molecule distances and low effective self-consistent barrier potentials for the conduction electron tunneling to the molecular site. If because of surface roughness or shape resonance there is the usual SPP excitation in addition, one will have a much larger enhancement of the Raman intensity for molecules in the first layer, with only the classical long range enhancement for molecules in subsequent layers. This is in qualitative agreement with the experiment of Sanda et al.[17] on a single-crystal of Ag with a sinusoidal grating on its well defined surface.

The short range enhancement is dominant for molecular vibrations normal to the local surface. To date, there is no detailed experimental proof for this selection rule. In any case, since the physics of adsorbed molecules is hidden in the additional short range polarizability $\partial \vec{\beta}/\partial \vec{Q}$, it would be an excellent idea to determine these quantities experimentally for different vibrations in various metal-molecule systems. For this, it may be necessary to do ultra-high vacuum experiments on smooth surfaces or measure the long range classical local field enhancement $|\vec{E}_L/\vec{E}_{inc}|$ separately to eliminate it from (43). Only then can we start testing different theoretical models in the literature and begin to use SERS more meaningfully in the study of surfaces.

REFERENCES

1. F. W. King, R. P. Van Duyne, and G. C. Schatz, Theory of Raman scattering by molecules adsorbed on electrode surfaces, J. Chem. Phys. 69:4472 (1978).

2. S. Efrima and H. Metiu, Light scattering by a molecule near a solid surface. II Model calculations, J. Chem. Phys. 70:2297 (1979).

3. E. Burstein, Y. J. Chen, C. Y. Chen, S. Lundquist, and E. Tosatti, "Giant" Raman scattering by adsorbed molecules on metal surfaces, Solid State Commun. 29:567 (1979).

4. M. Moskovits, Surface roughness and the enhanced intensity of Raman scattering by molecules adsorbed on metals, J. Chem. Phys. 69:4159 (1978); M. R. Philpott, Effect of surface plasmons on transitions in molecules, J. Chem. Phys. 62:1812 (1975).

5. J. R. Kirtley, S. S. Jha, and J. C. Tsang, Surface plasmon model of surface enhanced Raman scattering, Solid State Commun. 35:509 (1980).

6. J. Billman, G. Kovacs, and A. Otto, Enhanced Raman effect from cyanide adsorbed on a silver electrode, Surf. Sci. 92:153 (1980); S. H. Weber and G. W. Ford, Enhanced Raman scattering by adsorbates including the nonlocal response of the metal and the excitation of nonradiative modes, Phys. Rev. Lett. 44:1774 (1980).

7. S. S. Jha, J. R. Kirtley, and J. C. Tsang, Intensity of Raman scattering from molecules adsorbed on a metallic grating, Phys. Rev. B 22:3973 (1980).

8. S. L. McCall, P. M. Platzman, and P. A. Wolff, Surface enhanced Raman scattering, Phys. Lett. 77A:381 (1980); S. L. McCall and P. M. Platzman, Reflectivity modulation theory of enhanced surface Raman scattering, Phys. Rev. B 22:1660 (1980).

9. M. Kerker, D. S. Wang, and H. Chew, Surface enhanced Raman scattering (SERS) by molecules adsorbed at spherical particles: errata, Appl. Opt. 19:4159 (1980).

10. J. L. Gersten and A. Nitzan, Electromagnetic theory of enhanced Raman scattering by molecules adsorbed on rough surfaces, J. Chem. Phys. 73:3023 (1980).

11. C. Y. Chen and E. Burstein, Giant Raman scattering by molecules at metal island films, Phys. Rev. Lett. 45:1287 (1980).

12. J. C. Tsang, J. R. Kirtley, and J. A. Bradley, Surface enhanced Raman spectroscopy and surface plasmons, Phys. Rev. Lett. 43:772 (1979).

13. J. E. Rowe, C. V. Shank, D. A. Zwemer, and C. A. Murray, Ultra high vacuum studies of enhanced Raman scattering from pyridine on Ag surfaces, Phys. Rev. Lett. 44:1770 (1980).

14. J. C. Tsang, J. R. Kirtley, T. N. Theis, and S. S. Jha, Surface enhanced Raman scattering from molecules in tunneling junctions (to be published).

15. G. S. Agarwal, S. S. Jha, and J. C. Tsang, Surface-plasmon-polariton resonance factors in classical calculation of surface enhanced Raman scattering (to be published).

16. T. E. Furtak and J. Kester, Do metal alloys work as substrates for surface enhanced Raman spectroscopy?, Phys. Rev. Lett.

45:1652 (1980); J. C. Tsang, S. S. Jha, and J. R. Kirtley, Dependence of surface enhanced Raman scattering from Ag-Pd alloys on substrate dielectric function, Phys. Rev. Lett. 46:1044 (1981).

17. P. N. Sanda, J. M. Warlaumont, J. E. Demuth, J. C. Tsang, K. Christmann, and J. A. Bradley, Surface enhanced Raman scattering from pyridine on Ag(111), Phys. Rev. Lett. 45:1519 (1980).

THE "ADATOM MODEL": HOW IMPORTANT IS ATOMIC SCALE ROUGHNESS?

A. Otto, I. Pockrand, J. Billmann, and C. Pettenkofer

Physikalisches Institut III, Universität Düsseldorf

D-4000 Düsseldorf, Federal Republic of Germany

I.   INTRODUCTION

One might describe the situation in surface enhanced Raman spectroscopy (SERS) roughly as follows: For an adsorbate on a metal surface, the Stokes shifted Raman intensity for a particular vibration of the adsorbate is proportional to

$$\sum_i G_{EM}^{(i)}(\omega_L,\omega_L-\omega_{vi})\cdot\sigma_i(\omega_L,\omega_{vi}) \quad . \tag{I}$$

i denotes the adsorbed molecules, characterized by their adsorption site and the kind of bonding at this site, the index i at the vibrational frequency ω_v reflects the possibility that ω_v may depend on the kind of bonding to site i, ω_L is the frequency of the incident laser light, $G_{EM}^{(i)}(\omega_L,\omega_L-\omega_{vi})$ is the classical enhancement by electromagnetic resonances of the local fields at frequency ω_L and $\omega_L-\omega_{vi}$[1,2,3] and $\sigma_i(\omega_L,\omega_{vi})$ is the Stokes scattering cross section of the adsorbed molecule i. The dependence of the Stokes signal on polarizations, on angles of incidence and emission and any kind of coupling between the adsorbed molecules or between sites is neglected here, because it is not important for the following discussion.

For molecules not chemisorbed or physisorbed on the metal surface, but at some distance to it, the Stokes shifted intensity will be proportional to

$$\sum_i G_{EM}^{(j)}(\omega_L,\omega_L-\omega_v)\cdot\sigma_{free}(\omega_L,\omega_v) \quad . \tag{II}$$

j denotes the molecules not in immediate contact with the surface. They have a cross section $\sigma_{free}(\omega_L,\omega_v)$ which is not affected by the presence of the metal surface. Here we have neglected the possible van der Waals interactions between molecule and the surface, such as the so-called image field effect[4,5] and the "Raman reflectivity"[e.g.6,7]. Different opinions on SERS can be easily demonstrated with the help of formulae (I) and (II).

Proponents of "classical enhancement" assume that $\sigma_i(\omega_L,\omega_{vi}) \approx \sigma_{free}(\omega_L,\omega_v)$. In this case, the enhancement is due to the local field enhancement by surface plasmon polariton type resonances at a surface of supraatomic roughness on a scale of 50 - 5000 Å[1,10]. σ is not changed greatly by chemisorption and the enhancement is of long range nature and falls off gradually with the distance of the molecule from the surface[8].

Apart from "classical enhancement" there exists the possibility of a "chemical effect", namely an increase of $\sigma(\omega_L,\omega_v)$ after chemisorption by electronic interaction between the adsorbate and the metal[e.g.9,11,12,13]. The "chemical effect" would be roughly independent of the site of chemisorption i. For instance it would be the same for all CN^- ions chemisorbed on a silver electrode.

Finally, there is the hypothesis of special active sites i, for which $\sigma_i(\omega_i,\omega_{vi}) >> \sigma_{free}(\omega_L,\omega_v)$. These may be sites of atomic scale roughness[14] or of chemical bonds intimately associated with microroughness by bonding site availability[15], for instance silver adatoms in the case of CN^- ions chemisorbed on a silver electrode.

Since our publication[14] we have been fairly outspoken proponents of the idea of active sites provided by atomic scale roughness (which has become known under the name "adatom model") and of relatively strong photon-electron coupling at sites of atomic scale disorder. We have discussed the SERS experiments known to us before January 1981 with respect to the ideas of "classical enhancement" and the "adatom model" in a review article[16]. We have discussed in detail the theories of electromagnetic enhancement and the experiments which indicate contributions to SERS beyond that which is predicted by the electromagnetic local field enhancement theory[17].

The present article will be organized as follows: In section II we present experiments of our group which in our opinion indicate that the EM enhancement is not the strongest contribution to the overall enhancement, or at least not the only one. In section III we discuss the ideas of the hypothetical "adatom model" and in section IV we present our experiments indicating the importance of atomic scale roughness. Within this chapter, we will refer only to other work as it pertains to the discussion of our work. Finally in section V we will present our preliminary results for oxygen on silver. These results indicate that SERS may become a tool of catalytic research, even though the SERS process is not yet well understood. We summarize our results in section VI.

II. EXPERIMENTAL EVIDENCE FOR AN ENHANCEMENT MECHANISM BEYOND THE ELECTROMAGNETIC FIELD EFFECT

The electromagnetic enhancement G_{EM} of the Raman signal by surface plasmon type polariton resonances is a long range effect[8]. Direct contact of the molecule with the surface is not required. The enhancement for molecules in the second layer on top of the directly adsorbed first layer is of the same order as for molecules in the first layer. In order to reach values of G_{EM} 10^4-10^6, the intrinsic energy dissipation to the metal must be low, which means a high intrinsic reflectivity of the metal[17] (for instance greater than .95, as for silver in the visible spectral range). We have performed experiments, which raise doubts about both the points of long range and low energy dissipation.

a.) Enhancements of 10^6 are short range effects.

The average enhancement on "activated" silver electrodes is about 6 orders of magnitude for pyridine[18] and cyanide[14]. We have performed an experiment[19] where both pyridine and cyanide were present in the electrolyte. By variation of the potential, one could switch reversibly from the pure cyanide to the pure pyridine SERS spectra. This proved that the SERS signal was originating from the directly adsorbed CN^- or pyridine molecules. On the other hand, no enhancement of the Raman intensity of water was observed in these experiments, even though water is abundant in the "second layer" and most probably is also in direct contact with the silver electrode surface. Both observations rule out an enhancement of 10^6 for distances greater than 3 Å from the surface[19]. A long range enhancement of smaller than 10^4 cannot be ruled out.

b.) Is SERS compatible with high intrinsic energy dissipation?

In the following we will show that "SERS active", so-called "cold silver films" (see below) show an anomalously high bulk absorption. When a surface enhanced Raman signal is observed, it always occurs together with an inelastic background[20], which was first assigned to the continuum of low energy e-h pair excitations[6]. This background is present for "activated" samples without adsorbates[21], and is therefore an intrinsic property of the "SERS active" substrate. This is obvious from our experiments on "cold films" (under "cold film" we mean a silver film, evaporated in a vacuum in the low 10^{-10} Torr range on a cooled copper substrate of about 120 K, with evaporation rate ca. 1 nm/sec and about 2000 Å thickness, and kept at 120 K, following a first report by Wood and Klein[22]). When a cold film is exposed to 1 L

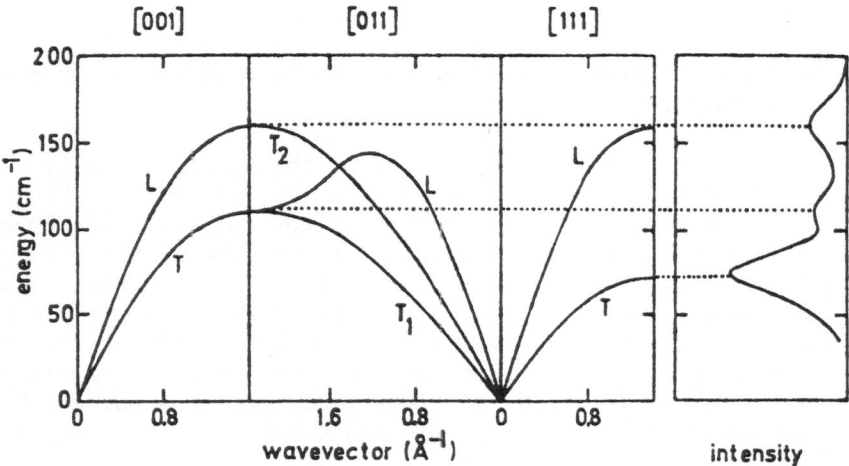

Fig. 1   Phonon dispersion in silver compared to the light scattering
         spectrum of a "cold silver film"[25].

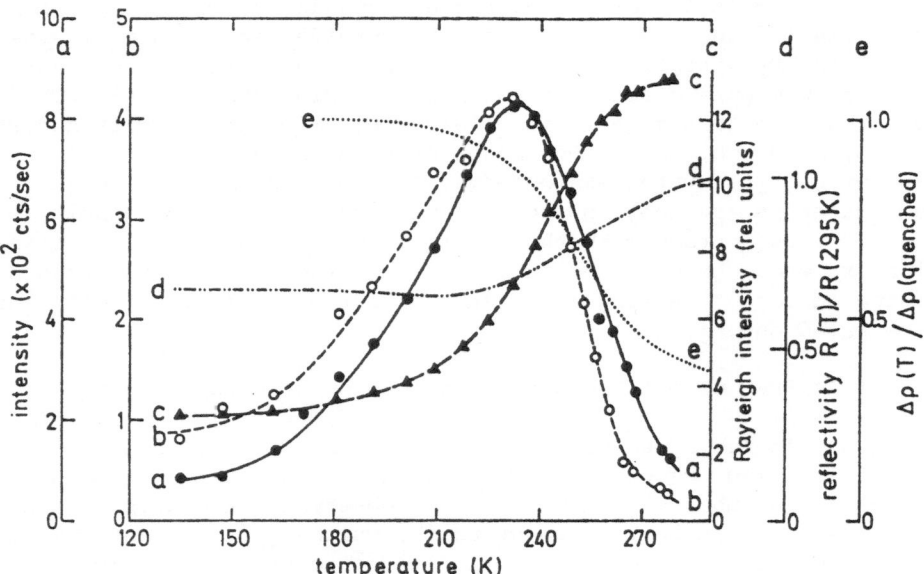

Fig. 2   Intensities of the background (a), of the peak at 161 cm^{-1} (b;
         fig. 1), of the Rayleigh scattered light (c), and of the rel-
         ative reflectivity at 514.5 nm wavelength (d) of a "cold sil-
         ver film" versus temperature. Isochronal annealing of excess
         resistivity[27] of silver quenched in at 1016 K (e).

pyridine, the Raman signal of pyridine is enhanced by 4 to 5 orders of magnitude[23]. For an unexposed "cold film" we find in the inelastic light spectrum a structureless background and a low frequency structure[24]. The latter is assigned to the density of phonon states in silver on the basis of the good agreement with neutron scattering data for the phonon dispersion[24,25] - see fig. 1. The assignment is corroborated by preliminary experiments on "cold copper films", which show a similar agreement between the low frequency structure in the inelastic light spectrum and the copper phonon dispersion. The Raman scattering by phonons is induced by point-like lattice defects in the bulk of the cold film (larger structures, like crystalline grains, would not provide the necessary momentum). If the "cold film" is warmed up to room temperature in about 2 to 3 hours (about 1 K per minute near 200 K), the intensities of the phonon signal and of the inelastic background exhibit the same variation, see fig. 2. This variation is irreversible (see fig. 9). The similar variation of phonon and background intensities indicates a common origin, namely electron-electron coupling and electron-phonon coupling at the same bulk point defects, whose concentration varies irreversibly with temperature. It is likely that point defects are sites of strong photon-electron coupling. In fact we are unable to understand the existence of disorder induced Raman scattering from phonons without this assumption. In the ideal silver lattice there is very little photon-electron coupling for photon energies below the threshold of interband transitions at 4 eV. Therefore, silver has a high intrinsic reflectivity in the visible spectral range.

The postulated strong photon-electron coupling at bulk defects should lead to additional optical absorption. Figure 3a shows indeed an irreversible annealing of an extra absorption (which is 1-R, where R is the reflectivity, diffuse elastic Rayleigh scattering is negligible). Because of the difficulty of measuring absolute reflectivities from samples in our Raman-UHV-system[26], we plot the reflectivity ratio [R(290 K)-R(T)]/R(290 K) versus λ. We thus neglect the possibility that the reflectivity R(290 K) of a cold film, warmed up to room temperature, might not yet have reached the ideal value. The irreversible variation of the reflectivity at a laser wavelength of 5145 Å is given in fig. 2, trace d. At 5145 Å the extra absorption disappears in the same temperature range (240 - 290 K) as the phonon and the inelastic background spectrum. However, it does not show the strong increase between 150 K and 230 K.

The reflectivity data can be understood with two assumptions. The first assumption is: The point defects in the cold film are vacancy clusters of various sizes which are subject to dissociation, migration and annealing (by reaching the surface or internal voids). Figure 4b shows how we envisage qualitatively the temperature dependence of the concentration of clusters of 5, 3, 2 vacancies and monovacancies. The available data of migration energies of divacancies and monovacancies

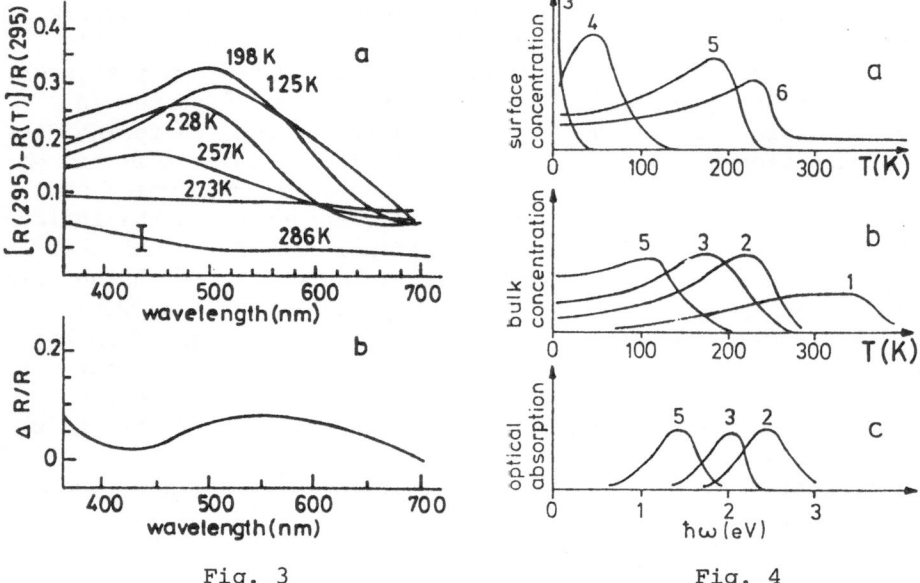

Fig. 3

Fig. 4

Fig. 3 a) Annealing of the relative optical absorption of a "cold sil-
        ver film". b) Relative loss of the reflectivity for a silver
        electrode after activation (see text).

Fig. 4 Hypothetical qualitative temperature variation of the concen-
        tration of defect sites at the surface (a) and in the bulk (b)
        of "cold silver film". Hypothetical absorption spectra of
        vacancy clusters of 5, 3, 2 vacancies in silver (c).

in silver are 0.58 and 0.99 eV[27]. The second hypothetical assumption
is: The bigger the number of vacancies in the cluster, the longer the
wavelength of maximum absorption induced by the cluster, as given in
fig. 4c (for clusters of 5, 3 and 2 vacancies). Both assumptions explain qualitatively the results in fig. 3a: The concentration of
bigger clusters decreases at lower temperatures than the concentration
of smaller vacancy clusters; therefore the extra absorption vanishes
at lower temperatures in the red than in the blue range. The optical
absorption at given wavelength is a superposition of absorption by
several types of vacancy clusters. When the temperature is raised, the
overall absorption will remain roughly constant, until the concentration of the smallest cluster contributing to the absorption starts to
decrease. This could explain the reflectivity variation in fig. 2,
trace d. A qualitative understanding of the temperature variation of
the intensity of the background and the phonon signal in fig. 2 is
possible under the following assumption: The coupling processes are
most pronounced for a particular vacancy cluster, namely for the

divacancy. The annealing of divacancies in silver is well known from the following experiments: Hot silver wires are quickly cooled to liquid nitrogen temperature, the annealing of the lattice defects upon warming up is monitored by DC resistivity measurements. A big decrease in resistivity (see curve e in fig. 2) occurs around 250 K[28,29,30] when the silver wires are warmed up at about the same rate as our "cold films". This decrease is assigned to the annealing of divacancies[27,28,30].

Alternatively, we have to discuss whether the optical absorption of the "cold film" and its temperature variation is caused by electromagnetic resonances of a bumpy surface and the annealing of bumps. Bumps of a dimension of 200 Å and greater are not annealed by warming to room temperature, as follows from ref. 31 (see section IV). However, according to Moskovits[32], the important bumps may be smaller, around 50 Å diameter. The dynamic dipole modes of the bumps are coupled, leading to a collective resonance (the so-called conduction resonance)[17,32]. The resonance frequency depends on the surface density of bumps; it shifts to the blue with decreasing surface density. So, one might tentatively assign the observed annealing of the extra absorption in fig. 3a to a transformation of a high surface density of very small bumps into a lower density of somewhat bigger bumps. These are still small enough that the electromagnetic modes beyond the dipole mode are negligible. The bigger bumps are eventually annealed as well. We do not favour this explanation for the following two reasons:

1.) According to Moskovits, the electromagnetic resonances in the hypothetical small bumps are the reasons for the SERS activity of a surface. Therefore, the extra absorption, shown in fig. 3, and the SERS activity should always occur together. The stronger the absorption, the stronger should be the Raman enhancement. This is not the case: The absorption of a silver electrode (redeposition of 56 Å of silver in 10^{-2} M KCl solution) which yields an average enhancement of 10^5 like the cold film is given in fig. 3b. It is evidently much weaker than the absorption of the cold film of fig. 3a. After an activation of a silver electrode with redeposition of 8 monolayers of silver in 0.1 M KCl and 0.05 M pyridine, Pettinger et al.[33] found $\Delta R/R < 0.05$.

In our opinion the SERS activity of a surface depends on the surface concentration of sites of atomic scale roughness, as described in section III. This surface defect concentration is not necessarily correlated with the bulk defect concentration. For instance, the bulk of a silver electrode may be well annealed (SERS spectra from adsorbates on electrodes are always taken at room temperature). On the other hand the defect concentration on the electrode surface may be quite high, because the Helmholtz layer (adsorbed ordered water layer and specifically adsorbed ions and counter ions) prevents the annealing of atomic scale surface roughness.

2.) It is not easy to understand why only very small bumps should
     yield the strong electromagnetic resonances necessary for SERS,
     whereas bumps of 200 Å diameter and the increasing more macro-
     scopic roughness indicated by the increasing Rayleigh scattering
     intensity (curve c in fig. 2) do not.

    Schultz et al.[63] concluded that roughness on a lateral scale
greater than 250 Å provides a maximum enhancement of 10^2. So we maintain, though the details are far from being clear, that the extra
absorption in fig. 3 is a bulk effect, caused by electron-photon
coupling at bulk defects. In this case, as discussed above, the surface of the "cold silver film" cannot support strong electromagnetic
resonances. But apparently electromagnetic resonances are not necessary, as follows from the occurrence of SERS on platinum, mercury and
maybe nickel[17].

    In summary, we interpret the data from the unexposed cold silver
film as point defect induced optical absorption (fig. 5a), point defect induced Raman scattering from phonons (fig. 5b) and an inelastic
background due to electron-photon interaction (fig. 5c). The point
in the vertices indicates that the coupling processes take place at
defects.

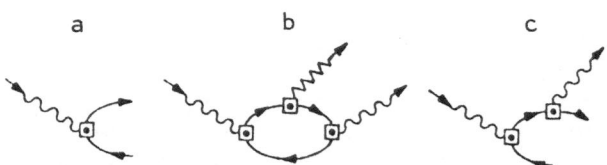

Fig. 5   Point defect induced optical absorption (a), Raman scattering
         by phonons (b) and electronic background (c).

III. THE "ADATOM MODEL", THE INFLUENCE OF ATOMIC SCALE ROUGHNESS.

    Figure 6 shows what we mean by atomic scale roughness, namely
adatoms, clusters of adatoms on plane terraces of crystallites, adatoms or adatom clusters at steps ("fuzzy steps"), kinksites, monoatomic
steps, surface vacancies, clusters of surface vacancies, but also
anchoring points of dislocations at the surface and grain boundaries
cutting through the surface. Compared to an ideal low index monocrys-
talline surface, these "surface defects" break the translational crystal symmetry parallel to the average surface.

    As explained in section II, the bulk point defects are sites of
photon-electron coupling. It is most likely that surface defects are
also sites of photon-electron coupling which is strong compared to
electron-photon coupling in the ideal crystalline metal[14,16]. The
variation of the coupling with the light frequency ω_L could be

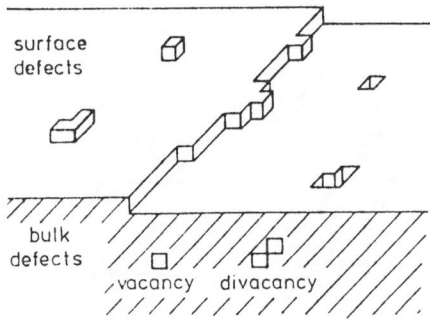

Fig. 6 Atomic scale surface roughness.

different for the different surface defects and could change after chemisorption of a molecule at the surface defect. Analogous to the discussion in section II for the case of bulk defects, surface defects should induce extra optical surface absorption and an inelastic background due to electron-photon coupling at these surface defects[16]. Those surface defects with significant electron-photon coupling for given ω_L are active sites in the following sense: when a molecule is chemisorbed on such a site, the electronic excitation may be transferred forth to the molecule and back to the metal. One of the postulated ways for energy transfer[9] is a transient negative ion state of the adsorbate[34]. The vibrational state of the adsorbate may be changed by excitation of the molecule. The resonant process characterized by the Feynman graph in fig. 7a is characterized schematically in real space in fig. 7b for the case of a resonant negative ion state. The electronic excitation of the active site is depicted schematically by an electron hole pair. The meaning of the middle vertex in the graph is depicted in fig. 7c. As discussed in[16] the Raman scattering process described by the graph in fig. 7a contains the fourth power of the electron-photon coupling constant. If this coupling constant is only 10 times stronger at a site of atomic scale roughness than at an adsorption site of a molecule on a plane terrace (for instance when a fraction of a monolayer is chemisorbed randomly on a smooth surface), this scattering process would be 10^4 times stronger for the molecule chemisorbed at the active site compared to the molecule chemisorbed to the terrace.

We are not able to calculate the strength of the scattering process. The first results on the electronic excited states of a smooth silver (110) surface have been recently presented[35]. The extension of these calculations to cases when translational symmetry is broken down is difficult. A promising scheme has been proposed by Herman[36]. He reintroduces translational symmetry by a regular spacing of adatoms. Even when one has the results on the excited states of an active site it is still a formidable task to calculate the Raman scattering process. The cross section for the postulated Raman

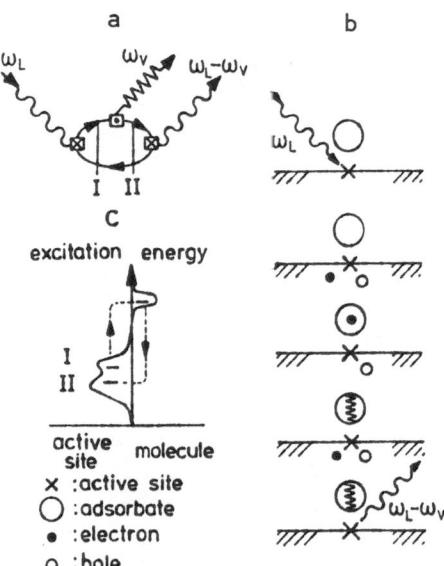

Fig. 7 The hypothetical SERS process at sites of atomic scale roughness, see text. I, II: Intermediate electronic states of the
active site.

scattering process will depend crucially on the lifetime of the excited electronic state of the active site (or of the system comprising
active site and adsorbate). The longer the lifetime the higher the
chance of energy transfer to the adsorbate and the higher the chance
of an excitation of a vibration. We will come back to this point below.

    In case that the cross section for the postulated Raman scat-
tering process $\sigma_i(\omega_L, \omega_{vi})$ is greater by several orders of magnitude
than the ordinary Raman cross section $\sigma_{free}(\omega_L, \omega_v)$ for the free molecule, this process makes a strong contribution to SERS even if the
surface concentration of active sites is small. The SERS signal would
be given approximately by

$$\sum_i \bar{G}_{EM}(\omega_L, \omega_{vi}) \cdot C_{Ai} \cdot \sigma_i(\omega_L, \omega_{vi}) \tag{III}$$

where i denotes the different kinds of active sites and $\bar{G}_{EM}$ is the
average electromagnetic enhancement. On a smooth surface without
active sites or a macroscopically rough surface with very small concentration of active sites no signal, or only a very weak signal, from
an adsorbate would be observed. If our conjecture is basically correct
one could explain the breakdown of Raman selection rules in SERS[32,37]
thus: The Raman selection rules are derived for direct photon-molecule

interaction under the condition that the direction of the incident field is homogeneous over the lateral extension of the molecule, and the scattered light is due to dipole emission from the molecule. These conditions are not fullfilled for the scattering process depicted in fig. 5. Here the molecule is excited by the highly inhomogeneous coulombic electron-electron interaction[16]. Strong overtones would only be expected when the energies of the excited states of active sites and of the adsorbed molecule coincide[16].

The main problem of our hypothesis is the following: The condition $\sigma_i(\omega_L,\omega_{vi}) \gg \sigma_{free}(\omega_L,\omega_v)$ implies an intermediate electronic excited state of the system active site and adsorbate with a lifetime much longer than 10^{-15} sec at the energy $\hbar\omega_L$ (in nonresonant Raman scattering the intermediate state is only populated for about 10^{-15} sec). There are two observations which should be mentioned in this context. Firstly the simultaneous presence of the disorder induced electronic background and the disorder induced scattering from phonons[24,25] discussed in section II. This indicates that the primary electronic excitations at or near point defects are coupled with comparable rates to both the continuum of the electronic excitations in the bulk and to the continuum of bulk phonon states. These coupling rates determine the lifetime of the primary excited state. A coupling rate of 10^{15} sec^{-1} to phonons corresponding to a lifetime of the intermediate state of 10^{-15} sec appears to be very high. Secondly, we should mention the surface picosecond Raman gain experiments by Heritage et al.[38] For an activated silver electrode surface, the reflectivity for the probe beam is increased by the pump beam up to a delay time of 200 psec after switching off the pump beam. One of the possible explanations given by Heritage et al. was that the effect is due to a change in the population of the metal electronic states by the pump beam. This would imply a surprisingly long relaxation time of the relevant electronic excitation of 10^{-10} sec. See also ref. 64.

IV. EXPERIMENTAL INDICATIONS FOR THE IMPORTANCE OF ATOMIC SCALE ROUGHNESS

Without any knowledge on the enhanced cross section $\sigma_i(\omega_L,\omega_{vi})$ in formula (III), the importance of atomic scale roughness in SERS can only be checked by variation of the concentration of active sites C_{Ai}. On activated polycrystalline surfaces, C_{Ai} is not known either. However, one may check whether variations of the observed intensity can be attributed to variations of C_{Ai}. The difficulty lies in the possible simultaneous variation of $\bar{G}_{EM}(\omega_L,\omega_{vi})$ and C_{Ai}. In the following, we will first describe our relevant experiments on silver-vacuum interfaces, and then our experiments on silver electrolyte interfaces.

a. SERS on silver-vacuum surfaces.

For a silver (110) surface, exposed to pyridine at 120 K, we found no indication for SERS[26]. In this case the Raman signal of pyridine grows linearly with exposure. Below an exposure of 30 L the signal was submerged in noise. Fig. 8c shows the unenhanced Raman spectrum of a solid layer of pyridine on Ag (110) after an exposure of 10^4 L. Crystal field (Davydov) splitting is observed for the 1037 - 1040 cm^{-1} vibration. On "cold silver films" we observed Raman enhancements for various adsorbates. Fig. 8a shows the SERS spectrum after exposure with 1 L pyridine. Comparable signal strength from an Ag (110) surface at 120 K is only observed after 10^4 to 10^5 L exposure[23].

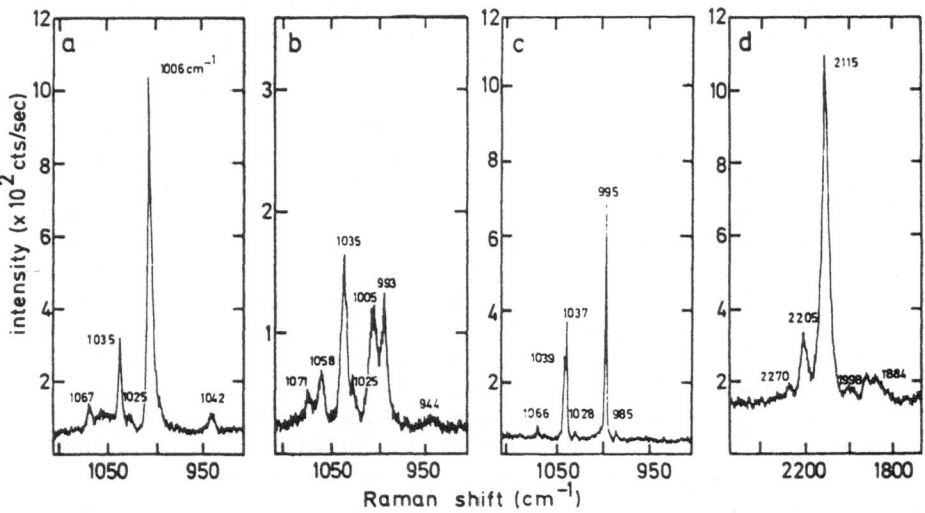

Fig. 8   SERS from a "cold silver film" exposed to 1 L (a) and 10^3 L (b) of pyridine and 3 x 10^4 L CO (d). c: Normal Raman scattering of an Ag (110) surface exposed to 10^4 L of pyridine.

This indicates an enhancement between four and five orders of magnitude. The differences of vibrational frequencies of the physisorbed pyridine (fig. 8c) and the SERS signal (fig. 8a) must be caused by chemisorption of those pyridine molecules involved in the enhancement mechanism. There exist several indications that the SERS signal in fig. 8a is a superposition of SERS from different kinds of adsorption sites (different active sites). We will discuss this elsewhere[39]. It is interesting to note, that Udagawa et al.[40] find a weak enhancement of about 250 for pyridine on Ag (100) surfaces. At the lowest exposure the frequencies coincide with the ones in fig. 8a. Therefore we think that this observed enhancement is not a "chemical effect" in the sense discussed in section I. Rather it reflects the low concentration c_{Ai} of active sites on the "smooth" surface, as discussed in section III. If the exposure of the "cold silver film" with pyridine is increased an extra line at 993 cm^{-1} is observed, see fig. 8b. The intensity variation of the 1006 cm^{-1} and the 993 cm^{-1} lines with exposure

is presented in ref. 23. For our "cold film" the enhancement of the
993 cm^{-1} line is only 2 orders of magnitude, when compared to the 995
cm^{-1} line of pysisorbed pyridine on Ag (110). The line is only seen
above 10 L exposure, but its intensity does not grow between 10 and
10^3 L exposure[23]. The dependence of the intensity of the 993 cm^{-1} line
upon exposure has been discussed as evidence for a long range enhancement mechanism[8]. According to this work, the 993 cm^{-1} line is from
molecules not in direct contact with the surface. On the other hand,
we have tentatively assigned the enhancement of the 993 cm^{-1} line to
Raman enhancement of pyridine in close contact with molecules directly
attached to active sites[23]. From a comparison of the available
data[8,23,31,41], one finds that the intensity of the line near 990 –
995 cm^{-1} compared to the line near 1006 cm^{-1} is higher, when the inelastic background (see chapter II) is higher. This leads to the open
question: Does the 993 cm^{-1} line originate from pyridine molecules on
atomically flat parts of the surface which interact with e-h excitations, created by electron-photon coupling at nearby bulk point defects a few lattice constants away?

Fig. 8d shows a SERS spectrum of adsorbed CO on a "cold film"
after exposure to 3×10^4 L CO, (note that one does not need a static
CO pressure, in agreement with the early reports of T. Wood and M.
Klein[22]). The line at 2115 cm^{-1} is a ^{12}CO stretch frequency, as verified by an analogous experiment with ^{13}CO , where this line was
shifted by 47 cm^{-1} to lower values in agreement with the increase in
the reduced mass (the origin of the other lines in fig. 8c is not
clear; catalytic reactions may be involved).

The left part of fig. 9 shows the variation of the intensity of
the 1006 cm^{-1} pyridine line on a cold film after exposure to 1 L at
120 K, when the film is warmed up to room temperature in about two
hours. The loss of intensity between 210 and 270 K is not caused by
thermal desorption of pyridine[25], as is shown by the right part of
fig. 9. After warming up the sample to about 240 K, it was recooled
to 120 K and only small changes in the intensity are recorded. At
120 K it was reexposed to 1 L of pyridine. This did not change the
signal intensity. When it was warmed up again to room temperature,
little change occurred below 240 K. Above 240 K the intensity varies as
in the uninterrupted cycle. Recooling after warming up to room temperature did not reestablish the SERS activity of the film (fig. 9, left).
These results show that the temperature variation of the signal is
caused by an annealing process of the surface topography. Temperature
dependent restructuring of the adsorbate or desorption does not
explain the results.

The annealing of the surface topography depends on the kind of
coverage of the surface; the relevant results are given in fig. 10.
It shows the intensity variation of the 1006 cm^{-1} and 993 cm^{-1} lines
after an initial exposure of the cold film to 10^3 L pyridine at 120 K.
The variation of the 1006 cm^{-1} line after an initial exposure of

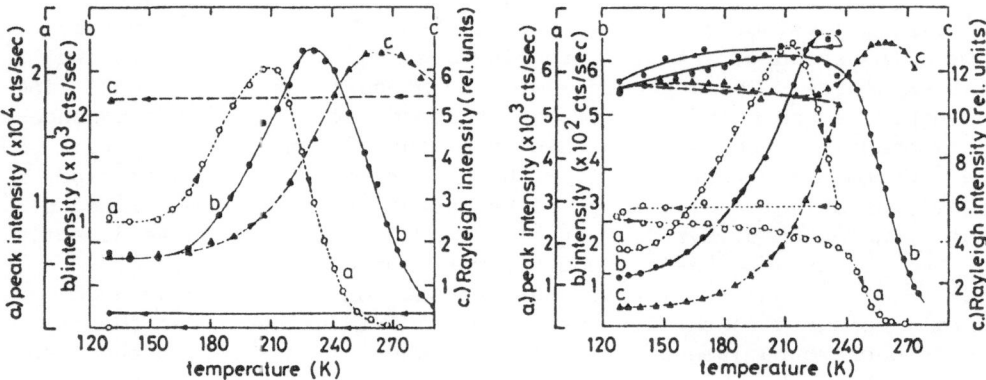

Fig. 9   Left: Temperature variation of the intensity of the background
(b), of the 1006 cm^{-1} pyridine signal above the background (a),
and of the Rayleigh light (c) of a "cold silver film" exposed
to 1 L of pyridine. Right: Recooling to 120 K and reexposing
to 1 L of pyridine. After[25].

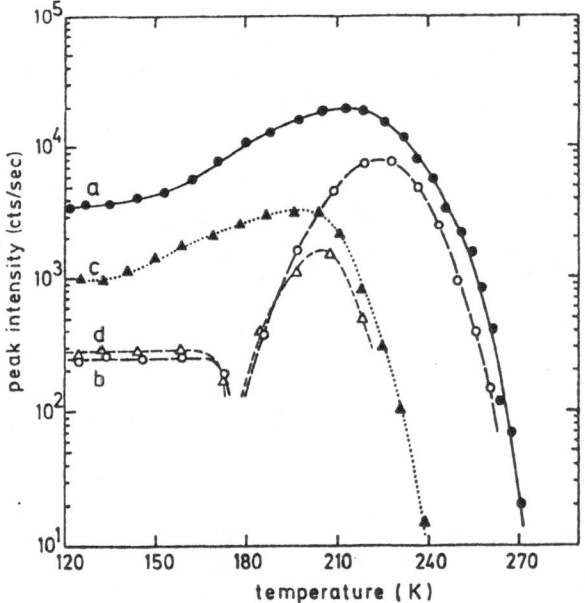

Fig. 10   Temperature variation of SERS of a "cold silver film" exposed
to 1 L of pyridine, 1006 cm^{-1} line (a); exposed to 10^3 L of
pyridine, 1006 cm^{-1} line (b) and 993 cm^{-1} line (d); and exposed to 3 x 10^4 L CO, 2115 cm^{-1} line (c).

only 1 L of pyridine at 120 K is different. Whereas the 1006 cm^{-1} intensity at 1 L exposure grows in the range between 120 and 170 K, it remains constant for the film with 10^3 L of pyridine exposure. The thickness of the pyridine film after 10^3 L is constant in this temperature range, the film only reevaporating near 175 K, within about one minute. This shows that the solid pyridine film prevents any annealing of the surface, in contrast to the surface with low pyridine coverage. When the solid pyridine film reevaporates near 175 K, we observe changes in the intensity of the 993 cm^{-1} and 1006 cm^{-1} lines. This may be related to the coverage dependent changes in the intensities of these lines, which we have reported[23]. Above 180 K, after evaporation of the bulk film, the intensity of the 1006 cm^{-1} line grows much faster with temperature than after an initial exposure of only 1 L. This indicates that the surface annealing between 120 K and 180 K, which was prevented by the bulk pyridine film, is now taking place much faster above 180 K. Eventually, above 230 K, the annealing no longer depends on the initial coverage. The film is "deactivated" around 270 K. The intensity of the 993 cm^{-1} line is lost at lower temperature. This is probably due to desorption of those molecules responsible for the 993 cm^{-1} line. The intensity of the CO stretch line at 2115 cm^{-1} after exposure of a "cold film" to 3 x 10^4 L CO disappears at lower temperatures than the intensity of the pyridine line at 1006 cm^{-1}. We have not yet checked whether this is caused by desorption of CO. It is possible that the difference between traces a and c in fig. 10 reflects the influence of different adsorbates on the surface annealing (e.g. the influence of adsorbates on the activation energy of the surface migration of silver adatoms). It is clear, that such an influence must exist. The irreversible "deactivation" of a "cold film" after annealing to room temperature is in contrast with the quasi permanent activation of an electrode-electrolyte interface at room temperature. In the latter case, the presence of the Helmholtz layer only allows a slow restructuring of the surface (see section II and below).

The crucial question is now: On what scale or to what extent is the silver surface restructured, what scale of roughness is annealed? One can safely exclude roughness on a lateral scale of 200 Å or greater: Eesley[31] cooled a mechanically polished and argon bombarded silver slug to 120 K. The combined signal of the 1006 cm^{-1} pyridine line and the background showed a temperature variation intermediate between curves a and b in fig. 9. After annealing, scanning electron microscope pictures were taken at room temperature. These showed a high surface concentration of bumps on the scale of 200 Å. We further think it unlikely that the reconstruction of small bumps of about 50 Å scale explains the data as discussed in section II.

We interpret our SERS data in fig. 9 as the temperature dependence of the surface concentration of a special active site, most probably adatoms at steps (see fig. 6). The arguments are as follows: For a cold film one would expect that the surface concentration of atoms with 3 nearest neighbours [e.g. single adatoms on (111) terraces], 4, 5 and 6 nearest neighbours [e.g. clusters of two adatoms, adatoms at steps, kinksites on (111) surfaces, respectively] varies irreversibly with temperature as given qualitatively in fig. 4a. The increase in the concentration of atoms with 5 nearest neighbours would be caused by decay of clusters of 2, 3 adatoms on terraces, which migrate separately to steps. The decrease in the concentration of atoms with 5 nearest neighbours is caused by the migration of the adatoms along steps to kinksites. The trends in surface concentrations of atoms in 4, 5, 6 nearest neighbours, sketched in fig. 4a, are the same as calculated by Schrammen[42] with a Monte Carlo method for a submonolayer coverage of Ni on an ideal Ni (100) surface. We think that all surface defects contribute to the increase of DC resistivity. Preliminary measurements on the DC conductivity of well annealed silver films after various coverage of additionally evaporated silver have been performed by Chauvineau[43], analogous to the work on gold[44]. The extra increase of resistivity after evaporation of less than a monolayer at 10 K is annealed between 10 K and room temperature in an approximately linear way. The increase in resistivity at 10 K is caused by diffuse scattering of conduction electrons at the silver adatoms. The annealing reflects the eventual incorporation of silver adatoms into the bulk of the silver crystals at kinksites. Because of the overall decrease in atomic scale surface roughness during warming up of the film, the extra resistivity of the film decreases monotonically with temperature.

As discussed in section III the hypothetical Raman cross section depends on the fourth power of the photon-electron coupling constant. In this way a particular kind of surface site with strong coupling could be "projected out" by SERS from the atomic scale surface roughness. In our case, we postulate this for adatoms at steps. The observation that the SERS activity of silver island films in UHV cannot be annealed by warming to room temperature[41] is not inconsistent with our point of view. It is inconceivable to destroy atomic scale roughness on a small particle - it will always have irregular steps, kinks, corners etc, because it cannot be formed into a shape with a flat surface. The most direct evidence for the adatom model would be the evaporation of Ag adatoms on flat Ag surfaces. In a preliminary experiment we have evaporated less than one monolayer of silver at 120 K on a "non active" silver film evaporated at room temperature. We found no Raman signal of pyridine, after exposure to 1 L of pyridine. Wood reported[45] that he needed 100 - 150 Å of additionally evaporated "cold silver" on a "non active" silver film before he could observe SERS. Does this indicate the failure of the "adatom model" and the validity of the model of electromagnetic resonances on a bumpy surface? Maybe. On the other hand, the two experiments might indicate that a sufficient surface concentration of active sites (e.g. adatoms at steps or surface atoms with 5 nearest neighbours) is only obtained after deposition of greater quantities of "cold silver", thus raising the surface concentration of steps sufficiently. We hope to prove (or falsify) our model by more refined experiments in this direction.

b. SERS on silver electrode surfaces.

At a silver vacuum interface the concentration of atomic scale roughness sites can only be changed by temperature variation. The separation of a kink site atom from the crystal surface can only be achieved by heating the crystal to high temperatures. In this separation process (see fig. 11a) about one-third of the cohesive energy of 3 eV would be needed to shift the atom from the kink site to an adatom site at the step, the second third of the cohesive energy would be needed to separate it from the step, thus transforming it into an adatom, and finally the last third of the cohesive energy would be needed to vaporize it.

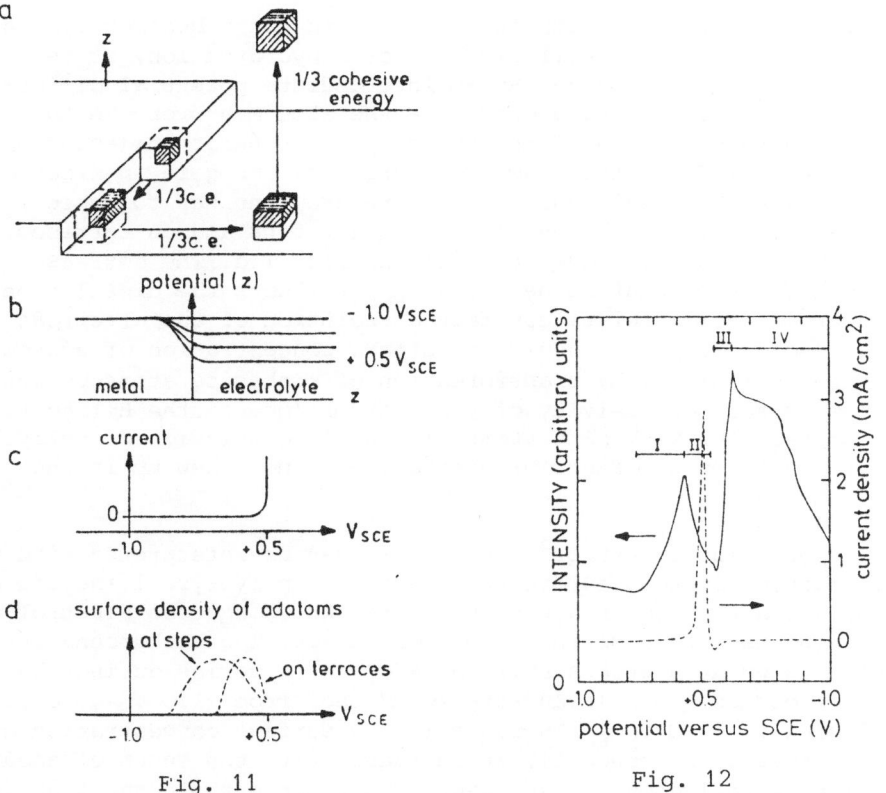

Fig. 11   Fig. 12

Fig. 11   Anodic dissolution of a silver electrode. Hatched parts in (a) characterize fractional positive charge. Potential variation at the silver electrode-electrolyte interface (b) and resulting current (c). Expected surface density of adatoms (d).

Fig. 12   Variation of the background intensity of an activated silver electrode with potential. See text.

The silver electrode-electrolyte interface has the advantage over
the metal vacuum interface that the separation described above can be
done at room temperature, because silver can be dissolved electrochemically. Silver goes into the electrolyte as a positively charged Ag^+
ion, when the potential difference between metal and electrolyte is
sufficiently large (about 1.5 V, see fig. 11b) that the electrostatic
energy and the hydration energy won by Ag^+ overcome the binding energy
of the Ag kink site atoms (which is equivalent to the cohesive energy
per atom). The separation of the Ag^+ ions from the electrode is reflected as the onset of current in the so-called voltammogram, see
fig. 11c. (For technical reasons, the current is not plotted versus
the potential difference electrode-electrolyte, but versus the potential difference V_{SCE} between the electrode and a SCE electrode,
immersed in the electrolyte. The SCE electrode has a fixed potential
difference with respect to the electrolyte.)

On the path of separation in fig. 11a, the atom becomes more and
more positively charged, until finally, as a hydrated ion, it is
completely positively charged. So at intermediate potential differences between electrode and electrolyte the atom may overcome the
activation energies for the first two steps of separation (about one
third and two thirds of the cohesive energy) by the gain in electrostatic energy and partial hydration. Therefore, one has to expect,
that the surface concentration of adatoms at steps and on terraces
varies with V_{SCE} as given qualitatively in fig. 11d. The decrease of
the concentration of adatoms near 0.5 V_{SCE} reflects the dissolution
of the adatoms. For a relatively fast dissolution of the electrode
(about 3 monolayers per second) the surface concentration of adatoms
will not be replenished by transformation of kink site atoms because
of the low surface diffusivity of the adatoms (due to the Helmholtz
layer). Rather, the kink site atoms will go into solution directly.
Note that the first two steps of separation do not show up in the
current.

Our experimental results[46] in fig. 12 can be interpreted with the
concentration of adatoms as discussed above. A polycrystalline silver
electrode was activated in a pure saturated $KClO_4$ aqueous electrolyte
by dissolving and redepositing 200 Å of silver. The background intensity (here taken at a Stokes shift of 4200 cm^{-1}) varies during the
triangular voltage sweep (scan rate 10 mV/sec) from -1.0 V_{SCE} to 0.5
V_{SCE} and back to -1.0 V_{SCE}. In range I, the surface concentration of
adatoms increases, in range II, it decreases with the onset of anodic
dissolution as indicated by the onset of the current. In the scan in
fig. 12, sixty monolayers are dissolved but the decrease in intensity
is still seen when less than a monolayer is dissolved during the scan.
It is hard to understand this decrease on the basis of electromagnetic
resonances of bumps, because bumps and hence the electromagnetic resonances would not be destroyed by dissolving less than one monolayer
of silver. In the cathodic back-sweep (from 0.5 to -1.0 V_{SCE} in range
II) two of the sixty dissolved monolayers are redeposited, as is

evident from the small negative current at about 0.4 V_{SCE}. At the end
of range II, when the discharging current of silver ions is very low,
the background intensity increases. This is assigned to the increase
of adatom concentration. The Ag^+ ions, which "arrive late" have a poor
chance of being incorporated into the lattice. It is again hard to
understand the increase of intensity in range III by electromagnetic
resonances of bumps on a surface. How can one form the necessary bumps
when only one or two monolayers are redeposited? In range IV, between
-0.1 V and -1.0 V_{SCE}, the adatoms eventually migrate to kink sites.
The slow migration of adatoms to kink sites in the range of small
potential differences between electrode and electrolyte leads to a
slow decay of the SERS signal at constant potential[47]. For further
details we refer the reader to ref. 46.

Besides pyridine, cyanide on silver electrodes is the best
studied system in SERS. All the different groups agree upon the
experimental results. We have discussed the results in detail[14] and
have reviewed them[16]. For reasons of limited space we simply restate
our conclusion with the help of fig. 13. The Raman signal of cyanide
chemisorbed on an atomically smooth silver surface, e.g. on a (111)
surface illustrated fig. 13a, is not enhanced, but the Raman signal
of $Ag(CN)_3^=$ and $Ag(CN)_2^-$ complexes at the surface are. A surface complex
$Ag(CN)_3^=$ is a partially positive silver adatom surrounded by 3 cyanide
groups - as depicted in fig. 13b. For steric reasons an adsorption of
3 cyanide groups to one silver atom at a smooth surface is not
possible.

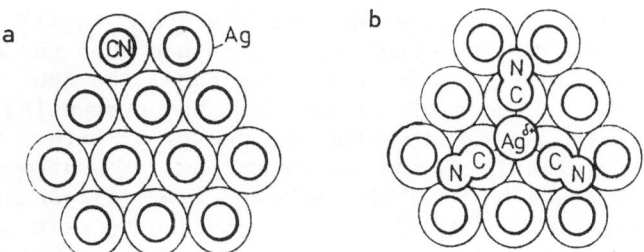

Fig. 13   a: "SERS inactive" chemisorption of CN^-.
          b: "SERS active" $Ag(CN)_3^=$ surface complex.

Last but not least, other experimental groups engaged in SERS
research have also interpreted their data by active sites composed of
adatom-adsorbate complexes[48,49,50].

V.   SERS, A METHOD FOR CATALYTICAL RESEARCH?

Catalysis is an important method in chemical industry but the catalytical processes are not well understood on an atomic basis. At least for certain reactions, catalytically active sites on the catalyst surface are important. For instance on single crystalline platinum surfaces atomic steps have been identified as the active sites for C-H and H-H bond breaking processes[51], and kinks in steps as active sites for C-C bond scissions[51]. Most likely for some special reactions, the activity of dispersed metal catalysts may also be attributed to active sites of a special atomic surface structure.

Silver catalysts are used in the glycol industry for selective oxidation of ethylene to ethylene oxide[52]. The nature of the oxygen species adsorbed on the silver surface is of special importance in this process. We cite the conclusion of the most recent review[52]: "It appears to be agreed generally that oxygen adsorbs on silver in more than one energy state and that such states are associated with several different adsorbed species. Whether these species include O_2^-, $O_2^=$, O^- or $O^=$ remains a matter of debate and interpretation of experimental data". In this situation the search for the different oxygen species by vibrational spectroscopy should be helpful. In electron energy loss spectroscopy on silver (110) surfaces a molecular oxygen species with an O_2-Ag stretch frequency of about 240 cm^{-1} and an O_2 stretch frequency of about 630 - 640 cm^{-1} was found below 185 K, and an atomic species with the Ag-O stretch frequency at about 320 cm^{-1} above 185 K and below 600 K[53,54]. In the SERS spectrum of oxygen on a polycrystalline silver film, evaporated on a cooled silver (111) face and exposed to 2000 L of oxygen, Wood et al.[55] found only low frequency lines of atomically adsorbed oxygen at 260 cm^{-1} and 324 cm^{-1}.

Figure 14 shows our[56] SERS spectra of a "cold Ag film" after exposure to 10^3 L O_2. The spectrum did not change when the oxygen gas was led to the cold film through a glass tube. The tube only terminated immediately in front of the sample, thus preventing reactions of the oxygen with the walls of the UHV vessel or within the ion getter pump before striking the silver surface. Measurements with $^{18}O_2$ showed all peaks between 600 cm^{-1} and 1300 cm^{-1} shifted to lower values with respect to $^{16}O_2$ (fig. 14) according to mass ratios between 1.03 and 1.06. One may therefore safely assign all vibrations between 600 and 1300 cm^{-1} in fig. 14 to oxygen, most probably to the stretch vibration of different kinds of dioxygen $O_2^{\sigma-}$ with various values of σ. The frequencies are different from the ones found by EELS on Ag (110). We do not yet know which of the oxygen species are the relevant ones in the oxidation of ethylene, or whether a form of oxygen not detected in SERS is the relevant one. SERS spectra of ethylene on cold silver have already been reported[57,58], although the spectra do not agree. There are reports of SERS on platinum[59,60,61,62] which is an important catalyst. These examples show that SERS can provide information from catalyst surfaces. However, the interpretation

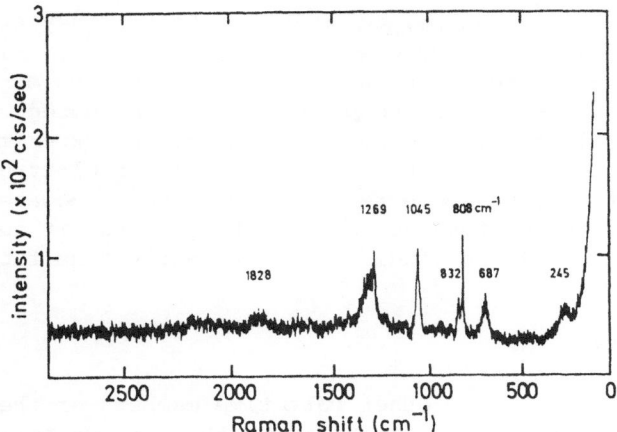

Fig. 14 SERS spectrum of a "cold silver film" exposed to 10^3 L of oxygen.

of the data is difficult - the final decision on the relevance of the concept of active sites in SERS will be important. If in fact SERS active sites do exist, then the question arises whether they are, or which of them are, identical with catalytically active sites.

VI. SUMMARY

The strong enhancement of 10^6 on silver electrode surfaces is a short range effect. The observation of disorder induced Raman scattering from phonons in SERS active "cold silver films" is caused by bulk point defects. We assign the low specular reflectivity of the "cold silver films" to bulk optical absorption at those defects. This would rule out strong electromagnetic enhancement in this case. Irreversible intensity variations at the reflectivity, background, phonon signals and of SERS intensity when the "cold films" are warmed up is assigned to annealing of bulk and surface point like defects. The details of surface annealing are not understood.

The potential dependence of the background intensity from an electrode surface reflects the surface concentration of adatoms. SERS intensity from CN^- on silver originates from CN^- ions adsorbed to silver adatoms.

In view of the open questions and speculations mentioned in our text and the difficulties of discriminating in some cases between atomic scale roughness and very small bumps (ca. 50 Å), we have to agree that the importance of atomic scale roughness is not yet established without doubt. In the case that small bumps are the most important roughness feature, one may ask whether they enhance the Raman spectrum via an electromagnetic resonance[8,32], or, via increased photon-electron coupling. In solid particles of very small dimensions, photon-electron coupling will be increased, because momentum is no longer a good quantum number. The experimental evidence against the opinion that SERS is merely an electromagnetic local field effect, has been recently compiled[17]. However, even without a final understanding of the SERS process, it seems worthwhile to start catalytic studies on "SERS-active" surfaces.

ACKNOWLEDGEMENT

Besides the authors, other part time members of the research group at Düsseldorf have contributed to the work described here: G. Kovacs, J. Timper, R. Zeller and A. Gadou and the technician J. Liebetrau. We thank P. E. Simmonds for a careful reading of the manuscript.

REFERENCES

1. E. Burstein, S. Lundqvist, and D. L. Mills, The roles of surface roughness, this volume.
2. J. I. Gersten and A. Nitzan, Electromagnetic theory: A spheroidal model, this volume.
3. M. Kerker, D.-S. Wang, H. Chew, O. Siiman, and L. A. Bumm, Enhanced Raman scattering by molecules adsorbed at the surface of colloidal particles, this volume.
4. G. C. Shatz, The image field effect: How important is it?, this volume.
5. T. K. Lee and J. L. Birman, Coupled excitation model and quantum test of image field effect, this volume.
6. A. Otto, Raman scattering from adsorbates on silver, in: "Proceedings of the International Conference on Vibrations in Adsorbed Layers, Jülich" (1978); unaltered reprint: Surf. Sci. 92:145 (1980).
7. H. Metiu, A survey of recent theoretical work, this volume.
8. C. A. Murray, Molecule-silver separation dependence, this volume.
9. E. Burstein, Y. J. Chen, C. Y. Chen, S. Lundquist, and E. Tossati, "Giant" Raman scattering by adsorbed molecules on metal surfaces, Solid State Commun. 29:567 (1979).
10. J. Gersten and A. Nitzan, Electromagnetic theory of enhanced Raman scattering by molecules adsorbed on rough surfaces, J. Chem. Phys. 73:3023 (1980).
11. H. Ueba, Induced resonance model, this volume.
12. S. S. Jha, J. R. Kirtley, and J. C. Tsang, Raman scattering from molecules adsorbed on a metal grating, Phys. Rev. B 22:3973 (1980).

13. R. K. Chang, R. E. Benner, R. Dornhaus, K. U. von Raben, and
     B. L. Laube, Enhanced Raman scattering of molecules adsorbed
     on silver, copper, and gold surfaces, in "Proceedings of the
     Sergio Porto Memorial Conference on Lasers and Applications,"
     W. O. N. Guimaraes, C. T. Lin, and A. Mooradian, eds.,
     Springer, Berlin, 1981, p. 55.

14. J. Billmann, G. Kovacs, and A. Otto, Enhanced Raman effect from
     cyanide adsorbed on a silver electrode, Surf. Sci. 92:153
     (1980).

15. T. E. Furtak, G. Trott, and B. H. Loo, Enhanced light scattering
     from the metal solution interface: Chemical origins, Surf.
     Sci. 101:374 (1980).

16. A. Otto, Surface enhanced Raman scattering: What do we know?,
     Appl. Surf. Sci. 6:309 (1980).

17. A. Otto, SERS: Only a local field effect?, in: "Light Scatter-
     ing in Solids," M. Cardona and G. Güntherodt, eds., Springer
     (1981).

18. D. L. Jeanmaire and R. P. Van Duyne, Surface Raman spectro-
     electro-chemistry, Part I, J. Electroanal. Chem. 84:1 (1977).

19. J. Billmann and A. Otto, Experimental evidence for a local
     mechanism of surface enhanced Raman scattering, Appl. Surf.
     Sci. 6:356 (1980).

20. A. Otto, Raman spectra of cyanide adsorbed at a silver surface,
     Surf. Sci. 75:L392 (1978).

21. A. Otto, J. Timper, J. Billmann, G. Kovacs, and I. Pockrand,
     Surface roughness induced electronic Raman scattering, Surf.
     Sci. 92:L55 (1980).

22. T. H. Wood and M. V. Klein, Enhanced Raman scattering from
     adsorbates on metal films in ultra high vacuum, Solid State
     Commun. 35:263 (1980).

23. I. Pockrand and A. Otto, Coverage dependence of Raman scatter-
     ing from pyridine adsorbed to silver vacuum interfaces,
     Solid State Commun. 35:861 (1980).

24. I. Pockrand and A. Otto, Surface enhanced and disorder induced
     Raman scattering from silver film, Solid State Commun. 37:109
     (1980).

25. I. Pockrand and A. Otto, Surface enhanced Raman scattering:
     Annealing the silver substrate, Solid State Commun., in
     press.

26. I. Pockrand and A. Otto, Raman scattering from silver vacuum
     interfaces, Appl. Surf. Sci. 6:362 (1980).

27. H. Mehren and A. Seeger, Interpretation of self-diffusion and
     vacancy properties in silver, Phys. Stat. Sol. 39:647 (1970).

28. L. J. Cuddy and E. S. Machlin, Quenching in and annealing of
     point defects in silver, Phil. Mag. 7:745 (1962).

29. L. M. Clarebrough, R. L. Segall, M. H. Loretto, and M. E.
     Hargreaves, The annealing of vacancies and vacancy aggregates
     in quenched gold, silver, and copper, Phil. Mag. 9:377 (1964).

30. M. Doyama and J. S. Koehler, Quenching and annealing of lattice
     vacancies in pure silver, Phys. Rev. 127:21 (1962).

31. G. L. Eesley, Coverage dependence of enhanced adsorbate Raman scattering, Phys. Lett. 81A:193 (1981).
32. M. Moskovits and D. P. DiLella, Vibrational spectroscopy of molecules adsorbed on vapor-deposited metals, this volume.
33. B. Pettinger, U. Wenning, and D. M. Kolb, Raman and reflectance spectroscopy of pyridine adsorbed on single crystal silver electrodes, Ber. Bunsenges Phys. Chem. 82:1326 (1978).
34. I. Nenner and G. J. Schulz, Temporary negative ions and electron affinities of benzene and N-heterocyclic molecules: Pyridine, pyridazine, pyrimidine, pyrazine, and s-triazine, J. Chem. Phys. 62:1747 (1975).
35. Kai Ming Ho, B. N. Harmon, and S. H. Liu, Surface state contribution to the electroreflectance of noble metals, Phys. Rev. Lett. 44:1531 (1980).
36. F. Herman, Interaction of cyanide molecules with smooth and microscopically rough silver surfaces, Bull. Am. Phys. Soc. 26:338 (1981).
37. R. Dornhaus, M. B. Long, R. E. Benner, and R. K. Chang, Time development of SERS from pyridine, pyrimidine, pyrizine, and cyanide adsorbed on silver electrodes during an oxidation-reduction cycle, Surf. Sci. 93:240 (1980).
38. J. P. Heritage, J. G. Bergman, A. Pinczuk, and J. M. Worlock, Surface picosecond Raman gain spectroscopy of a cyanide monolayer on silver, Chem. Phys. Lett. 67:229 (1979).
39. A. Otto and I. Pockrand, Proceedings of the International School, "Dynamics of Gas Surface Interactions," in preparation.
40. M. Udagawa, Chih-Cong Chou, J. C. Hemminger, and S. Ushioda, Raman scattering cross section of adsorbed pyridine molecules on a smooth silver surface, UCI Technical Report 80-96.
41. H. Seki, Surface enhanced Raman scattering of pyridines on different silver substrates, Bull. Am. Phys. Soc. 26:338 (1981).
42. P. Schrammen, Dissertation, Kassel, 1981.
43. J. P. Chauvineau, private communication.
44. J. P. Chauvineau, Diffusion superficielle des adatomes observee basse temperature par la variation de resistance electrique de films minces d'or et de bismuth, Surf. Sci. 93:471 (1980).
45. T. H. Wood and D. A. Zwemer, The dependence of surface enhanced Raman scattering on surface preparation, Bull. Am. Phys. Soc. 26:339 (1981).
46. A. Otto, J. Timper, J. Billmann, and I. Pockrand, Enhanced inelastic light scattering from metal electrodes caused by adatoms, Phys. Rev. Lett. 45:46 (1980).
47. J. Timper, J. Billmann, A. Otto, and I. Pockrand, Surface enhanced light scattering from silver electrodes: Background and cyanide stretch vibration, Surf. Sci. 101:348 (1980).
48. V. V. Marinyuk, R. M. Lazararko-Manevich, and Ya. M. Kolotyrkin, Nature of the interaction of adsorbate molecules with metal adatoms, J. Electroanal. Chem. 110:111 (1980).
49. K. A. Bunding, R. L. Birke, and J. R. Lombardi, The surface

enhanced Raman spectrum of 2, 6-lutidine, Chem. Phys. 54:115 (1980); J. R. Lombardi, E. A. Shields Knight, and R. L. Birke, Evidence for a Ag adatom-molecule complex in surface enhanced Raman scattering, Chem. Phys. Lett. 79:214 (1981).

50. H. Wetzel, H. Gerischer, and B. Pettinger, Surface enhanced Raman scattering from silver halide and silver pyridine vibrations and the role of silver adatoms, Chem. Phys. Lett. 78:392 (1981).

51. D. W. Blakely and G. A. Somorjai, The dehydrogenation and hydrogenolysis of cyclohexane and cyclohexane on stepped (high miller index) platinum surfaces, J. Catalysis 42:181 (1976).

52. E. Verykios, F. P. Stein, and R. W. Coughlin, Oxidation of ethylene over silver: Adsorption, kinetics, catalyst, Catal. Rev. Sci. Eng. 22:197 (1980).

53. B. A. Sexton and R. J. Madix, Vibrational spectra of molecular and atomic oxygen on silver (110), Chem. Phys. Lett. 76:294 (1980).

54. C. Backx, C. P. M. de Groot, and P. Biloen, Adsorption of oxygen on Ag(110) studied by high resolution ELS and TPD, Surf. Sci. 104:300 (1980).

55. T. H. Wood, M. V. Klein, and D. A. Zwemer, Enhanced Raman scattering from adsorbates on metal films in ultra high vacuum, preprint.

56. C. Pettenkofer, I. Pockrand, and A. Otto, to be published.

57. T. H. Wood and P. A. Zwemer, Surface reaction studied on roughened silver using enhanced Raman scattering, preprint.

58. M. Moskovits and D. P. DiLella, Enhanced Raman spectra of ethylene adsorbed on silver, Chem. Phys. Lett. 73:500 (1980).

59. J. Heitbaum, XPS, Raman, and IR-analysis of a surface product formed with the electrooxidation of phenylhyrazine, Z. Physik. Chem. 105:307 (1977).

60. R. P. Cooney, M. Fleischmann, and P. J. Hendra, Raman spectrum of CO on a platinum electrode surface, J. Chem. Soc. Chem. Commun. (1977), p. 235.

61. R. P. Cooney, E. S. Reid, P. J. Hendra, and M. Fleischmann, Thiocyanate adsorption and corrosion at silver electrodes: A Raman spectroscopic study, J. Am. Chem. Soc. 99:2002 (1977).

62. H. Yamada and Y. Yamamoto, Surface enhanced Raman spectra of pyridine adsorbed on silver, gold, nickel, and platinum metals, Chem. Phys. Lett. 77:520 (1981).

63. S. G. Schultz, M. Janik-Czachor, and R. P. Van Duyne, Surface enhanced Raman spectroscopy: A re-examination of the role of surface roughness and electrochemical anodization, Surf. Sci. 104:419 (1981).

64. C. K. Chen, A. R. B. de Castro, and Y. R. Shen, Surface enhanced second harmonic generation, Phys. Rev. Lett. 46:145 (1981). These authors find for activated electrodes (bulk silver and films of silver) an "exceptionally strong"

antistokes background beyond the second harmonic of 1.06 μ,
extending in wavelength up to about 1/3 of 1.06 μ. In our
opinion, this implies a continuum of rather long living excited electronic states. The authors state: "Furthermore,
the temporal behavior of the rough film background showed
a clear tail, several pulse widths (of 10 nsec) long. We
do not understand this observed anomaly on the films. The
result, however, makes us believe that the broadband background is of luminescence origin."

INDUCED RESONANCE MODEL

Hiromu Ueba

Department of Electronics
Toyama University
Takaoka, Toyama, Japan

INTRODUCTION

Mounting experimental observation of SERS has led to considerable theoretical efforts towards the understanding of various aspects of this phenomenon. The existing theories should be tested by their ability to explain the following important properties of SERS: (1) The Raman intensity of adsorbed molecules is enhanced by a factor of 10^5-10^6 compared to that of free molecules. (2) The degree of the enhancement depends on the excitation energy and also on the properties of the substrate metal. (3) The enhanced Raman scattering spectra are accompanied by a broad continuum spectrum. (4) The roughness of a metal surface plays a crucial role.

Since a critical analysis of theories has been given by Furtak and Reyes,[1] several models have also been added, which focus their primary attention to the secondary electromagnetic field near a rough metal surface. The various types of roughness models which differ with each other in the structure of the roughness are explained in this book.

In a course of theoretical progress, a class of theories which do not require any surface roughness includes the classical image dipole model of adsorbed molecule,[2,3] and the quantum mechanical treatment of the interaction between adsorbed molecule and surface plasmon.[4] Another model for a flat surface claims an importance of the electron transfer between adsorbed molecule and metal, which we name here the induced resonance model.

Although it is now widely accepted that the surface roughness is responsible for a large factor of the enhancement, there has

been no clear demonstration which forces us to conclude that a
flat surface does not generate any enhancement. Recently, Udagawa
et al.[5] have observed the Raman scattering of pyridine molecules
adsorbed on a "smooth" surface of Ag and reported the enhancement
factor of 4.4×10^2 when the 514.5 nm excitation is used. A stronger
enhancement ($\sim 10^3$) is then expected even on a smooth surface at
longer wavelength excitations. Furthermore, it is worth mentioning
that they also noted the presence of the background continuum,
even though the intensity is weaker than is observed for rough
surfaces.

Contradictory interpretations have been proposed for the origin
of the continuum on which the enhanced Raman spectra are superimposed. Those are the coupling of surface plasmon polaritons to
light through the grating roughness,[6] the resonant electronic
Raman scattering due to the adatom roughness,[7] the superposition
of innumerable very weak SERS lines arising from vibrational modes
of the adsorbed species,[8] and the luminescence involving charge
transfer recombination predicted in the induced resonance model.

We are now at the stage where each model seems to be supported
by some experimental evidence, but no unique mechanism is able to
explain all of the properties associated with SERS of various
adsorbed molecule-metal systems. It is, therefore, of importance
to find the general effect of all systems and to classify the
specific component of each system. In what follows, the induced
resonance model is explained on the basis of previous works[9-11]
and its possibilities and limitations are discussed.

RENORMALIZED MOLECULAR STATE AND INDUCED RESONANCE

We consider a system consisting of a molecule with two discrete
electronic levels interacting with both vibrational modes and a
metal. The molecule-metal interaction is assumed to take place only
between the molecular excited state and metal via the s-d mixing
interaction with the constant coupling strength V. Such a mixing
allows the electron to shuttle between the molecule and metal.
Furthermore, the photon field is assumed to couple only with the
molecular states. In this sense, the Raman scattering of the adsorbed molecule is considered through its intermediate state, which is
"dressed" by electron transfer between the molecule and the metal.

The model Hamiltonian of the coupled molecule-metal system is
then written as

$$H = H_g + H_e \tag{1}$$

$$H_g = a_g^+ a_g (\varepsilon_g + H_{ph}) + H_m \tag{2}$$

$$H_e = a_e^+ a_e (\varepsilon_e + H_{ph}) + H_m + H'_{ph} + H'_m \tag{3}$$

$$H_m = \sum_k \varepsilon_k c_k^+ c_k \tag{4}$$

$$H'_m = \sum_k (V c_k^+ a_e + V^* a_e^+ c_k) \tag{5}$$

where $a_g(a_g^+)$ and $a_e(a_e^+)$ are the annihilation (creation) operators of the molecular electronic ground and excited states with energy ε_g and ε_e, respectively. The electronic state in a metal is expressed by its operator $c_k(c_k^+)$. In Eqs. (2) and (3) H_{ph} and H'_{ph} are the kinetic energy of phonons and the electron-phonon interaction, respectively.

A general formula of the second order optical process supplemented with a damping constant γ of the excited state is given by

$$I(\omega_1,\omega_2) = \sum_{f} \text{Av.} \sum_{i} <f|H_I \frac{1}{\omega_1+E_i+i\gamma-H_e} H_I|i>^2 \delta(\omega_1-\omega_2+E_i-E_f) \tag{6}$$

where ω_1 and ω_2 are the energies of incident and scattered photons, respectively, and i and f refer to the initial and final states with energies E_i and E_f. The electron-radiation field interaction H_I takes a form of the dipole allowed optical transition, which is decomposed into an excitation part $M_\uparrow$ and a deexcitation part $M_\downarrow$:

$$H_I = M_\uparrow + M_\downarrow = M^* a_e^+ a_g + M a_g^+ a_e \quad . \tag{7}$$

Equation (6) can be rewritten as the perturbation expansion with respect to $H'=H'_{ph}+H'_m$ as

$$I(\omega_1,\omega_2) = \text{Av.} \sum_{i} \sum_{m=0}^{\infty} \sum_{n=0}^{\infty} <i|M_\downarrow \left[\frac{1}{Z^*-H_e^0}H'\right]^m \frac{1}{Z^*-H_e^0} M\delta(\omega_1-\omega_2+E_i-H_g)$$

$$\times M_\downarrow \frac{1}{Z-H_e^0}\left[H'\frac{1}{Z-H_e^0}\right]^n M_\uparrow|i> \tag{8}$$

where $Z=\omega_1+E_i+i\gamma$ and H_e^0 is the unperturbed part of Eq. (3). In Eq. (8) the following quantity

$$G(\omega_1) = \sum_{n=0}^{\infty} M_\downarrow \frac{1}{Z-H_e^0}\left[H'\frac{1}{Z-H_e^0}\right]^n M_\uparrow \tag{9}$$

is the polarizability operator, since its average

$$P(\omega_1) = \text{Av.} <i|G(\omega_1)|i> \tag{10}$$
$$\quad\quad i$$

gives the polarizability of the system. Because of the linear interaction of H', the perturbation expansion of Eq.(8) is restricted to the case n+m=2j (j=0,1,2,$\cdots$). A renormalization for the diagonal propagators can be embodied by replacing H_e^0 by $H_e^0+\Sigma$ and taking the n, m summation as n=m=0,1,2$\cdots$ in Eq.(8). Here, the self-energy Σ of the excited molecular state is given by:[10]

$$\Sigma(\omega) = \Delta_{ph} + \Delta_m(\omega) -i\{\Gamma_{ph} + \Gamma_m(\omega)\} \tag{11}$$

where the self-energy due to the linear electron-phonon interaction is calculated as

$$\Delta_{ph} = \int d\omega \frac{\Gamma_{ph}(\omega)}{\omega} \tag{12}$$

$$\Gamma_{ph} = \int d\omega \, \Gamma_{ph}(\omega)\delta(\omega) = 0 \tag{13}$$

$$\Gamma_{ph}(\omega) = \sum_j \omega_j^2 \, \eta_j^2 \, \delta(\omega-\omega_j) \tag{14}$$

at the zero temperature case, where ω_j and η_j are the phonon energy and the expansion coefficient of the electron-phonon interaction, respectively. Since we are implicitly concerned with weakly adsorbed molecules on the metal surface the self-energy due to the molecule-metal interaction is evaluated in the lowest order as

$$\Sigma_m(\omega) = \frac{1}{|M|^2} \text{Av.} <i|M \, H_m' \frac{1}{\omega+E_i-H_e^0+i\delta} H_m'M |i> \tag{15}$$
$$\quad\quad i$$

$$= |V|^2 \sum_k \frac{1}{\omega+\varepsilon_g-\varepsilon_k+i\delta} \quad (\delta\to +0)$$

$$= \rho|V|^2 \left[\ln\left|\frac{\tilde{\omega}}{\tilde{\omega}-D}\right| - i\{\theta(\tilde{\omega})- \theta(\tilde{\omega} - D)\} \right] \tag{16}$$

where $\rho=\rho(\varepsilon_f)$ is the density of states of a metal at the Fermi energy ε_f, and D is the half bandwidth, θ is the usual step function, and $\tilde{\omega}=\omega+\varepsilon_g-\varepsilon_f$. Thus the polarizability defined by Eq.(10) is obtained as

$$p(\omega_1) = \{ (\omega_1-\tilde{\varepsilon}- \Delta_m) + i(\gamma + \Gamma_m)\}^{-1}. \tag{17}$$

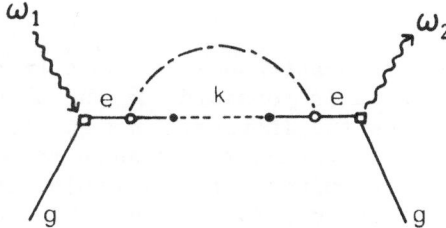

Figure 1. Feynman diagram of the Raman scattering of adsorbed
molecules. The solid, wavy and dot-dashed lines represent the
electron in the molecular states, photon and phonon, respectively.
The dashed line denotes the electron in a metal.

Now, we come to a stage to calculate the light scattering
spectrum of adsorbed molecules. The first term $n=m=0$ in Eq.(8)
gives the Rayleigh scattering, and a choice of $n=m=1$ after the
treatment mentioned below Eq.(10) gives the Raman scattering
spectrum

$$I(\omega_1,\omega_2) = |M|^4 \, F(\omega_1)F(\omega_2)\{\Gamma_{ph}(\omega_1-\omega_2) + \Gamma_m(\omega_1-\omega_2+\varepsilon_e-\varepsilon_g)\} \quad (18)$$

where

$$F(\omega) = \frac{1}{\{\omega-\hat{\varepsilon}-\Delta_m(\omega)\}^2 + \{\gamma+\Gamma_m(\omega)\}^2} \quad (19)$$

$$\hat{\varepsilon} = \varepsilon_e - \varepsilon_g + \Delta_{ph} . \quad (20)$$

The Feynman diagram leading to this result is shown in Fig.1.

The interesting characters of the Raman spectrum of adsorbed
molecules manifest themselves in Eq.(18). The first term of Eq.
(18) represents the one phonon Raman spectrum in the sense that
ω_2 is correlated with ω_1 through $\Gamma_{ph}(\omega_1-\omega_2)$ defined by Eq.(14).
The explicit form of $\Gamma_{ph}(\omega)$ is obtained by assuming the density of
states in a simple form $\rho_{ph}(\omega)=\delta(\omega-\omega_{ph})$, ω_{ph} being the
characteristic phonon energy, and is given by

$$\Gamma_{ph}(\omega) = g\delta(\omega-\omega_{ph}) \quad (21)$$

where g is the electron-phonon coupling strength. As we can see
from Eq.(19), $F(\omega)$ becomes quite large when the following
condition is satisfied:

$$\omega - \hat{\epsilon} - \Gamma_m(\omega) = 0 \quad , \tag{22}$$

whereas the intrinsic resonant condition of a free molecule is given by $\omega - \hat{\epsilon} = 0$. Thus it is possible for the adsorbed molecules to be in resonance below the intrinsic one due to the molecule-metal interaction, which we call the induced resonance. In Fig. 2 the shift and the broadening of the molecular excited state are shown as functions of the excitation energy ω_1, where all the energies are measured in units of $\hat{\epsilon}$ relative to ϵ_g. In the case when $D \gg \omega + \epsilon_g - \epsilon_f$, the solution of Eq.(22) is given by

$$\omega + \epsilon_g = \epsilon_f - D \exp\left[- \frac{\epsilon_e + \Delta_{ph} - \epsilon_f}{\rho|V|^2} \right] , \tag{23}$$

which suggests to us that the induced resonance is realized when the electron in the molecular ground state is virtually excited to slightly below the Fermi level of the metal. We then expect the induced resonance Raman scattering of adsorbed molecules, even with the off-resonant excitation of a free molecule.

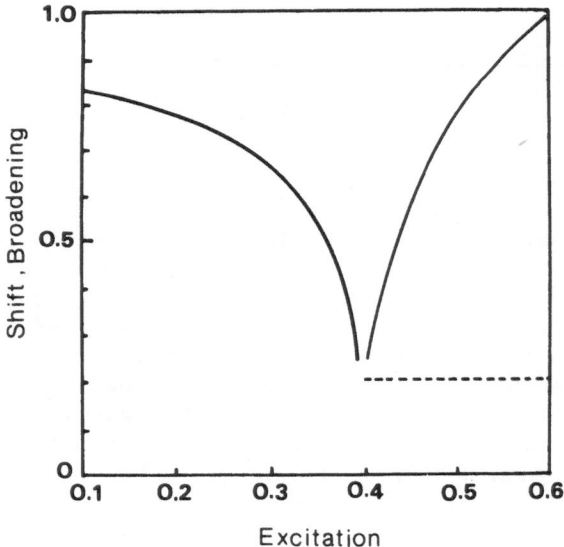

Figure 2. The shift (solid line) and the broadening (dashed line) of the molecular excited state as a function of the excitation $\omega_1 + \epsilon_g$, where $\rho|V|^2 = 0.2$ and $\epsilon_f = 0.4$.

The exact solution of Eq.(22) indicates that a resonance is realized near the threshhold for the photoinjection of electrons from the molecule to the conduction band of the metal, $\omega_1 + \varepsilon_g \approx \varepsilon_f$. This was first conceived by Gersten et al.[9] When $\omega_1 + \varepsilon_g > \varepsilon_f$, the photoionization of the molecule results in the electron transfer from the molecule to the metal. Such a transfer causes a rise in the Fermi level of the metal relative to the molecular ground state. The shift of the Fermi level continues until the condition $\omega_1 + \varepsilon_g = \varepsilon_f$ is realized, where the one-photon photoionization process is cut off. Since the electron transfer between molecule and metal plays an important role in the establishment of the induced resonance, it can also be related with an applied voltage dependence of SERS. Gersten et al.[9] suggested that any adsorbed molecule having a stable positive ion radical can be promoted to the virtual excited state near the Fermi level and should exhibit SERS. In a similar fashion, a molecule having a negative ion radical shows the same effect when the electrons in the metal are excited to the lowest unoccupied state of the molecule. Such a situation will be destroyed when the electron transfer is shut off by an interfacial applied voltage.

Using Eq.(18), the Raman scattering intensity of the adsorbed molecule is compared with that of a free molecule. Since the Raman scattering of a free molecule is simply obtained from Eq.(18) by neglecting Σ_m, the degree of the enhancement is given by

$$R(\omega_1, \omega_2) = \frac{F(\omega_1) F(\omega_2)}{F^0(\omega_1) F^0(\omega_2)} \left[1 + \frac{\Gamma_m(\omega_1 - \omega_2 + \varepsilon_e - \varepsilon_g)}{\Gamma_{ph}(\omega_1 - \omega_2)} \right] . \quad (24)$$

At the induced resonance situation where $\omega_1 + \varepsilon_g < \varepsilon_f$ and $\omega_2 = \omega_1 - \omega_{ph}$, we obtained

$$R = \frac{(\varepsilon_f - \hat{\varepsilon}_e)^2 + \gamma^2}{\gamma^2} \frac{(\varepsilon_f - \hat{\varepsilon}_e - \omega_{ph})^2 + \gamma^2}{(\varepsilon_f - \hat{\varepsilon}_e - \omega_{ph})^2 + (\gamma + \rho|v|^2)^2}$$

$$\times \left[1 + \frac{\rho|v|^2}{g\omega_{ph}} \right] ; \quad \hat{\varepsilon}_e = \varepsilon_e + \Delta_{ph} . \quad (25)$$

Since the damping constant γ is a few orders of magnitude less than $\hat{\varepsilon}_e - \varepsilon_g$, the strong enhancement can be expected from Eq.(25). It is, however, needless to point out that when the induced resonance condition is satisfied at $\omega_1 + \varepsilon_g > \varepsilon_f \approx \hat{\varepsilon}_e$, the enhancement will be diminished because such an excitation produces the resonant Raman scattering of a free molecule. Furthermore, it is strongly dependent on the value of $\rho|v|^2$.

It should also be mentioned that the integrated intensity of the
enhanced Raman band will, however, be less than that of a free
molecule. The enhancement factor calculated from Eq.(24) is shown
in Fig.3 in the log-scale, where $\gamma=0.005$ and $\omega_{ph}=0.01$ are used in
the unit of $\hat{\varepsilon}_e-\varepsilon_g$. The enhancement factor is clearly dependent on
the excitation and shows the strong peak due to the induced
resonance.

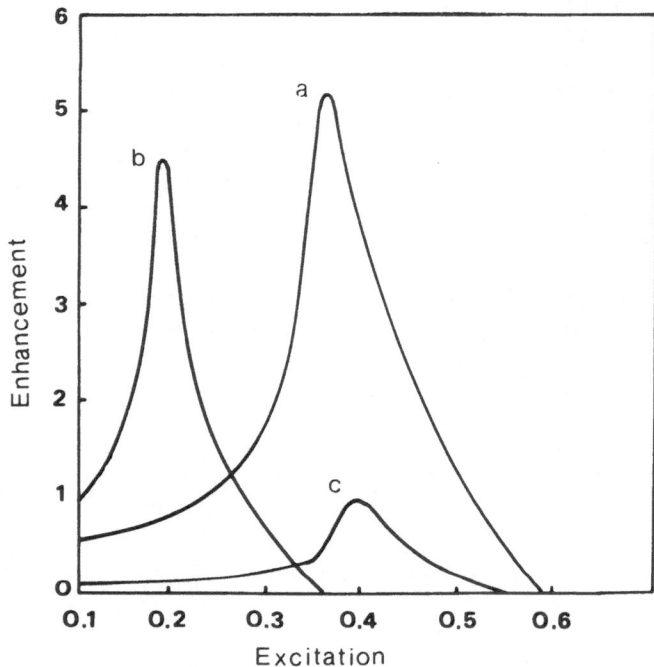

Figure 3. Intensity of the enhanced Raman scattering at $\omega_2=\omega_1-\omega_{ph}$
as a function of the excitation ω_1, where $\rho|V|^2=0.2$ in (a) and (b),
0.05 in (c) and $\varepsilon_f=0.4$ in (a) and (c), 0.2 in(b). The logarithmic
singularity due to Δ_m is not shown in these figures.

Since we have found the resonant character of the Raman
scattering of adsorbed molecules, such an effect will manifest
itself in the excitation energy dependence of the Raman intensity.
By using Eq. (18), the spectral intensity is given by

$$S(\omega_1) = g\omega_{ph}(\omega_1-\omega_{ph})^4 F(\omega_1)F(\omega_1-\omega_{ph}) \quad . \tag{26}$$

When the excitation energy approaches the induced resonant situation, the Raman intensity $S(\omega_1)$ starts to depart from an ω^4 law and shows the peak.

Let us turn our attention to the effect of substrate metals on SERS. A comprehensive comparison of the results for pyridine adsorbed on Ag, Cu, and Au has been done by Wenning et al.,[12] and also on Ag and Cu sols by Creighton et al.[13] Although the excitation profile and a degree of enhancement show a different behavior dependent on the nature of metal surfaces on which pyridine is adsorbed, it has been confirmed that Ag is a better enhancer than

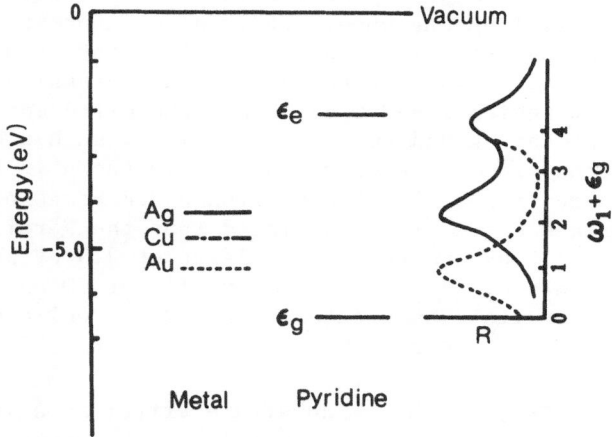

Figure 4. Tentative energy diagram of pyridine-Ag, Cu, and Au metals. The vacuum level is chosen as a common energy origin. The excitation profiles R estimated from the diagram are also shown for Ag (solid line) and Au (dashed line). The peak around $\omega_1+\varepsilon_g \approx 4.3$. eV is due to the intrinsic resonant Raman scattering whose intensity is expected to be weaker than that of a free molecule.

Cu or Au for a variety of excitation energies. Such a difference is very important in gaining a better understanding of the origin of SERS.

According to the present model, the molecule-metal interaction has moved the resonance to the vicinity of the Fermi level. The substrate metal effect, therefore, appears in the position of the Fermi level relative to the molecular ground state. Several pieces of experimental results[12,13] have been used in proposing a tentative energy diagram of the pyridine-Ag, Cu, and Au[11] as

shown in Fig.4. Since the precise determination of the metal work
function with adsorbed molecules has not yet been reported, the values
without a molecule are used in Fig.4. In spite of this uncertainty,
the relative position of the Fermi level of these metals shown in
Fig.4 would not be altered with adsorbed molecules. The energy diagram enables us to estimate a qualitative trend of the excitation
profiles of adsorbed pyridines on these metals. The schematic illus-
trations of the excitation profile are also shown in Fig.4. The
additional peak around 4.3 eV is due to the intrinsic resonant Raman
scattering. (The actual intensity ratio between SERS and the intrinsic one is not known.) The induced resonance model predicts that at
the excitation energy ω_1, which can produce a strong SERS of pyridine
on Ag, i.e., $\omega_1 + \varepsilon_g \simeq \varepsilon_f(Ag)$, will be unable to produce a strong enhancement on Au because $\omega_1 + \varepsilon_g \gg \varepsilon_f(Au)$. It is, however, possible for
pyridine–Au to exhibit a strong SERS at lower excitation energies
where SERS on Ag will be quenched. It is also interesting to mention
the case of CO on Ag and Au.[14] The excitation profile of CO on Ag is
an increasing function, while that on Au is a decreasing one in the
excitation energy region 1.9–2.6 eV. Since the difference in the
work function between Ag and Au is about 1.1 eV, such a behavior
will be understood from the excitation profile shown in Fig.4,
where the solid and dashed lines are crossing in certain energy
regions. In a case of CO it is estimated that the Fermi level of Au
is located about 1.8 eV and that of Au is about 2.9 eV above the
molecular ground state. The excitation profile of CO on Au should
decrease for the excitations greater than 1.8 eV, while on Ag it is
still increasing up to around 2.9 eV excitation.

The present analyses on the substrate effect as a function of
the excitation energy rely on the position of the Fermi level. The
experimental determination of a metal work function with adsorbed
molecules is desirable for a crucial check of the induced
resonance model. However, the validity of this model seems to be
so limited that it is unable to explain why pyridine on roughened
Ag exhibits a different Raman excitation profile from that it
exhibits when adsorbed on Ag colloids.

So far we have discussed the Raman component of Eq.(18). In
addition to the enhanced Raman part, we obtained another spectrum
given by the second term of Eq.(18). This spectrum appears in the
energy region

$$\omega_2 < \omega_1 + \hat{\varepsilon}_e - \varepsilon_f \quad . \tag{27}$$

Consequently, this gives rise to the continuum background on which
the enhanced Raman scattering spectra are superimposed and is
interpreted as the characteristic luminescence associated with
adsorbed molecules, not as a broad Raman scattering. The origin of
the broad background continuum has been a subject of great interest

and controversy. Particularly, a question has been raised as to
whether the background continuum is related to SERS or not. Otto
et al.[15] observed the continuum for the unexposed surface (without
adsorbed molecules) of the polycrystalline Ag. They measured the
potential dependence of that intensity from a Ag electrode in an
electrolyte and its change was attributed to the change of the
atomic-scale roughness due to Ag adatoms. On the other hand, support
of the luminescence hypothesis has been reported by Birke et al.,[16]
who monitored the intensity of enhanced Raman spectra and the continuum spectrum as a function of the electrode potential. Both signals were found to behave in a similar fashion. Especially, the
intensity decrease at negative potentials was assigned due to desorption of adsorbed molecules, not due to the desorption of adatoms.
The continuum, therefore, has been considered to be related to the
presence of the adsorbed molecules.

Although there have been pieces of experimental results to
support each interpretation, it is not easy to find the above
mentioned discrepancy including the differences in the experimental
condition. One of the diverging points to investigate the origin of
the continuum spectrum is closely connected with how the surface
roughness contributes to it. Most of the predictions towards the
continuum are concerned with the roughness, whatever a microscopic
mechanism is. As mentioned at the beginning of this article,
Udagawa et al.[5] have detected the continuum spectrum even on a
smooth surface of Ag on which pyridine is adsorbed. Their careful
treatment of the Ag surface enabled them to define the clean and
smooth surface. They also noted that the intensity of the background
is lower than is observed for rougher surfaces. To my knowledge, it
is still not clear whether the Ag surface they used is optically
flat or contains the atomic-scale roughness.

Even though various theories of SERS exist none of them
include the theoretical explanation for the continuum. Burstein
et al.[17] proposed the hot luminescence process involving the
radiative recombination of electron-hole pairs created by the
incident photon through the surface roughness. However, no calculation has been made to verify the plausibility of this theory. The
significance of the electron-hole excitation in a metal has been
extensively investigated by Persson[18] on the damping of excited
molecules above a metal surface. He argued that the radiative decay
of the excited molecule is a dominant channel for a large separation
between molecule and metal. At a small distance, within a few hundred
angstroms, nonradiative energy transfer to the metal due to the
excitation of the electron-hole becomes important in explaining the
life time variation of the excited state of adsorbed molecules.
Such a pair excitation process, nevertheless remains a possible
source for the background and exploring its functional form is
worthwhile.[19]

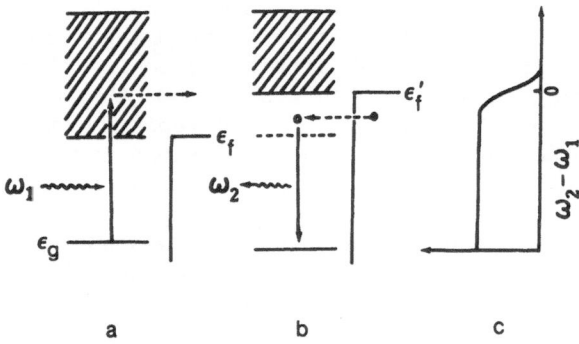

a   b   c

Figure 5. Radiative recombination process due to electron transfer
between adsorbed molecule and metal. The hatched part is the
broadened molecular excited state due to the molecule-metal
interaction. (See the text for detailed explanation.)

     The induced resonance model explicitly predicts the radiative
recombination as an explanation of the continuum spectrum. This is
illustrated in Fig.5. The photoionization process leading to SERS
drives the electrons from the adsorbed molecule to the metal when
$\omega_1 + \epsilon_g > \epsilon_f$, (a) in Fig.5. The photoinjection results in a shift of
Fermi level (ϵ_f'). The change of the Fermi level due to electron
transfer has to be determined from the charge conservation. The
injected electrons return to the vacant molecular electronic state
leading to the luminescence over a wide energy region, (b) in Fig.
5. Since the electrons below the Fermi level are responsible for
such a tunneling process, the spectrum arising from this mechanism
would reflect the Fermi distribution function, (c) in Fig.5. Birke
et al.[16] have observed that the intensity of the continuum
spectrum at the anti-Stokes side decreases along the Fermi
distribution function. The overall agreement between the shape of
the continuum spectrum and the Fermi distribution function supports
the luminescence hypothesis. In fact Eq.(27) tells us that the
continuum appears at energies below the excitation energy, i.e.,
at the Stokes side, [ϵ_f in Eq.(27) should be considered as ϵ_f'].
Since the present analysis does not require any roughness, a part
of the strong continuum spectrum is still attributed to the
radiative recombination which can occur at a smooth surface of a
metal. Furthermore, it strongly contributes to the enhancement
factor given by Eq.(25), since $\rho|V|^2/g\omega_{ph} \gg 1$. Finally, the effect
derived from the induced resonance model would occur even for
a rough surface, which does not destroy any part of the model.

DISCUSSION

Analyses of SERS by the induced resonant model rely on unknown
quantities associated with a coupled molecule-metal system. It is,
therefore, difficult to make a quantitative comparison between
experimental and theoretical results. It would be fair to remark that
more information about the electronic properties of the adsorbed
molecule-metal system helps to establish or to disprove the present
model as a part of a general effect over a wide range of systems.
One of the most important pieces of information is the precise position of the Fermi level measured from the molecular ground state.
Once such an experiment is performed, a detailed excitation energy
dependence of SERS or an optical absorption spectrum [which is given
by $F(\omega_1)$ itself defined by Eq.(19)] will provide a more crucial
test of the model.

In the present treatment the k-dependence of the molecule-
metal coupling V_k is not taken into account. When the electronic
properties of a metal are incorporated with the model through the
k-dependence, the density of states of an adsorbed molecule
contains more information about the substrate metal. The structure
of the continuum spectrum, for instance, reflects the density of
states of a metal. It is interesting to note here that the densities
of states of Ag, Cu, and Au[20] are almost flat at a few eV below
and above the Fermi level, being mainly due to the s-electrons.
Accordingly, the replacement of $\rho(\varepsilon)$ by $\rho(\varepsilon_f)$ used here seems to be
qualitatively consistent with the energy band calculation of these
metals. This would be a possible reason why the continuum is
structureless over a wide range of the relevant energies. In
addition to Ag, Cu, and Au, SERS of pyridine has been also observed
on Ni and Pt.[21] Since the Fermi levels of these transition metals
are located inside the sharp d-band region, it is conjectured that
the continuum spectrum would mimic such a band structure of Ni or
Pt if the spectrum were due to the radiative recombination involving
the electron transfer.

The study of SERS has progressed so widely and extensively
that a variety of information has become available to us. Among
the important experimental results to examine the validity and the
possibility of theories is the molecule-metal distance dependence
of the SERS intensity,[22,23] which provides the range of the
effect where SERS is observed. Although the present model is, in
essence, proposed on the basis of the short range s-d mixing
interaction, it is powerless to discuss such a coverage or distance
effect within the present treatment.

From the angular dependence of SERS[6,24,25] it is now clear
that adsorbed molecules are influenced by the strong field arising
from the direct excitation of a surface plasmon polariton by the

incident photon via surface roughness. The strength of this secondary field depends on the dielectric function of the metal which characterizes the change of the excitation profile of SERS for different metals. Thus, a model including such a role is one of the promising theories of SERS when it appears to be successful in explaining other primary aspects of SERS. A calculation of the light scattering spectra from rough surfaces with or without adsorbed molecules[26] is imposed on any model in this category. It is, however, doubtful to use the long range electromagnetic field to obtain the anomalous enhancement for the adsorbed molecule within several tens of angstroms from a metal surface. This can be also pointed out for the image dipole model which has been forced to assume a few angstroms separation between molecule and metal. In fact, the decrease in the life time of an excited molecule within a few angstroms away from a surface is not fully explained by the image dipole model.[27]

We have now been convinced that both the long range electromagnetic field of the surface plasmon and the short range force involving the molecular interaction with the metal give rise to the strong enhancement. The degree of each contribution to SERS varies depending on the properties of the metal and also on the environment in which the coupled molecule-metal system is placed.

ACKNOWLEDGMENTS

A main part of the work described here was completed while the author was visiting the University of Waterloo. He would like to express his sincere thanks to Professor S. G. Davison for a very pleasant and stimulating time. Thanks are also extended to Professor S. Ichimura for his interest and encouragement.

REFERENCES

1. T. E. Furtak and J. Reyes, A critical analysis of theoretical models for the giant Raman effect from adsorbed molecules, Surf. Sci. 93:351 (1980).
2. F. W. King, R. P. Van Duyne and G. C. Schatz, Theory of Raman scattering by molecules adsorbed on electrode surfaces, J. Chem. Phys. 69:4472 (1978).
3. S. Efrima and H. Metiu, Classical theory of light scattering by an adsorbed molecule I. Theory, J. Chem. Phys. 70:1602 (1979); II. Model calculations, J. Chem. Phys. 70:2297 (1979); Surface induced resonant Raman scattering (SERS), Surf. Sci. 92:417 (1980).
4. T. K. Lee and J. L. Birman, Molecule adsorbed on plane metal surface: Coupled system eigenstates, Phys. Rev. B 22:5953 (1980);

Quantum theory of enhanced Raman scattering by molecules on metals: Surface-plasmon mechanism for plane metal surface, Phys. Rev. B 22:5961 (1980).

5. M. Udagawa, C. C. Chou, J. C. Hemminger and S. Ushioda, Raman scattering cross section of adsorbed pyridine molecules on a smooth silver surface, preprint.

6. J. C. Tsang, J. R. Kirtley and T. N. Theis, Surface plasmon polariton contributions to Stokes emission from molecular monolayers on periodic Ag surfaces, Solid State Commun. 35:667 (1980).

7. A. Otto, Surface enhanced Raman scattering (SERS): What do we know?, in Proceedings of 6th Solid Vacuum Interface Conference, Delft, The Netherlands, 1980, to be published.

8. B. Pettinger, Surface enhanced Raman spectroscopy (SERS) of pyridine on Ag electrodes. Part II: Evidence for overtones, Chem. Phys. Lett. 78:404 (1981).

9. J. I. Gersten, R. L. Birke and J. R. Lombardi, Theory of enhanced light scattering from molecules adsorbed at the metal-solution interface, Phys. Rev. Lett. 43:147 (1979).

10. H. Ueba, Effective resonant light scattering from adsorbed molecules, J. Chem. Phys. 73:725 (1980).

11. H. Ueba and S. Ichimura, Raman scattering of adsorbed molecules, J. Chem. Phys. 74:3070 (1981).

12. U. Wenning, B. Pettinger and H. Wetzel, Angular-resolved Raman spectroscopy of pyridine on copper and gold electrodes, Chem. Phys. Lett. 70:49 (1980).

13. J. A. Creighton, C. G. Blatchford and M. G. Albrecht, Plasma resonance enhancement of Raman scattering by pyridine adsorbed on silver or gold sol particles of size comparable to the excitation wavelength, J. Chem. Soc. Faraday II 75:790 (1979).

14. T. M. Wood and M. V. Klein, Studies of the mechanism of enhanced Raman scattering in ultrahigh vacuum, Solid State Commun. 35:263 (1980).

15. A. Otto, J. Timper, J. Billmann and I. Pockrand, Enhanced inelastic light scattering from metal electrodes caused by adatoms, Phys. Rev. Lett. 45:46 (1980).

16. R. L. Birke, J. R. Lombardi and J. I. Gersten, Observation of a continuum in enhanced Raman scattering from a metal-solution interface, Phys. Rev. Lett. 43:71 (1979).

17. E. Burstein, Y.J. Chen, C. Y. Chen, S. Lundquist and E. Tosatti, Giant Raman scattering by adsorbed molecules on metal surfaces, Solid State Commun. 29:567 (1979).

18. B. N. J. Persson, Theory of the damping of excited molecules located above a metal surface, J. Phys. C (Solid State Phys.) 11:4251 (1978).

19. Furtak and Reyes (Ref. 1) mentioned about the work by R. Fuchs, Enhanced Raman scattering by molecules adsorbed on metals, Bull. Am. Phys. Soc. 24:339 (1979). However, its final form has not yet been available to us.

20. For example, Ag: N. E. Christensen, The band structure of silver and optical interband transitions, Phys. Status Solidi (b) 54: 551 (1972), Cu: G. A. Burdick, Energy band structure of copper, Phys. Rev. 129:138 (1963), Au: N. E. Christensen and B. O. Seraphin, Relativistic band calculation and the optical properties of gold, Phys. Rev. B 4:3321 (1971).

21. H. Yamada and Y. Yamamoto, Surface enhanced Raman spectra of pyridine adsorbed on silver, gold, nickel and platinum metals, Chem. Phys. Lett. 77:520 (1981).

22. J. W. Rowe, C. V. Shank, D. A. Zwemer and C. A. Murray, Ultra-high-vacuum studies of enhanced Raman scattering from pyridine on Ag surfaces, Phys. Rev. Lett. 44:1770 (1980).

23. P. N. Sanda, J. M. Warlaumont, J. E. Demuth, J. C. Tsang, K. Christmann and J. A. Bradley, Surface enhanced Raman scattering from pyridine on Ag (111), Phys. Rev. Lett. 45:1519 (1980).

24. R. Dornhaus, R. E. Benner and R. K. Chang, Surface plasmon contribution to SERS, Surf. Sci. 101:367 (1980).

25. B. Pettinger, U. Wenning and H. Wetzel, Surface plasmon enhanced Raman scattering frequency and angular resonance of Raman scattered light from pyridine on Au, Ag and Cu electrodes, Surf. Sci. 101:409 (1980).

26. V. Celli, A. Marvin and F. Toigo, Light scattering from rough surfaces, Phys. Rev. B 11:1779 (1975).

27. R. R. Chance, A. Prock and R. Silbey, Molecular fluorescence and energy transfer near interfaces, in Advances in Chemical Physics, Vol. 37, ed. by I. Prigogine and S. A. Rice (John Wiley and Sons, New York, 1978), p. 1.

COVERAGE DEPENDENCE

P.N. Sanda, J.E. Demuth, J.C. Tsang,
and J.M. Warlaumont

IBM Thomas J. Watson Research Center
Yorktown Heights, NY 10598

1. INTRODUCTION

The determination of the detailed coverage-dependence of surface-enhanced Raman scattering (SERS) is of fundamental importance in understanding the enhancement process. The dependence of the enhancement factor on the distance away from the metal surface is an experimentally measurable quantity which can be compared with theoretical predictions. Also, in the submonolayer regime, comparisons of the SERS spectra with coverage-dependent bonding behavior can yield information on effects due to chemical bonding.

Although SERS is observed in a large variety of experimental systems as discussed in other articles in this volume, most current experiments share a common problem which makes it difficult to gain an understanding of SERS: there is yet very little known about the detailed nature of the metal substrate and of the adsorbed molecules. In many of the experimental systems for which surface Raman scattering has been observed, it is not possible to start off with a bare surface, continuously vary the molecular coverage, or understand the nature of the molecular bonding sites.

The coverage-dependence of SERS is ideally studied using samples prepared in ultra-high vacuum (UHV). The UHV environment allows the use of a clean substrate, cleaned in-situ through ion-bombardment and annealing cycles or by utilizing a freshly evaporated film. A low base pressure ($\sim 10^{-10}$ Torr) allows controlled dosing of the substrate with gas phase molecules to adsorb known quantities of molecules onto the metal surface. Direct distance-dependent measurements of the enhancement factor can thus be performed by observing changes in the intensity of the Raman features as molecules are adsorbed layer by layer on the metal surface. This capability also allows one to look in the submonolayer regime to observe additional effects specifically related to the bonding of the molecules to the metal surface.

189

Several SERS experiments have recently been reported which utilize UHV capabilities[1-3,5-10]. Rowe et. al.[2] have used the photoreaction of iodine on Ag to form silver balls. They measured SERS spectra as a function of pyridine exposure to this substrate and determined their coverages using the carbon densities derived from Auger electron spectroscopy (AES). They conclude that their enhancement can be explained in terms of an electromagnetic field enhancement, based on the observation that substantial growth in the SERS signals is observed as coverage beyond the first layer is increased. This behavior was also observed in continuing experiments by Zwemer et. al.[3] in which deuterated pyridine was adsorbed on top of a spacer layer of pyridine, and also in the experiments using polymer spacers by Murray et. al.[4] which are discussed elsewhere in this volume. Eesley and coworkers[5], on the other hand, claim that the SERS signal is produced mainly by the first layer of molecules on their sputter-etched polycrystalline surfaces. Seki et. al.[6] have reported coverage-dependent SERS measurements for pyridine adsorbed on silver island films using a quartz crystal microbalance to monitor the pyridine coverage. They propose that the enhancement only occurs for some fraction of the first molecular layer and suggest that only molecules which are adsorbed on certain special sites, termed "active sites" produce the major portion of the enhanced Raman signal. This view of "active sites" as being responsible for all, or at least a large part of the enhancement is shared by Pockrand et. al..[7] Their coverage-dependent measurements using silver films evaporated at low temperature lead them to postulate that atomic scale roughness due to "adatoms" provide special "active" adsorption sites.

As one can observe through the aforementioned examples, there is presently controversy about the nature of the enhancement mechanism: Interpretations range from the Bell labs workers[2-4] who claim that it can be completely attributed to a field enhancement caused by electromagnetic resonances of metallic particles on the scale of $\gtrsim 100\text{Å}$, to the interpretation of Otto and coworkers that it is predominantly due to atomic scale surface roughness. A major difficulty in resolving this controversy is that the types of surfaces used vary from laboratory to laboratory thereby making detailed comparisons impossible. Also, in most current experiments, one has very little knowledge of the surface profile or the detailed nature of the molecular bonding sites. In this chapter, we will describe our SERS measurements from known quantities of pyridine on a well-defined, single crystal silver surface and discuss these results in view of related experiments by other workers. Our experiments are intended to elucidate the enhancement process for the particular case of pyridine adsorption on a well-defined single crystal Ag(111) surface with a gentle periodic surface modulation, for which we have characterized the detailed structure of the surface and the nature of the molecular bonding. UV-photoemission spectroscopy (UPS) has provided a coverage calibration by which a clear delineation between the chemisorbed and condensed phases was possible. This allowed the interpretation of the Raman scattering signals as a function of coverage, for molecular multilayers, as well as for the chemisorbed phase. By examining the incremental changes in intensity of the Raman signals for varying coverages and thicknesses of pyridine on Ag(111), we have separated two types of enhancements. The enhancement for physisorbed pyridine ($\sim 10^2$) can be attributed to the increased field strength at the molecular site associated with the excitation of substrate conduction electron resonances. The enhancement for chemisorbed pyridine ($\sim 10^4$) shows a further contribution in addition to this field enhancement. For submonolayer coverages of pyridine on Ag(111), our Raman measurements are correlated to the coverage-dependent phase transition[11] which was determined with the aid of high-resolution electron energy loss spectroscopy (EELS).

2. EXPERIMENTAL PROCEDURES

The experimental measurements were performed in three separate UHV systems (typical base pressures of $\leq 1 \times 10^{-10}$ Torr). The first (ion- and titanium sublimation-pumped) system is mobile, has facilities for low-energy-electron diffraction (LEED), AES, argon-ion sputtering, and optical windows to allow light scattering measurements. The second (turbomolecular pumped) system described elsewhere[12] allows UPS, LEED, AES and thermal desorption spectroscopy to be performed. The third (ion- and titanium sublimation-pumped) system permits EELS and work function change measurements.

The Ag crystal was first spark cut, mechanically polished, and subsequently chemically polished[13] with a chromic-acid and HCl solution. Then several UHV argon sputtering/thermal annealing cycles were used to remove C and S impurities from the bulk and to segregate dislocation defects to the surface. A final chemical polish left the crystal with a mirror-like finish, free of etch pits. The 10,000Å periodic surface modulation was then fabricated into a $4 \times 4 \text{mm}^2$ area of the $\sim 8 \times 6 \text{mm}^2$ face of the crystal, with the modulation wave vector K_s oriented along the (110) direction. This structure was prepared by first creating a polymer stencil (PMMA) on the sample with the use of x-ray lithography techniques followed by chemical polishing to remove about 3000Å of material in the unmasked regions ($\sim 50\%$). The PMMA was then dissolved in acetone and the crystal was repeatedly argon sputter-etched and annealed in UHV (T$\sim$500K). This preparation also served to reduce the higher-order Fourier components of the profile. The final result was a portion of nearly-sinusoidal surface (valleys slightly wider than the peaks) with a 10,000Å wavelength and $\sim$1000Å height as estimated by the LEED beam profiles, surrounded by the flat surface which served as a "control" region for comparison. The modulated region of the sample showed a well-defined, low-background LEED pattern, comparable to the control region of the sample. However, satellite lobes were observed in the beam profiles, which indicated a distribution of steps and terraces parallel to K_s. The peak in this distribution corresponded to a terrace width to step height ratio of about 10 to 1. The intensity of the main peak relative to the side lobes indicated that roughly 90% of the surface is of (111) orientation.

Most of the Raman spectra were measured with the sample at $\sim$80K with $\sim$150mW radiation from the 5145Å Ar$^+$ line, although measurements were also made with the 4880Å Ar$^+$ line and the 5309Å line from a Kr$^+$ laser. Standard backscattering geometry was used, with light collected over a solid angle of 45°. The scattered light was analysed by a conventional double-grating monochromator operating at $\sim$6 cm^{-1} resolution. Given our operating conditions, the threshold enhancement for detecting a monolayer is estimated to be $\sim 5 \times 10^2$.

3. COVERAGE CALIBRATION

Although AES was used to monitor and characterize surface cleanliness there was no attempt made to use AES to determine a pyridine coverage calibration, as done by Rowe et. al.[2,3] and other workers.[10] Due to uncertainties induced by possible desorption or dissociation of the molecular adsorbate, it is estimated that the error margin for the coverage using this method can be as much as 50%. Instead, UPS measurements were used to calibrate relative coverages of pyridine from the adsorbate-derived ionization features and to allow a clear delineation between the first molecular layer (chemisorbed) and subsequent layers (physisorbed). These measurements were performed in a separate vacuum system with use of a differentially pumped He-resonance lamp ($h\nu$=21.2eV) and a double-pass cylindrical mirror analyser.[12] Comparison of the coverage-dependent UPS spectra for the modulated Ag(111) and flat Ag(111) surfaces revealed no measurable differences.

UPS difference curves, $\Delta N(E)$, in Fig. 1 show the adsorbate-induced changes in emission for consecutive pyridine exposures on the modulated Ag(111) surface.

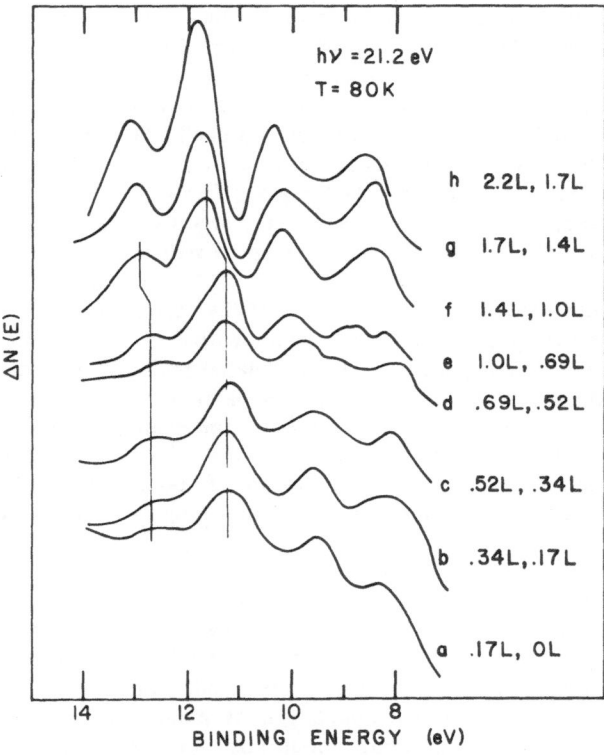

Fig. 1 UPS difference spectra $\Delta N(E)$ for pairs of successive pyridine exposures, as indicated to the right (1L=10^{-6} Torr-sec), where the ion-gauge pressure reading has been divided by 5.8 to account for the gauge correction (Ref. 14). Spectra e-h have been divided by 2.

The ionization features in Fig. 1 demonstrate a pronounced energy shift starting at coverages above ~1.2L. This increase in binding energy for physisorbed pyridine compared to chemisorbed pyridine is expected. It is attributed to an increase in the relaxation effects[15,16] for the first molecular layer. This shift can be most reliably seen (light vertical lines) in the lower-lying levels (binding energy ~11.5 and 13eV) which should be least affected by initial-state chemical-bonding effects. The absolute coverage calibration was obtained by taking the total adsorbate-induced intensity of the large peak at about 11.5eV (combination of a^1 and b^2 orbitals)[17] for a 1.2L exposure to correspond to the saturation of one monolayer.

From thermal desorption measurements we find that layer by layer growth occurs for at least the first two monolayers with condensation thereafter. We thereby find no evidence for "clustering" of adsorbed pyridine and assume layer by layer growth for all coverages. Fig. 2 shows the coverage vs. exposure curve, derived from the intensity of the 11.5eV peak and scaled to 1 monolayer at 1.2 L as described above. The change in slope of the curve at ~.6 L corresponds to the occurrence of a phase transition from π-bonding to lone-pair-bonding, which will be described further in Sec. 4. Above this phase transition, no change in slope was observed and we assume a constant sticking coefficient above 1L.

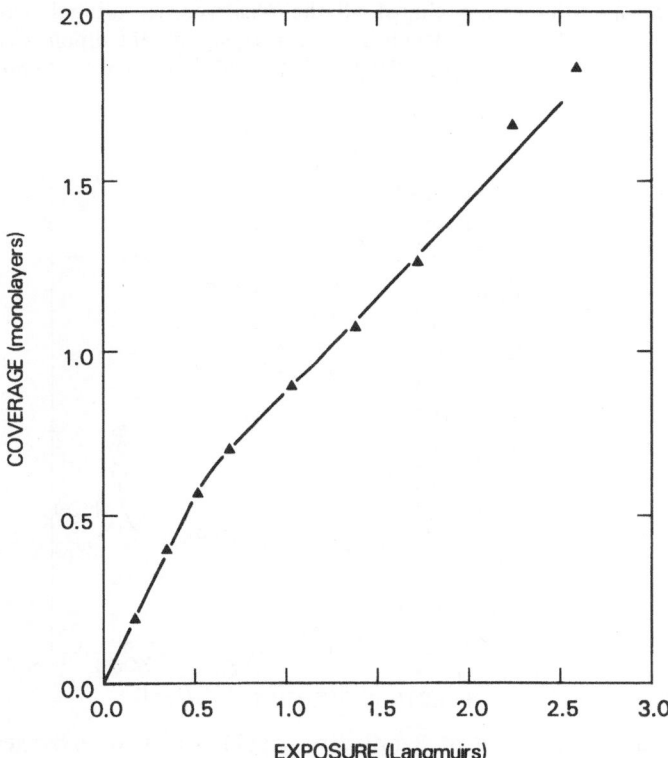

Fig. 2 Coverage as a function of pyridine exposure determined using UPS (see text).

4. EELS RESULTS: BONDING INFORMATION FOR CHEMISORBED PYRIDINE

The detailed bonding behavior of pyridine chemisorbed on a flat Ag(111)surface was determined by studying the vibrational energy loss features observed in EELS as a function of coverage. The electron optics for EELS consist of two sets of 2.5 cm hemispherical deflection analyzers with associated focusing elements[18] so as to allow monochromatization, reflection from a sample (total scattering angle of 90°) and energy analysis of a well-defined (<0.2 mm dia.), collimated (<1°) low-energy 2-100 eV electron beam. Our results indicate that pyridine chemisorbed on Ag(111) exists in two phases with the transition coverage occurring at ~0.5L. Fig. 3 shows the EELS vibrational loss features before (solid line) and after (dotted line) this transition. Strong changes in the relative intensities of the C-H deformation modes between 400 - 850 cm⁻¹ are clearly observed, especially when compared to the relative infra-red absorbances, also shown in Fig. 3. Such IR absorbances directly reflect the dipole scattering intensities of a randomly oriented pyridine molecule and can be compared to EELS results for specular scattering. The transition is more clearly observed in the coverage dependent EELS spectra shown in Fig. 4.

In order to obtain structural information, the vibrational losses must be assigned to the vibrational modes of the molecule. Such an assignment would seem formidable based upon the relatively poor resolution of EELS and the fact that pyridine is of low symmetry and has 27 IR-active modes[21]. Fortunately, we find that we can straightforwardly assign almost all the observed vibrational losses since (a) only a fraction of the 27 free-molecule modes have significant dipole scattering cross sections (ie: IR absorbances, see Fig. 3); and (b) pyridine is weakly chemisorbed and

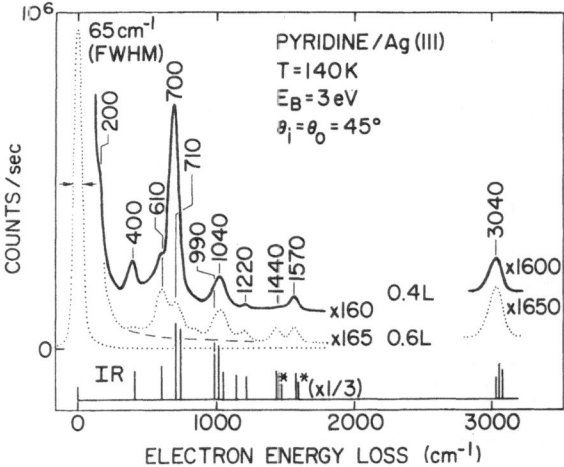

Fig. 3 Vibrational loss spectra of pyridine on Ag(111) at two coverages above and below the compressional phase transition. The IR absorbances (Ref. 19 and 20) are shown below for comparison where the * levels are reduced by 1/3.

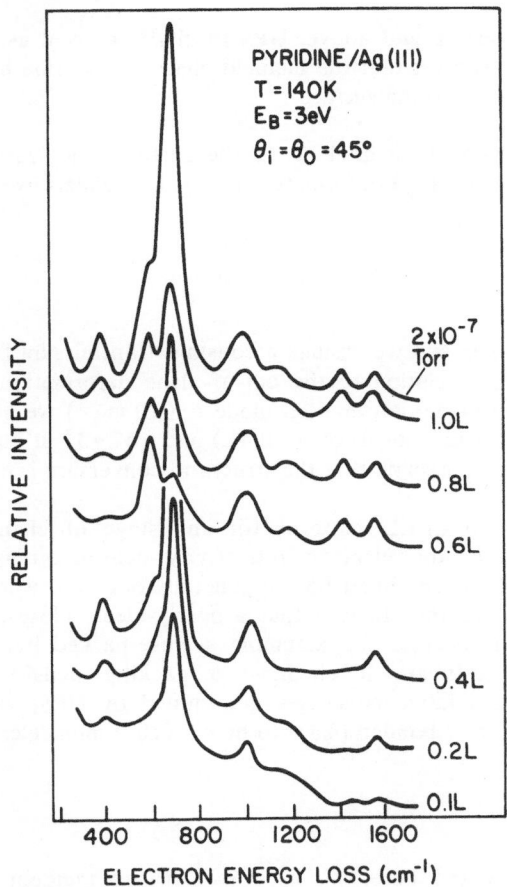

Fig. 4 Coverage dependence of the vibrational spectra of chemisorbed pyridine (<1.0L exposure). The spectra taken at a 2×10^{-7} Torr ambient pressure correspond to the condensation and the formation of the first physisorbed layers.

leads to weakly perturbed vibrational frequencies ($\Delta \nu_{avg}= 6cm^{-1}$) relative to liquid pyridine. These assignments are tabulated and described elsewhere[11].

Based upon the mode assignments we consider the (dipole-derived) scattering intensities of in-plane and out-of-plane vibrations relative to the IR-absorbances to obtain structural information. The low coverage phase shows intense, out-of-plane, CH deformation vibrations at 700 cm^{-1} and 400 cm^{-1} which become suppressed after the phase change. With this phase change the in-plane, CH-deformation vibration at 610 cm^{-1} becomes more intense. From our angle-dependent measurements we find that these aforementioned in-plane and out-of-plane modes are dominated by dipole scattering in the specular direction which thereby permits their use in a structural analysis. As a result of the surface selection rule[22] the relative intensities of these in-plane and out-of-plane vibrations indicate that the molecule becomes more inclined to the surface at 0.6L exposure and above. UPS results[11], as well as other chemical arguments and results[23,24] suggest that the inclined phase of pyridine has the nitrogen end of the molecule directed into the surface.

For pyridine inclined at an angle θ to the surface, the loss intensities for in-plane (I_{in}) and out-of plane (I_{out}) deformation modes are related by

$$(A_{in}/A_{out})\tan\theta = I_{in}/I_{out}$$

where A is the IR absorbance and we assume a constant transmission function for our spectrometer. Applying this relation to the out-of-plane deformation modes ($\sim$400 and $\sim$700 cm^{-1}) and the in-plane deformation mode ($\sim$610 cm^{-1}) we have determined averaged angles of $\sim$5$\pm$3° at low coverages ($<$0.4L) and $\sim$57$\pm$3° at higher coverages. For these higher coverages we assume that the structural conversion is complete.

Although we see no LEED patterns for any stage of chemisorption[25], a dramatic increase occurs in the electron reflectivity nearing completion of the π-bonded phase. Such phenomena in EELS are generally observed when well-ordered superstructures form.[26,27] We thus believe that a nearly-ideal close-packed ordered array of π-bonded pyridine occurs; and assuming a close-packed but commensurate pyridine monolayer, we estimate it to have a packing density of $\sim$3$\times$10^{14} molecules/cm^2. From the relative coverages determined by UPS, we estimate the packing density of the lone-pair bonded phase to be $\sim$5$\times$10^{14} molecules/cm^2.

5. RAMAN RESULTS

For the data presented here, we used p-polarized incident radiation and oriented the sample so that the surface wavevector K_s was in the plane of incidence. The angle of incidence was set to the minimum in intensity of the direct reflected beam, corresponding to surface plasmon-polariton excitation. The Raman scattered signal for pyridine adsorbed on the modulated portion of the sample was observed as the incident angle was brought to within 5° of this condition. As expected, there was no Raman scattered signal observed from the flat (control) portion of the sample. The features found to be observable in the Raman spectrum for chemisorbed pyridine on our modulated Ag(111) surface occurred between 950 and 1050 cm^{-1} and are shown for several exposure levels in Fig. 5. In the liquid phase, pyridine shows two strong features corresponding to the symmetric (991 cm^{-1}) and asymmetric (1030 cm^{-1}) ring-breathing modes. Our SERS spectra show selective enhancement of the symmetric ring breathing mode for chemisorbed pyridine. It is not until thick condensed layers are obtained (exposure $>$20L) that the asymmetric ring-breathing mode starts to be observed and continues to grow with increasing exposure. No C-H modes were seen and the carbon-ring deformation modes in the 1300-1600 cm^{-1} range were also not detected. These latter features are probably masked by the broad peaks at 1350 and 1550 cm^{-1} which arise from trace amounts of amorphous carbon.[28,29] We found these peaks to persist even at carbon levels undetectable by AES ($<$1/10 monolayer). Also, prior to complete annealing of the modulated sample, we observed an additional peak at 986 cm^{-1} which we associate with pyridine bound to step sites.

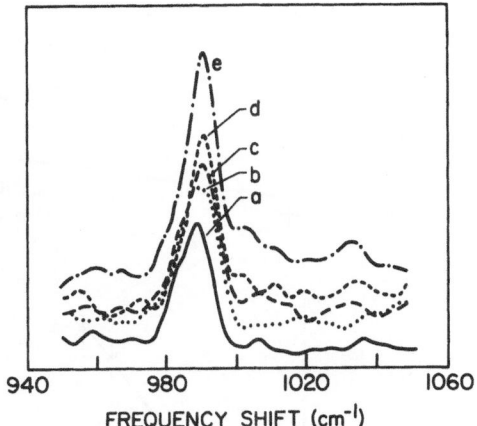

Fig. 5 Raman spectra for increasing pyridine exposures: curve a, 1.7L; curve b, 3.4L; curve c, 6.9L; curve d, 19.3L; curve e, 44L. The incident laser power is 150 mW.

Using the coverage calibration as determined by UPS (see section 3), we determine the coverage dependence of the 990 cm^{-1} peak. This is plotted in Fig. 6 which shows the behavior of the signal as we increase the number of pyridine layers. Relative coverages, on the horizontal axis, are expressed in terms of monolayer equivalents - the coverage at which adsorption into the first (chemisorbed) layer is complete and condensation begins. The growth of the signal within one monolayer coverage is shown in the inset in Fig. 6. We observe a signal starting at about ~.6 monolayers which rises rapidly until 1 monolayer coverage, and continues to grow at a much slower rate at higher coverages. The enhancement factor at 1 monolayer coverage is ~10^4, as determined from comparative Raman measurements with liquid pyridine using the same optics and operating conditions and assuming the liquid phase packing density for chemisorbed pyridine.

The behavior of the enhancement factor is more readily observed by taking the derivative of the data in Fig. 6. $dI/d\Theta$ vs. Θ shown by the solid points in Fig. 7 thus give the distance dependence of the enhancement factor, as a function of the number of layers away from our silver surface. $dI/d\Theta$ drops off quite rapidly in the region from $\Theta = 1$-2 monolayers, and for larger values of Θ, approaches an asymptotic value of ~3% of the monolayer value.

6. DISCUSSION

The similarities in the chemisorption behavior on a flat Ag(111) crystal and on the modulated Ag(111) surface encourage us to believe that the terrace edge atoms present on the latter surface do not significantly alter the adsorption properties. Namely, pyridine shows ionization levels and coverage dependent features in UPS which are similar on both surfaces. We therefore compare the coverage-dependence

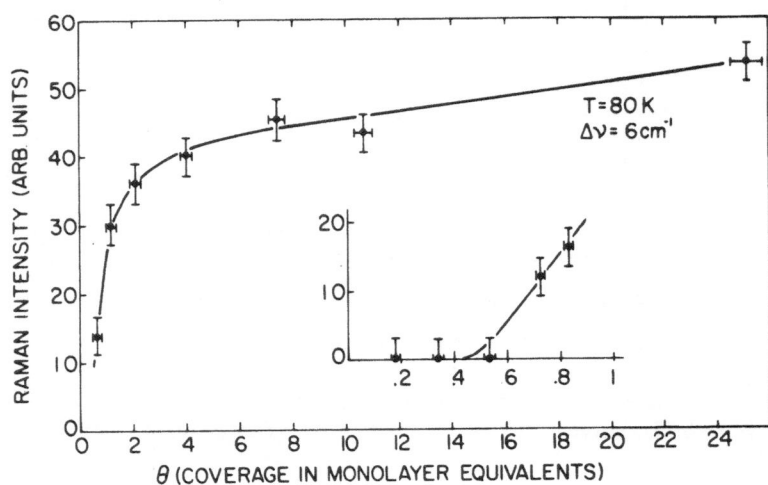

Fig. 6 Raman scattering intensity for the 990 cm⁻¹ peak as a function of coverage (in
monolayer equivalents - see text). The insert shows the detailed low coverage
behavior. (λ=5145Å).

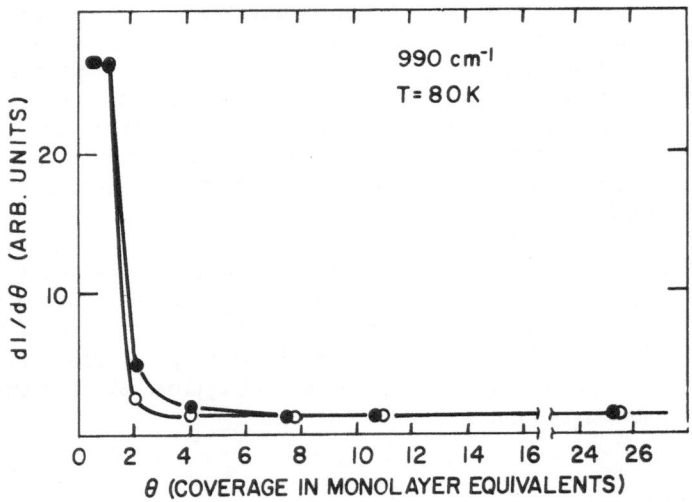

Fig. 7 Incremental enhancement factor dI/dΘ, as a function of coverage as determined from Fig. 6 (solid circles) and predicted by theory (open circles Ref.
30).

of the 990 cm^{-1} SERS feature with the bonding behavior on Ag(111) as understood from EELS. Below 0.5 monolayers coverage, where pyridine lies nearly flat and is π-bonded, the SERS signal is below the limits of detectability ($< 5 \times 10^2$ enhancement). However, at the onset of lone-pair bonding, where pyridine is inclined to the surface, the 990 cm^{-1} peak is observed. We note that if our SERS spectra were dominated by pyridine adsorbed on terrace edge sites, we expect these sites to become occupied first so as to produce an onset of the signal at lower exposures.

Another important feature of our enhanced Raman scattering for chemisorbed pyridine is the mode selectivity. Namely, the symmetric ring-breathing mode is selectively enhanced while for liquid or thick condensed layers of pyridine the symmetric and asymmetric modes are of comparable intensities. Any model of the enhancement must account for such mode selectivity.

The distance-dependence of the SERS enhancement factor shown in Fig. 7 (filled circles) demonstrates two distinct mechanisms contributing to the total enhancement for the case of pyridine on our modulated Ag(111) surface: one which applies to the first, chemisorbed layer, and the other which extends many molecular layers away from the surface. The long-range contribution is attributed to the enhancement of the electromagnetic fields when surface plasmon-polaritons are excited (or Mie resonances, when the substrate is composed of spheroidal particles). Molecules chemisorbed on the surface may demonstrate another factor in addition to the aforementioned field enhancement, since the polarizability of the molecule plus substrate system is modified from that of the free molecule. Namely, additional polarizability may arise from interactions between the substrate conduction electrons and the molecular vibrations to give rise to an enhanced polarizability.

The open circles shown in Fig. 7 correspond to values derived from a theoretical model by Jha, Kirtley, and Tsang[30] which is described in a preceding chapter. This theory has been chosen for comparison with our experiment because Jha et. al. explicitly include both a field enhancement (10^2-10^4) and polarizability enhancement for Raman scattering from a molecule near a sinusoidal grating. They calculate the enhancement by modeling the system in terms of a dielectric function which varies across the interface, and a substrate-molecular layer barrier height for tunneling. The molecular vibrations may modulate the surface charge density to produce an added surface dipole moment contribution to the system polarizability. For modeling the theoretical distance-dependence we used a pyridine interlayer spacing[31] of 5 Å and a Ag-pyridine barrier height[30] of 0.25V. In order to allow a qualitative comparison, we have scaled the short-range and long-range contributions in the theory to the experimental values at 1 monolayer and 24 layers, respectively.

Our results show a strong decrease in the SERS enhancement between the 1st and 2nd layers whereas the results by Rowe et. al. show a weaker change. This difference is reasonable considering the different surface profiles. Namely, their iodine-roughened surfaces have a high local radius of curvature which will produce larger field gradients compared to our low-amplitude, gently modulated surfaces. In our case, since the fields are expected to be longer-ranged and attenuate less rapidly away from the interface, the first layer, chemically-derived enhancement can be more readily distinguished from the field-derived enhancement.

On the other hand, a much more abrupt distance-dependence between chemisorption and physisorption coverages has been observed by Pockrand and Otto for the case of pyridine adsorbed on silver surfaces which have been evaporated onto cooled copper substrates at T~130K. They observe the strongest peak at 1006 cm^{-1} which can readily be detected after exposure to only 3×10^{-2}L, corresponding to coverages on the scale of 1/100th of a monolayer. Pockrand and Otto have proposed that increased electron-hole pair production from the reduced symmetry caused by such atomic scale roughness leads to an enhanced polarizability for the first molecular layer. Such a large polarizability enhancement for the first layer could mask other possible contributions from electromagnetic field enhancement. In fact their coverage-dependent measurements suggest a strong, short-range enhancement which dominates their spectra.

In conclusion, there are both short-range (first molecular layer) and long-range enhancements contributing to the total SERS intensity. The exact coverage-dependence would be expected to vary depending on the relative strength of these components. For our experimental system, the first layer enhancement and the electromagnetic field enhancement contributed roughly equal amounts ($\sim 10^2$) to the total ($\sim 10^4$) enhancement for chemisorbed pyridine. For submonolayer coverages, we have also shown that SERS is sensitive to the details of the chemisorption bonding. Such coverage-dependent studies performed at well-defined interfaces can provide fundamental information on the nature of SERS which may help to understand the more complicated cases such as electrochemical cell systems and layered structures.

ACKNOWLEDGEMENTS

We thank J.A. Bradley, K. Christmann, S.S. Jha, and J.R. Kirtley for helpful discussions and valuable assistance during the course of this work. This work was supported in part by the U.S. Office of Naval Research and the Cornell University Materials Science Center.

REFERENCES

1.   P.N. Sanda, J.M. Warlaumont, J.E. Demuth, J.C. Tsang, K. Christmann, and J.A. Bradley, Surface-enhanced Raman scattering from pyridine on Ag(111), *Phys. Rev. Lett.* 45:1519 (1980).

2.   J.E. Rowe, C.V. Shank, D.A. Zwemer, and C.A. Murray, Ultrahigh-vacuum studies of enhanced Raman scattering from pyridine on Ag surfaces, *Phys. Rev. Lett.* 44:1770 (1980).

3.   D.A. Zwemer, C.V. Shank, and J.E. Rowe, Surface-enhanced Raman scattering as a function of molecule-surface separation, *Chem. Phys. Lett.* 73:201 (1980).

4.   C.A. Murray, D.L. Allara, and M. Rhinewine, Silver-molecule separation dependence of surface enhanced Raman scattering, *Phys. Rev. Lett.* 46:57 (1980).

5.   G.L. Eesley, Coverage dependence of enhanced adsorbate Raman scattering, *Phys. Lett.* 81A:193 (1981); G.L. Eesley, J.M. Burkstrand, and D.L. Simon,

X-ray photoemission and Raman spectroscopy investigation of pyridine on Ag, preprint.

6.   H. Seki, SERS of pyridine on Ag island films prepared on a sapphire substrate, *J. Vac. Sci. Technol.* 18: (1980); H. Seki and M.R. Philpott, Surface enhanced Raman scattering by pyridine on silver island films in an ultra-high vacuum, *J. Chem. Phys.* 73:5376 (1980).

7.   I. Pockrand and A. Otto, Coverage dependence of Raman scattering from pyridine adsorbed to silver/vacuum interfaces, *Solid State Commun.* 35:861 (1980).

8.   T.H. Wood and M.V. Klein, Raman scattering from carbon monoxide adsorbed on evaporated silver films, *J. Vac. Sci. Technol.* 16:459 (1979); T.H. Wood, M.V. Klein, and D.A. Zwemer, Enhanced Raman scattering from adsorbates on evaporated silver films, *Surf. Sci.,* in press.

9.   R.R. Smardzewski, R.J. Colton, and J.S. Murday, Enhanced Raman scattering by pyridine physisorbed on a clean silver surface in ultra-high vacuum, *Chem. Phys. Lett.* 68:53 (1979).

10.   M. Udagawa, Chih-Cong Chou, J.C. Hemminger, and S. Ushioda, Raman scattering cross-section of adsorbed pyridine molecules on a smooth silver surface, preprint.

11.   J.E. Demuth, K. Christmann, and P.N. Sanda, The vibrations and structure of pyridine chemisorbed on Ag(111): the occurrence of a compressional phase transformation, *Chem. Phys. Lett.* 76:201 (1980).

12.   J.E. Demuth, The interaction of acetylene with Ni(111), chemisorbed oxygen on Ni(111), and NiO(111); the formation of CH species on chemically modified Ni(111) surfaces, *Surf. Sci.* 69:365 (1977).

13.   H.J. Levinstein and W.H. Robinson, Etch pits at dislocations in silver single crystals, *J. Appl. Phys.* 33:3149 (1962).

14.   This corresponds to the ionization-gauge correction for benzene as obtained from the gas sensitivity tables supplied with our Varian ion gauge.

15.   G. Kaindl, T.-C. Chiang, D.E. Eastman, and F.J. Himpsel., Distance-dependent relaxation shifts of photoemission and auger energies for Xe on Pd(001), *Phys. Rev. Lett.* 45:1808 (1980).

16.   J.E. Demuth and D.E. Eastman, Photoemission observations of π-d bonding and surface reactions of adsorbed hydrocarbons on Ni(111), *Phys. Rev. Lett.* 32:1123 (1974).

17.   W. von Niesen, G.H.F. Diercksen, and L.S. Cederbaum, The electronic structure of molecules by a many-body approach, *Chem. Phys.* 10:345 (1975).

18.   J.A. Simpson and C.E. Kuyatt, Electron monochromator design, *Rev. Sci. Instr.* 38:103 (1967).

19.   L. Corrsin, B. Fox, and R.C.Lord, The vibrational spectra of pyridine and pyridine-d_5, *J. Chem. Phys.* 21:1170 (1953).

20.   D.J. Pouchart, Aldrich Library of Infrared Spectra, Aldrich Chem. Co., Wisconsin, 1975.

21.   D.A. Long and E.L. Thomas, Spectroscopic and thermodynamic studies of pyridine compounds, *Trans. Far. Soc.* 59:783 (1962).

22.   D. Sokcevic, Z. Lenac, R. Brako, and M. Sunjic, Excitation of adsorbed molecule vibrations in low energy electron scattering, *Z. Physik B* 28:273 (1977).

23.   B.J. Bandy, D.R. Lloyd, and N.V. Richardson, Selection rules in photoemission from adsorbates: pyridine adsorbed on copper, *Surf. Sci.* 89:344 (1979).

24.   F.P. Netzer, E. Bertel, and J.A.D. Matthew, Sensitivity of electronic transitions to molecule-surface orientation: ELS of benzene and pyridine on Ir(111), *Surf. Sci.* 92:43 (1980).

25.   We use a conventional LEED display apparatus with ~ 10^{-6} A beam currents which are well recognized to disrupt, desorb or disorder weakly adsorbed molecular species. For example, see J.C. Tracy, Structural influences on adsorption energy II. CO on Ni(111), *J. Chem. Phys.* 56:2736 (1972).

26.   H. Ibach, private communication.

27.   J.E. Demuth and H. Ibach, Experimental study of the vibrations of acetylene chemisorbed on Ni(111), *Surf. Sci.* 85:365 (1979).

28.   J.C. Tsang, J.E. Demuth, P.N. Sanda, and J.R. Kirtley, Enhanced Raman scattering from carbon layers on silver, *Chem. Phys. Lett.* 76:54 (1980).

29.   M.R. Mahony, M.W. Howard, and R.P. Cooney, Carbon dioxide conversion to hydrocarbons at silver electrode surfaces. Raman spectroscopic evidence for surface carbon intermediates. *Chem. Phys. Lett.* 71:59 (1980).

30.   S.S. Jha, J.R. Kirtley, and J.C. Tsang, Intensity of Raman scattering from molecules adsorbed on a metallic grating, *Phys. Rev. B* 22:3973 (1980).

31.   C.S.G. Biswas, X-ray analysis of frozen pyridine and its solution at -180°C, *Indian J. Phys.* 32:13 (1958).

MOLECULE-SILVER SEPARATION DEPENDENCE

Cherry A. Murray

Bell Laboratories
Murray Hill, N.J. 07974

INTRODUCTION

In this chapter I will discuss some of the surface enhanced Raman scattering (SERS) experiments with which I have been involved, focusing on one aspect of the results: the determination of the molecule-surface separation dependence of the Raman scattering enhancement of a molecule placed near a silver enhancing surface. This aspect of the SERS problem is important for both conceptual and practical reasons: conceptual, as the knowledge of the spatial range of SERS is crucial for determining its mechanisms; and practical, as one needs to know just how surface sensitive a technique it is for potential applications. We find that different silver surfaces exhibit spatial ranges of Raman enhancement which vary considerably from roughly 5 to 50 Å. How can this large variation be explained? Is this observed behavior consistent with proposed mechanisms for SERS? Can we make use of the information we have gained in the experiments for the development of a new surface spectroscopy? I believe that the different spatial ranges of Raman enhancement found for various surfaces can be explained by differences in the surface roughness features from sample to sample and are qualitatively consistent with proposed electromagnetic enhancement mechanisms. Our results are useful for developing SERS as a more universal surface probe in two ways: first, these experiments have led to qualitative understanding of much of the physics of the enhancement, and second, they have given us the insight necessary to design appropriate silver enhancers in order to use SERS on other interfaces.

THEORETICAL PREDICTIONS

Before describing the experimental results, I will briefly review the predictions of various theories for the molecule-surface separation dependence of SERS.

Chemical

By the term "chemical enhancement mechanisms", I mean those in which the actual Raman matrix element of an adsorbed molecule is altered by the presence of the surface. These include direct charge transfer between the molecule and metal[1], induced modulation of the surface tunneling barrier for electrons[2], excitation of electron-hole pairs[3] requiring atomic scale roughness, or chemisorption induced resonance[4] theories. The Raman enhancement caused by chemical mechanisms is generally limited to those molecules in the first monolayer in direct contact with the surface or on specific active sites.

Electromagnetic

In the term "electromagnetic enhancement mechanisms", I include all local electromagnetic field enhancements on a surface, as well as any additional enhancement of the Raman shifted radiation due to electromagnetic coupling.

Smooth Surfaces. On an atomically smooth metal surface in vacuum, classical Fresnel reflection effects[5] can cause an increase in intensity (mean squared field strength) by no more than a factor of four of a correctly polarized incident plane light wave. The intensity increase is, of course, due to interference and occurs on the scale of a wavelength of the incident light. An induced dipole near a metallic surface will induce in turn an image dipole of screening electrons in the metal which will affect the Raman scattering from the molecule.[6] If this problem is solved self-consistently using realistic models for both the radiating dipole[7] and the metal,[8] the enhancement of Raman scattering from a molecule located about 1Å from the surface is estimated to be from 1 to 10^3. The enhancement of a radiating molecule farther from the surface due to interaction with its image should decay rapidly on an atomic length scale.

Gratings and small roughness. Any addition of roughness to the surface (including, perhaps, an adsorbed molecule itself) can induce additional local field enhancements. If the surface roughness of the metal can be treated as a small perturbation on an otherwise flat surface, then it can induce previously forbidden coupling of light to longer wavelength surface plasmon-polaritons of the metal providing the roughness has the correct spatial Fourier components to conserve both wavevector and energy in the process.[9] The radiation of a dipole near a slightly rough surface should be enhanced for similar reasons.[10] The maximum field enhancement for coupling to surface plasmons on a randomly rough surface is expected to be small, of the order of unity.[11] The most efficient way to couple light to surface plasmon-polaritons is to use prism momentum couplers on a very smooth surface[12,13] or to make the surface itself into a perfect sinusoidal grating.[12,14] Then one can couple all of the incident radiation in a monochromatic plane wave in a narrow cone about a specific angle into a narrow wavevector band of surface plasmon-polaritons and in doing so theoretically increase the local field intensity by as much as 10^2 - 10^3 over that of the incident light.[11] The surface plasmon-polaritons decay exponentially into the vacuum region with decay lengths on the order of the wavelength of light.[12] Thus any electromagnetic SERS due to this electric field intensity enhancement will exhibit a decay length on the order of 10^3Å for the molecule-surface separation dependence. The Raman scattered light from a molecule situated on a grating is not efficiently coupled out by plasmon-polaritons because the necessary momentum matching condition can only be satisfied in a few directions in the near field of the radiating dipole.[15] Any extra random roughness added to the small amplitude sinusoidal grating on the surface will cause damping of the excited surface plasmon-polaritons, broadening of the narrow resonance, and intensity loss in the local surface field.

Extremely rough surfaces. All of the Raman enhancing surfaces which we have studied are extremely rough on a distance scale smaller than or comparable to the wavelength of the exciting light. If a surface becomes extremely rough, so that it is highly undulating or it appears to be strewn with boulders, its optical properties are no longer well described by a model with small perturbations on a flat surface. The optical properties of very rough silver films, silver island films and colloid suspensions are well known[16] to be governed by optical conduction resonances,[17] or collective electron excitations[18] of localized protrusions of metal. These collective electron resonances in silver can exist throughout the visible region of the spectrum.[19] Their resonant frequencies depend on particular surface's morphology or colloid's aggregation properties.

The essential physics of the electromagnetic SERS models[18,20] on such rough surfaces is schematically depicted in Fig. 1. A classical point dipole molecule is situated near a tip of a surface protrusion, here modeled as an ellipsoid of dielectric constant $\epsilon = \epsilon_1 + i\epsilon_2$ and of a size much smaller than the wavelength of the incident light, λ. When the latter approximation is valid, the problem is reduced to one of classical electrostatics: an ellipsoid in a uniform applied electric field. The

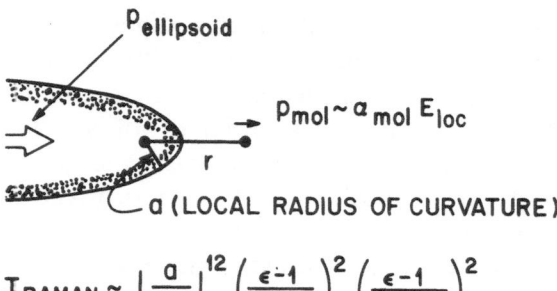

$$I_{RAMAN} \sim \left| \frac{a}{r} \right|^{12} \left(\frac{\epsilon -1}{\epsilon + \epsilon_R} \right)^2 \left(\frac{\epsilon -1}{\epsilon + \epsilon_{R^I}} \right)^2$$

Fig. 1.   Schematic drawing of a molecule located a distance r from the center of the local radius of curvature, a, of the tip of a surface protrusion. In the simplest electromagnetic resonance theory described in the text, the protrusion is modeled as an ellipsoid of energy dependent dielectric constant ϵ. If incident light is resonant with the dipole collective electron oscillation of the ellipsoid, or $\epsilon = - \epsilon_R$, then the local field E_{loc} at the molecule is enhanced tremendously. The local field induces a dipole moment in the molecule. If the Raman scattered light, resonant at $\epsilon = - \epsilon_{R'}$, is within the dipole resonance of the protrusion, it will also be amplified by the ellipsoid 'antenna'. The total Raman scattered intensity is proportional to the expression above, neglecting image and chemical enhancement effects. It is proportional to the normal Raman cross section of the molecule.

relevant collective electron resonance with resonant frequency in the visible region of the spectrum for silver is the Mie dipole resonance parallel to the long axis of the ellipsoid. The frequency of this confined plasma oscillation is determined only by the shape of the surface protrusion (i.e. the ratio of major to minor axis of the ellipsoid) if it is small compared to the wavelength of light.[21] When the incident light frequency ω is resonant with the dipole plasma oscillation, $\epsilon_1(\omega) = -\epsilon_R$ where ϵ_R determines the dipole resonance of the ellipsoid. In this case the local electric field immediately at the tip of the protrusion is enhanced by the factor $\left| \frac{\epsilon -1}{\epsilon + \epsilon_R} \right|$ or $\sim \left| \frac{\epsilon -1}{\epsilon_2} \right|$. For silver at $\lambda = 5000$ Å, this field enhancement is roughly a factor of 30, so that the local intensity at the surface of the tip is about 900 times larger than that of the applied field in free space. The local field on the side of the protrusion far from the tip is reduced in amplitude by a factor of about $|\epsilon|$ (about 10 for silver at $\lambda = 5000$ Å), as can be easily discerned by considering boundary conditions for Maxwell's equations in the electrostatic limit.

The Raman scattering of a molecule on the surface of the protrusion is not only enhanced from its free space value by the local electric field intensity increase of $\left|\dfrac{\epsilon-1}{\epsilon_2}\right|^2$ at the tip, $\left(\dfrac{1}{|\epsilon|^2}\left|\dfrac{\epsilon-1}{\epsilon_2}\right|^2\right.$ at the side), but is further enhanced by the now very efficient coupling out of the Raman scattered radiation by the nearby resonant "antenna". If the Raman shifted radiation is still at a frequency within the width of the dipole resonance of the protrusion, the enhancement of the outgoing radiation from a molecule exactly at the tip will be increased by an additional multiplicative factor $\left|\dfrac{\epsilon-1}{\epsilon+\epsilon_{R'}}\right|^2$ or $\sim\left|\dfrac{\epsilon-1}{\epsilon_2}\right|^2$, where $\epsilon_{R'} = -\epsilon_1(\omega-\Omega) \simeq \epsilon_R$ is then the frequency dependent dielectric constant of the metal at the Raman shifted frequency $\omega - \Omega$. Molecules located on the side of the protrusion will have an overall Raman enhancement which is lower than that of molecules at the tip by a factor of $|\epsilon|^{-4}$. For silver protrusions and incident light with wavelength 5000 Å, this factor is 10^{-4}, and thus SERS will only be observed from molecules at the tip.

If the induced and Raman radiating dipoles of the molecule are modeled as point dipoles situated in space a certain distance r along the long axis of the ellipsoidal protrusion from the center of the local radius of curvature, a, of the tip, then the molecule-surface separation dependence of the SERS is proportional to $(a/r)^{12}$. This is just the spatial dependence of simple dipole-dipole coupling, squared in order to take into account both the coupling in of the incident and the coupling out of the Raman shifted radiated light. The scaling parameter of the enhancement is the local radius of curvature of the tip, a, and not the average size (major or minor axis) of the protrusion. This is an important point as for example the local radius of curvature can be much smaller than the length of a dendrite on the surface, or the width of an island. The tip radius of curvature and not its average size determines the range of SERS enhancement on silver in the visible because only the molecules at the tip contribute to the observed SERS.

Obviously, a detailed knowledge of the morphology of the Raman enhancing surfaces will be necessary for a quantitative comparison with experimental results. Of course, protrusions on real surfaces are poorly modeled by isolated small ellipsoids in vacuum. Both the size of the protrusions and the interaction effects[22] of the resonances of nearby protrusions and of a possible metallic ground plane need to be taken into account in the theory. This will necessitate complete Mie scattering cal-

with blue light to form small silver particles in a photochemical process. A scanning electron micrograph of one of these surfaces, taken after the Raman measurements, is shown in Fig. 2. The morphology of this surface consisted of 500-1000 Å diameter balls of silver separated by roughly 1500-3000 Å. Some iodine remained on the surface, as determined by Auger spectroscopy; however its presence did not affect the results I will discuss for the coverage dependence of pyridine. Similar spectra were obtained in our vacuum chamber on lightly ion sputtered electrochemically roughened silver surfaces which had similar size (but more interconnected) roughness features.

We determined relative pyridine coverage by Auger electron spectroscopy peak height ratios of carbon, nitrogen and silver. Well before pyridine exposures for which appreciable charging of the surface was noticeable during the Auger measurements, we observed a sharp break in the peak heights with increasing exposure. This break always occurred at roughly 7 Langmuirs exposure, uncorrected for the better ionization efficiency of the gauge for pyridine compared to N_2 ($1L = 10^{-6}$ torr-sec). This break corresponds to a decrease of the sticking coefficient for the pyridine by roughly a factor of two after the coverage attained at 7L exposure. We assumed that this sticking or condensation coefficient change occurs at the completion of the first monolayer. Subsequent photoemission[24,25,26] and Auger[26,27] measurements by other groups on different silver surfaces confirm monolayer completion at exposures from 5-7 L (uncorrected gauge readings). Similar changes in sticking coefficient near that exposure have also been found,[24,27] in excellent agreement with our results.

The numerical value of coverage for one completed monolayer which we estimated both from quantitative comparisons of Auger peak heights on clean and pyridine covered surfaces and also the measured value of the pyridine sticking coefficient at low coverages is $(3\pm1) \times 10^{14}$ molecules/cm^2. This is in good agreement with photoemission[24,26] estimates. However, the actual number of molecules on the surface is not as important for separation dependence SERS measurements as the knowledge of the completion of one monolayer and the condensation coefficient change afterwards. If we assume that the thickness of one molecular layer is roughly 5 Å, and that pyridine coats the surface more or less uniformly, then the saturation of the SERS intensity with coverage in the physisorbed condensed layers after the completion of the first monolayer provides quantitative information on the spatial range of the enhancement.

Our Raman spectra on the surface photographed in Fig. 2 exhibit two

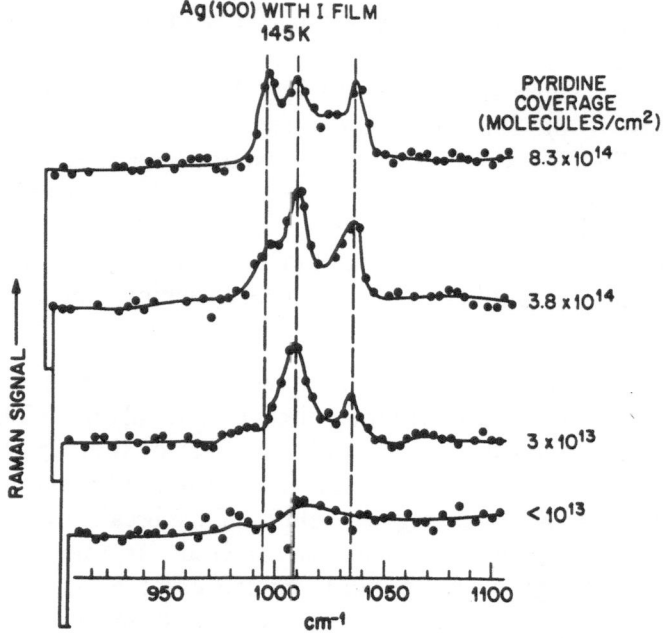

Fig. 3.   Raman spectra of various coverages of pyridine adsorbed at 150K on an I-roughened Ag surface, taken with ~100mW of 4880 Å incident light. The two lower frequency peaks at 991 cm^{-1} and 1003 cm^{-1} are associated with the symmetric breathing modes of two different surface species. The higher frequency, 1030 cm^{-1}, mode is associated with both species.

different symmetric ring breathing mode frequencies of pyridine, Fig. 3. We interpret these two modes as belonging to two separate surface species. The species which first appears at low coverages has a breathing mode that is shifted in frequency slightly higher than that of the liquid, to 1003 cm^{-1}, a shift that is consistent with nitrogen coordination to the surface.[26,28] The intensity of this 1003 cm^{-1} mode in the spectrum saturates at nearly a monolayer coverage. This pyridine species is coordinated to sites on the silver surface in the first layer. At about a monolayer coverage, a shoulder on the 1003 cm^{-1} peak appears at 991 cm^{-1}, a frequency unshifted from that of liquid (or solid) pyridine. The intensities of these modes are plotted in Fig. 4. The peak intensity at 991 cm^{-1} grows considerably after the completion of one monolayer. Therefore we definitely observe scattering from the condensed pyridine in multilayers on the surface. The first layer 1003 cm^{-1} species desorbs at a higher temperature than the 991 cm^{-1} species and exhibits at most an order of magnitude larger SERS for the symmetric stretching mode. The 991 cm^{-1} pyridine peak may include contributions from π bonded or vanderWaals bonded pyridine in the first monolayer[24] as well as physisorbed condensed species in the upper layers. The saturation of Raman intensity for this mode occurs at about 7 layers, or ~35 Å. We conclude that both pyridine species are enhanced by long range electromagnetic effects on this surface. In addition, the nitrogen coordinated 1003 cm^{-1} species in the first layer may have a multiplicative short range, perhaps chemical, enhancement of at most a factor of ten. If one makes a comparison with the Mie dipole resonance model described in the last section, one obtains an average effective local radius of curvature for this surface of 250-300 Å. The exact numbers should not be taken too seriously as the model is extremely crude, but nevertheless, this is in good agreement with the size of the particles in the micrograph in Fig. 2. The pyridine exposure dependence of SERS in recent experiments on thick silver films evaporated on cold (~100K) substrates by Pockrand and Otto[29] and Wood et al.[30] shows dramatically different behavior compared to our results on I-roughened silver. On the cold evaporated films the 1003 cm^{-1} peak associated with the nitrogen coordinated pyridine species exhibits coverage dependence essentially identical to that shown in Fig. 4. The 991 cm^{-1} peak associated with physisorbed pyridine, however, is extremely weak and shows up as a distinct shoulder only at high coverages. A com-

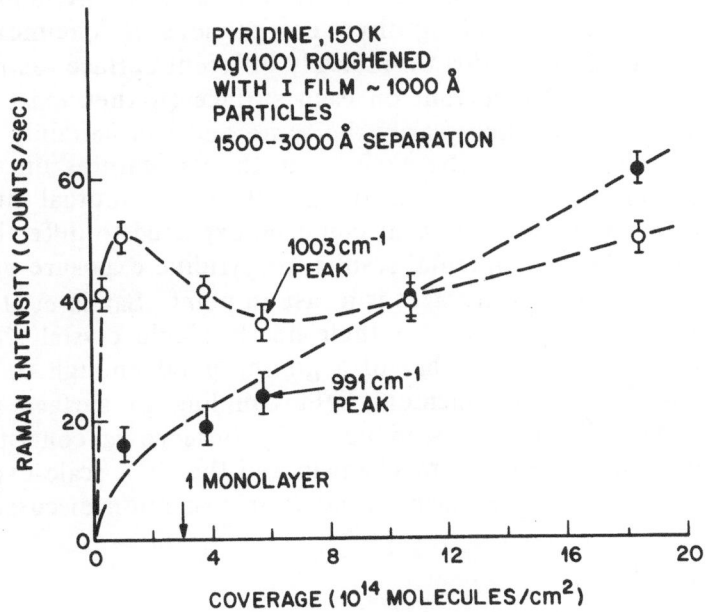

Fig. 4.   Peak intensity versus pyridine coverage for the symmetric ring breathing modes of the two surface species on I-roughened silver. The lines are guides to the eye.

parison of the 991 cm^{-1} shoulder intensity of Wood's data to the Mie model yields an effective local radius of curvature of from 30-50 Å, about a factor of ten smaller than that on the I-roughened silver. The morphology of these cold evaporated films is unknown as by the time they are heated to room temperature the surfaces anneal and the SERS is destroyed.[31,32] One would expect that the roughness features are considerably smaller on these surfaces compared to those on surfaces which are stable at room temperature. Therefore the difference in saturation dependence of SERS on these cold evaporated film surfaces compared to the I-roughened surfaces is not inconsistent with the electromagnetic Mie enhancement mechanism.

In his experiments on silver island films in ultrahigh vacuum, Seki[33] found a saturation of SERS for pyridine at coverages very close to a monolayer. His results, which show a very short range enhancement, are in agreement with the distance dependence for polymers we have found on similar island films as described in the following section. In other pyridine/silver vacuum interface experiments using ion bombarded cold surfaces[34] and mechanically abraded rough surfaces,[26,35] only very short range enhancements are found. Between all of these pyridine/silver-vacuum interface results there are large variations in the numbers and relative concentrations of different surface species and the intensity ratios of different Raman modes for the same pyridine species. One must be careful in making direct comparisons of "chemical" versus "electromagnetic" SERS ratios on radically different surfaces as the active sites for chemical enhancement on each surface (if they exist at all on some surfaces) are not necessarily the same and will certainly not exist in similar numbers. Since the surface roughness features on each surface are also certain to be different in size, shape and mutual interaction, their electromagnetic enhancement could be expected to differ by orders of magnitude. The experimental results for pyridine exposure on a silver grating with single crystal terraces is a case in point. Sanda et al.[36] do not observe the 1003 cm^{-1} mode on their nearly single crystal silver, and their surface morphology is that of a grating good enough to see field enhancement due to the incident light coupling to surface plasmon-polaritons. The distance dependence they observe is consistent with both a short range "chemical" mechanism and the 10^3 Å scale exponential decay expected for surface plasmon-polariton excitation discussed in the last section.

Silver Island Films

Silver island films are good SERS enhancers for adsorbed monolayers.[18] We[37] prepared island films by slow .1 Å/sec. evaporation of silver onto room temperature glass substrates. These films ranged in thickness from 50-200 Å and enhanced the Raman scattering from benzoic acid and CN^- monolayers by from 10^4–10^5. In order to study the spatial range of SERS on island films, we prepared a number of identical films and measured the Raman scattering from different thicknesses of polystyrene spun from chlorobenzene solution on top of the films. As polystyrene is not expected to interact strongly with the silver we hoped to avoid any strong "chemical" enhancement mechanisms.

The polymer layer thicknesses were measured by ellipsometry at 6328 Å[38] assuming that the refractive index of the thin layers remains constant with thickness. It is known[39] that a monolayer of polystyrene forms on gold in the presence of a nonpolar solvent in a layer of about 40 Å thickness with roughly half the bulk density. The polymer chains are vanderWaals bonded to the surface in a few spots along the chain with the remainder coiled about surface voids. The structure of polystyrene adsorbed to silver has not been studied in such detail, but we can expect it will be similar to that on gold. The change in morphology in the polymer network on the surface is unknown when the solvent is removed by spinning, as in our experiments. However, one can guess that it will collapse more against the surface as the polymer-surface interactions can only increase when the solvent is out of the way. We used polystyrene thicknesses larger than 20 Å in order to insure that the average polymer density very close to the surface would not change dramatically as the layer thickness was increased.

The intensity of the polystyrene 1000 cm^{-1} ring breathing mode in the Raman spectrum of samples in air taken with 4880 Å incident light is plotted versus polystyrene layer thickness in Fig. 5. This particular experiment was performed on a 100 Å mass thickness (thickness if uniform with bulk density) silver island film, but we obtained essentially identical distance dependence results for other silver thicknesses from 50-150 Å. We find that the SERS from polystyrene saturates at a thickness smaller than 20 Å, implying that the electromagnetic range of enhancement is very short on these island films.

The islands for the 100 Å film on glass were roughly 300 Å in diameter and a little less than 100 Å in thickness under scanning electron microscope inspection. They resemble oblate ellipsoids crowded together with their centers separated by roughly a diameter. For incident blue light, the dipole resonance of a silver ellipsoid of this shape is polarized in the plane of the pancake, parallel to the substrate surface. Thus following the discussion relating to Fig. 1 of this chapter, one would expect that Raman enhancement will only take place at the edges of the pan-

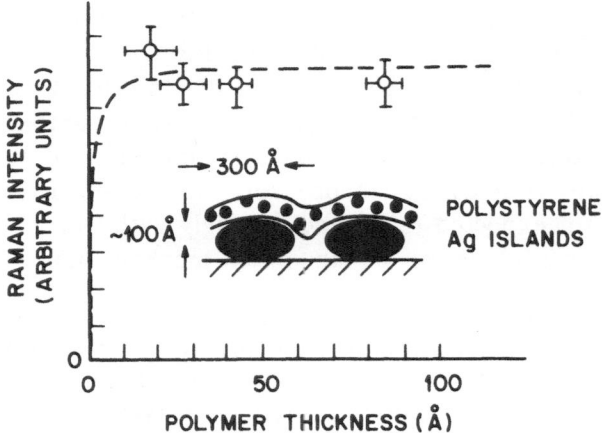

Fig. 5.   Raman intensity for the 1000 cm^{-1} ring breathing mode of polystyrene deposited on Ag islands versus polystyrene thickness. The
line is a guide to the eye.

cakes between the islands. The local radius of curvature of the oblate
ellipsoid relevant for the measurements of the range of SERS normal to
the surface is the small one. For a semimajor to semiminor axis ratio of
3:1, as is roughly the case for our 100 Å island films, the small radius of
curvature of the islands is a factor 0.06 smaller than the semimajor axis.
Our pancakes are 300 Å in diameter, and so the saturation in SERS
intensity which we find on these films is quite consistent with the 9 Å
local average radius of curvature and the predictions of the Mie resonance model. We found that dipping any of the samples into an index
matching medium (mineral oil in which the polystyrene is insoluable)
affected the SERS by less than 10 percent. From this we can conclude
that the additional thickness polymer films did not shift the resonant frequencies of the islands from those on the 20 Å thickness polystyrene
samples. Such a shift could have affected our thickness dependence
SERS measurements made using a single frequency of incident light.

CaF$_2$-roughened Silver in Multilayered Structures

Our experiments[40] on the separation dependence of SERS from
monolayers of p-nitrobenzoic acid in multilayer structures using CaF$_2$
roughened silver films as enhancers have been reported in detail elsewhere.[40,41] The structure of the samples used for a particular spacer
experiment is shown in the inset of Fig. 6. As the preparation of the
samples has been described at length,[41] I will only touch on the points
relevant to the understanding of the separation dependence measure-

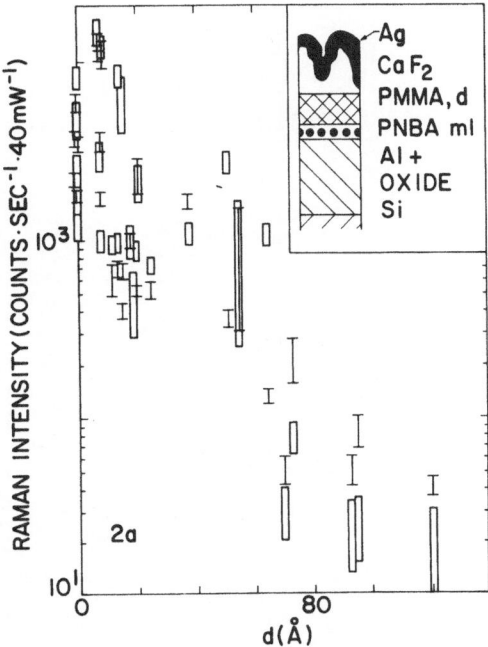

Fig. 6.   Intensity of two Raman modes of a PNBA monolayer, adsorbed in
the multilayer structure shown in the inset, versus PMMA spacer
thickness d. (Open symbols: 1100 cm^{-1} mode, and closed symbols:
1597 cm^{-1} mode.) The size of the symbols represents the variation
of intensity over the ~5cm^2 sample area.

ments. The p-nitrobenzoic acid (PNBA) molecule has a very strong
solution phase Raman cross section and is deposited as a monolayer to
serve as a marker for the SERS distance dependence measurements. It
is strongly chemisorbed on a native oxide layer on the aluminum and
forms the p-nitrobenzoate ion. The 2500 Å aluminum layer under the
PNBA is smooth (~200 Å surface ripple) and serves as a reflector for
infrared reflection measurements, used to insure that a monolayer exists
on the oxide.

On top of the PNBA monolayer a polymer spacer layer of thickness
$0 < d < 150$ Å is spun from chlorobenzene solution. The spacer used is
polymethylmethacrylate (PMMA) which is known to make a uniform
layer, smooth on at least a few hundred Angstrom scale,[42] with an
extremely low Raman cross section. Ellipsometry[38] is used to determined the spacer thickness before the completion of the sample fabrication. We have made measurements of the diffusion and dissolution of

cation of the uniformity of both the monolayer density and the SERS due to variations in the surface roughness across the sample.

We find a long range SERS effect on these CaF_2 roughened surfaces with the enhancement dropping off by an order of magnitude in roughly 50 Å spacer thickness. We make a very rough estimate of about 20 Å for the thickness of the CaF_2 in the valleys of the surface undulations where the silver is closest to the monolayer and assume that the electromagnetic Raman enhancement of the PNBA is taking place by means of resonances of the silver protrusions into these hollows. A comparison of the data of Fig. 6 with the Mie resonance theory yields an effective radius of curvature for the silver of from between 125-200 Å. Again, this is in reasonable agreement with the size of the structures observed on these surfaces (Fig. 7).

CONCLUSIONS

In conclusion, we have observed SERS silver-molecule separation dependences on several surfaces of very different random surface morphologies. The surfaces are extremely rough on a distance scale below 1000 Å. The separation dependences we observe for SERS on these surfaces are consistent with the sizes and shapes of the surface roughness features seen in electron micrographs of the samples and a simple electromagnetic model using Mie resonances of protrusions as enhancing "antennas" for the incident and Raman scattered light.

Electromagnetic Raman enhancement mechanisms should exist on all silver surfaces but their nature, distance dependences, resonance energies and magnitudes depend critically on the specific roughness features on each surface. These field enhancement mechanisms can be used in a number of interesting experiments; for example, surface enhanced non-linear optics.[44] They will be better understood with the help of more realistic model calculations taking into account the effects of retardation and interactions of resonances, and cleaner experiments on surfaces with more regular surface roughness.[45] Silver layers with long range enhancements can be used as enhancers to study potentially more interesting non noble-metal interfaces.

In addition to the electromagnetic SERS mechanisms, there are indications that chemical enhancement mechanisms play a role on some surfaces. Once the electromagnetic enhancements are understood quantitatively, we will be in a better position to study the "chemical" mechanisms which may tell us a great deal about adsorption and active sites for catalysis.

ACKNOWLEDGEMENTS

I am pleased to acknowledge the collaboration of J. E. Rowe, C. V. Shank and D. A. Zwemer in the ultra-high vacuum experiments, and D. L. Allara in the multilayer and island film work. I wish to thank M. Rhinewine, S. Christman and S. Bodoff for technical assistance; and T. H. Wood, P. M. Platzman, S. M. McCall and R. E. Slusher for useful discussions.

REFERENCES

1. S. L. McCall and P. M. Platzman, Raman scattering from chemisorbed molecules at surfaces, *Phys. Rev. B 22*: 1660 (1980); J. I. Gersten, R. L. Birke and J. R. Lombardi, Theory of enhanced light scattering from molecules adsorbed at the metal-solution interface, *Phys. Rev. Lett. 43*: 147 (1979); F. R. Aussenegg and M. E. Lippitch, On Raman scattering in molecular complexes involving charge transfer, *Chem. Phys. Lett. 59*: 214 (1978).

2. J. R. Kirtley, S. S. Jha, and J. C. Tsang, Surface plasmon model of surface enhanced Raman scattering, *Solid State Commun. 35*: 509 (1980).

3. E. Burstein, Y. J. Chen, C. Y. Chen, S. Lundquist and E. Tosatti, Giant Raman scattering by adsorbed molecules on metal surfaces, *Solid State Commun. 29*: 565 (1979); J. Billman, G. Kovacs, A. Otto, Enhanced Raman effect from cyanide adsorbed on a silver electrode, *Surf. Science 92*: 153 (1980); and A. Otto, I. Pockrand, J. Billman, and C. Pettenkofer, chapter in this volume.

4. M. R. Philpott, Effect of surface plasmons on transitions in molecules, *J. Chem. Phys. 62*: 1812 (1975), S. Effrima and H. Metiu, Resonant Raman scattering by adsorbed molecules, *J. Chem. Phys. 70*: 2297 (1979); R. M. Hexter and M. G. Albrecht, Metal surface Raman spectroscopy: Theory, *Spectrochemica Acta 35A* [3], (1979); D. P. Dilella, A. Gohin, R. H. Lipson, P. McBreen, and M. Moskovits, Enhanced Raman spectroscopy of CO adsorbed on vapor-deposited silver, *J. Chem. Phys. 73*: 4282 (1980); H. Ueba, Effective resonant light scattering from adsorbed molecules, *J. Chem. Phys. 73*: 725 (1980); and H. Ueba, chapter in this volume.

5. J. D. E. McIntyre, Optical reflection spectroscopy of chemisorbed monolayers, in: "Optical Properties of Solids-New Developments", B. O. Seraphim, ed., North Holland, Amsterdam (1976); and R. G. Greenler and T. L. Slager, Method for obtaining the Raman spectrum of a thin film on a metal surface, *Spectrochim. Acta 29A*: 193 (1973).

6. F. W. King, R. P. VanDuyne, and G. C. Schatz, Theory of Raman scattering by molecules adsorbed on electrode surfaces, *J. Chem. Phys. 69*: 4472 (1978); G. L. Easley and J. R. Smith, Enhanced Raman scattering on metal surfaces, *Solid State Commun. 31*: 815 (1979) S. Effrima and H. Metiu, Classical theory of light scattering by an adsorbed molecule. I. Theory, and Resonant Raman scattering by adsorbed molecules, *J. Chem. Phys. 70*: 1602, 2297 (1979); and G. C. Schatz and T. K. Lee and J. L. Birman, chapters in this volume.

7. P. R. Hilton and D. W. Oxtoby, Surface enhanced Raman spectra: A critical review of the image dipole description, *J. Chem. Phys. 72*: 6346 (1980); G. W. Ford and W. H. Weber, Electonmagnetic effects of a molecule at a metal surface. I-Effects

of nonlocality and finite molecular size, to be published.

8. N. D. Lang and A. R. Williams, Theory of atomic chemisorption on simple metals, *Phys. Rev. B 18*: 616 (1978); W. H. Weber and G. W. Ford, Enhanced Raman scattering by adsorbates including the nonlocal response of the metal and the excitation of nonradiative modes, *Phys. Rev. Lett. 44*: 1774 (1980); G. Korzeniewski, T. Maniv, and H. Metiu, Electrodynamics at metal surfaces. IV. The Electric fields caused by the polarization of a metal surface by an oscillating dipole, to be published.

9. J. C. Tsang, J. R. Kirtley and T. N. Theis, Surface plasmon polariton contributions to Stokes emission from molecular monolayers on periodic Ag surfaces, *Solid State Commun. 35*: 667 (1980).

10. P. K. Aravind and H. Metiu, The enhancement of Raman and fluorescent intensity by small surface roughness I. The change of dipole emission, *Chem. Phys. Lett. 74*: 301, (1980).

11. W. H. Weber and G. W. Ford, Optical electric-field enhancement at a metal surface arising from surface plasmon excitation, *Opt. Lett. 6*: 122 (1981).

12. H. Raether, Surface plasma oscillations and their applications, in: "Physics of Thin Films," eds. G. Hass, M. Francombe and R. Hoffman, Academic Press, New York (1977); and references therein.

13. Y. J. Chen, W. P. Chen and E. Burstein, Surface-electronmagnetic-wave-enhanced Raman scattering by overlayers on metals, *Phys. Rev. Lett. 36*: 1207 (1976).

14. S. S. Jha, J. R. Kirtley and J. C. Tsang, Intensity of Raman scattering from molecules adsorbed on a metallic grating, *Phys. Rev. B 22*: 3973 (1980); and S. S. Jha, chapter in this volume.

15. P. K. Aravind, E. Hood and H. Metiu, Angular resonances in the emission from a dipole located near a grating, to be published.

16. For example, see: J. G. Endriz and W. E. Spicer, Study of Al films. I. Optical Studies of reflectance drops and surface oscillations on controlled-roughness films, *Phys. Rev. B 4*: 4144 (1971); and J. A. Creighton, chapter in this volume.

17. J. P. Marton and J. R. Lemon, Optical properties of aggregated metal systems, *Phys. Rev. B 4*: 271 (1971).

18. C. Y. Chen and E. Burstein, Giant Raman scattering by molecules at metal-island films, *Phys. Rev. Lett. 45*: 1287 (1980); and E. Burstein, C. Y. Chen and S. Lundquist, Giant Raman scattering by molecules adsorbed on metals: An overview, in: "Proceedings of the Second Joint U.S.-U.S.S.R. Symposium on Inelastic Light Scattering in Condensed Matter," eds. J. L. Birman, H. V. Cummins, and K. K. Rebane, Plenum, New York (1979), p. 429; and E. Burstein and D. L. Mills, chapter in this volume.

19. See, for example: R. S. Sennett and G. D. Scott, The structure of evaporated metal films and their optical properties, *J. Opt. Soc. Am. 40*: 203 (1950).

20. S. L. McCall, P. M. Platzman, and P. A. Wolff, Surface enhanced Raman scattering, *Phys. Lett. 77A*: 381 (1980). D. S. Wang, M. Kerker, and H. Chew, Surface enhanced Raman scattering (SERS) by molecules adsorbed at spherical particles:

errata *Appl. Opt. 19*: 4159 (1980); J. L. Gersten and A. Nitzan, Electromagnetic theory of enhanced Raman scattering by molecules adsorbed on rough surfaces, *J. Chem. Phys. 73*: 3023 (1980); F. J. Adrian, Surface enhanced Raman scattering by surface plasmon enhancement of electromagnetic fields near spheriodal particles on a roughened metal surface, *Chem. Phys. Lett. 78*: 45 (1981); and chapters in this volume by J. I. Gersten and A. Nitzan, and M. Kerker, D. S. Wang, H. Chew, O. Siiman and L. A. Bumm.

21. E. C. Stoner, The demagnetizing factors for ellipsoids, *Phil. Mag. 36*: 803 (1945).

22. J. E. Sansonetti and J. K. Furdyna, Depolarization effects in arrays of spheres, *Phys. Rev. B 22*: 2866 (1980); T. Yamagudi, S. Yoshida and A. Kinbara, Effect of retarded dipole-dipole interactions between island particles on the optical plasma-resonance absorption of a silver-island film, *J. Opt. Soc. Am. 64*: 1563 (1974); and references therein.

23. J. E. Rowe, C. V. Shank, D. A. Zwemer and C. A. Murray, Ultrahigh-vacuum studies of enhanced Raman scattering from pyridine on Ag surfaces, *Phys. Rev. Lett. 44*: 1770 (1980); also see D. A. Zwemer, C. V. Shank and J. E. Rowe, Surface enhanced Raman scattering as a function of molecule-surface separation, *Chem. Phys. Lett. 73*: 201 (1980).

24. J. E. Demuth, K. Christmann and P. N. Sanda, The vibrations and structure of pyridine chemisorbed on Ag(111): The occurrence of a compressional phase transformation, *Chem. Phys. Lett. 76*: 201 (1980).

25. S. R. Kelemen and A. Kaldor, Pyridine adsorption on Ag(110), *Chem. Phys. Lett. 73*: 205 (1980).

26. G. L. Easley, J. M. Burkstrand and D. L. Simon, X-Ray photoemission and Raman spectroscopy investigation of pyridine on Ag, to be published.

27. M. Udagawa, C. C. Chow, J. C. Hemminger and S. Ushioda, Raman scattering cross-section of adsorbed pyridine molecules on a smooth silver surface, to be published.

28. W. Schindler and H. Posch, Rayleigh and Raman light scattering from pyridine-water mixtures, *Chem. Phys. 43*: 9 (1979).

29. I. Pockrand and A. Otto, Coverage dependence of Raman scattering from pyridine adsorbed to silver/vacuum interfaces, *Solid State Commun. 35*: 861 (1980).

30. T. H. Wood, D. A. Zwemer, C. V. Shank and J. E. Rowe, The dependence of surface-enhanced Raman scattering on surface preparation: Evidence for an electromagnetic mechanism, to be published; and T. H. Wood, private communication.

31. T. H. Wood and M. V. Klein, Studies of the mechanism of enhanced Raman scattering in ultra-high vacuum, *Solid State Commun. 35*: 263 (1980).

32. I. Pockrand and A. Otto, Surface enhanced Raman scattering (SERS): Annealing the silver substrate, to be published.

33. H. Seki and M. R. Philpott, Surface enhanced Raman scattering by pyridine on silver island films in ultra-high vacuum, *J. Chem. Phys. 73*: 5376 (1980); and H. Seki, SERS of pyridine on Ag island films prepared on a sapphire substrate, Proceedings of Am. Vac. Soc. Symp. Oct. 1980, *J. Vac. Sci. Tech*, to be published.

34. R. R. Smardewski, R. J. Colton and J. S. Murday, Enhanced Raman scattering by pyridine physisorbed on a clean silver surface in ultra-high vacuum, *Chem. Phys. Lett.* *68*: 53 (1979).

35. G. L. Easley, Coverage dependence of enhanced adsorbate Raman scattering, *Phys. Lett.* *81*: 193 (1981).

36. P. N. Sanda, J. M. Warlaumont, J. E. Demuth, J. C. Tsang, K. Christmann, and J. A. Bradley, Surface enhanced Raman scattering from pyridine on Ag(111), *Phys. Rev. Lett.* *45*: 1519 (1980); also see P. N. Sanda, J. E. Demuth, J. C. Tsang, and J. M. Warlaumont, chapter in this volume.

37. In collaboration with D. L. Allara.

38. D. L. Allara, A. Baca, and C. A. Pryde, Distortions of band shapes in external reflection infra-red spectra of thin polymer films on metal substrates, *Macromolecules 11*: 1215 (1978).

39. H. Gebhard and E. Killmann, Ellipsometric investigation of the adsorption of polystyrene and polymethylmethacrylate on metal surfaces, *Angew Makromole Chem. 53*: 171 (1976).

40. C. A. Murray, D. L. Allara and M. Rhinewine, Silver-molecule separation dependence of surface-enhanced Raman scattering, *Phys. Rev. Lett. 46*: 57 (1981).

41. C. A. Murray and D. L. Allara, Measurement of the molecule-silver separation dependence of surface enhanced Raman scattering in multilayered structures, to be published.

42. R. F. Roberts, D. L. Allara, C. A. Pryde, D. N. E. Buchanan and N. D. Hobbins, Mean-free path for inelastic scattering of 1.2 keV electrons in thin polymethylmethacrylate films, *Surf. and Interface Anal. 2*: 5 (1980).

43. In collaboration with D. L. Allara and A. H. Hebard.

44. For example: C. K. Chen, A. R. B. deCastro, Y. R. Shen, Surface enhanced second harmonic generation, *Phys. Rev. Lett. 46*: 145 (1981); and J. P. Heritage and A. M. Glass, chapter in this volume.

45. For example: P. F. Liao, J. G. Bergman, D. S. Chemla, A. Wokaun, J. Melugailis, A. M. Hawryluk and N. P. Economou, Surface enhanced Raman scattering from microlithographic silver surfaces, to be published; and P. F. Liao, chapter in this volume.

TUNNEL JUNCTION STRUCTURES

J.R. Kirtley, J.C. Tsang, and T.N. Theis

IBM Thomas J. Watson Research Center

Yorktown Heights, New York, 10598

We have observed Raman scattering from molecular layers absorbed on the oxide layer in Al-Al$_2$O$_3$-metal tunnel junctions[1-6]. The observation of these signals was remarkable, since the molecular layers were present only in monolayer coverages, and were separated from the incident laser beam by a typically 20 nm thick metal film. In fact, the Raman scattering could be measured because the observed signal levels were up to ~ 5x10^4 times more intense per molecule than expected from similar measurements on bulk specimens. Therefore, surface enhanced Raman scattering, analogous to that observed in electrochemical cells and in vacuum (and discussed in other Chapters in this volume), can be present in tunnel junction structures as well. Our experiments conclusively demonstrated the role of electromagnetic resonances in surface enhanced Raman scattering: 1) Junctions with Ag second metal electrodes had the most intense spectra, with Cu and Au ~ 100 times weaker, and no signals were observed for Pb, Sn, In, or Al second metal electrodes. This is consistent with the fact that low imaginary parts of the dielectric response of the metals were required for strong electromagnetic resonances. 2) We found that the observed spectra were most intense for tunnel junctions evaporated on highly roughened substrates. Surface roughness was required to couple the surface electromagnetic resonances to external radiation. 3) The Raman intensity from junctions on sinusoidal profile, holographically produced gratings became much stronger under conditions in which the incident light was coupled to collective electromagnetic resonances of the junction structures, and also under conditions in which these resonances coupled out to light. However, the field enhancements associated with the electromagnetic resonances were not large enough to explain the observed signals. Also, coupling out of the Stokes scattered radiation through electromagnetic resonances probably did not greatly increase the observed signal strengths: The outcoupling resonances from junctions on gratings were not large, and observations of sharp light emission peaks with little residual background when tunnel junctions on gratings were biased to voltages of 2-3 volts showed that localized resonances due to uncontrolled roughness did not dominate the emission properties of these structures. In addition, certain molecular layers had much stronger Raman signals in the tunneling junction geometry than others, even when the different scattering cross sections for the molecules were taken into account. We conclude that an enhancement mechanism in addition to electromagnetic resonances must be present in these devices. We discuss one possible enhancement mechanism.

Our samples were fabricated in a standard tunneling junction geometry[7]: 40 nm thick Al films were evaporated onto the substrates in an LN_2 baffled, 10^{-7} torr base pressure, oil diffusion pumped vacuum evaporator. The geometry of these films were defined by evaporation through mechanical masks. The Al films were oxidized by exposure to room air for a few minutes, doped with the organic monolayer to be studied, then returned to the evaporator for completion with a second metal electrode. The molecular monolayers were prepared by dissolving the organic acids or aldehydes in water or ethanol (with solution concentrations typically 0.5 mg/ml), dropping the solutions onto the oxidized aluminum surface with a medicine dropper, and then removing the excess by spinning the sample in a photoresist type spinner at 3000 rpm. When doping with an ethanol solution, it was often necessary to wash the surface with a non-polar solvent to remove excess physisorbed layers of the dopant. The second metal electrode was evaporated while the sample was cooled by mechanical contact with a stainless steel block through which LN_2 was passed. We found that cooling the substrate for the second metal evaporation was necessary to obtain high quality inelastic electron tunneling spectra from these junctions[8]. Cooling of the substrate was not necessary to obtain surface enhanced Raman spectra from these samples; several samples prepared with Ag evaporated at room temperature had Raman spectra comparable to those with cold evaporated Ag.

The tunnel junction geometry had several advantages: 1) Molecular monolayers could be " glued " to the alumina and coated with the top metal electrode, making a sample which could be transported in air and which degraded only slowly over a period of weeks or months. 2) The junction topology could be controlled on a scale of 10-1000 nm by evaporating the films on substrates with controlled roughness. 3) The vibrational spectra of the absorbed layer and to some extent the interaction between the top metal and the molecular layer could be independently determined by measuring the current-voltage characteristics of the tunneling junctions.

Leads were attached to the crossed films, the junctions were cooled to 4.2 K or colder, and the second derivatives of the junction current-voltage characteristics were obtained using standard constant current modulation techniques. The current through the barrier region in a tunneling junction has two components: an elastic component, in which no energy is lost in the transition, and an inelastic component, in which energy is lost to a vibrational mode of the barrier region. The inelastic tunneling represents an additional channel that can open up only when the bias voltage eV is greater than the energy lost to a particular vibrational mode $\hbar\omega_i$. Therefore the tunneling conductance has a step up at $eV = \hbar\omega_i$. The step in conductance is small, typically less than 1%, so the derivative of the conductance is taken to eliminate the large background conductance[7]. The vibrational spectra of the normal modes of vibration ω_i of the impurities in the junction region appear as peaks in d^2V/dI^2 at voltages $V = \hbar\omega_i$. Cooling of the sample is required since the width of the inelastic tunneling step is limited by thermal broadening of the Fermi surfaces of the two metals. The instrumental bandwidth contributed by thermal broadening is 5.4kT: about 1.8 meV at 4.2°K[7-9].

A tunneling spectrum of an Al-Al$_2$O$_3$-Ag tunnel junction doped with 4-pyridine-carboxylic acid appears in Fig. 1a. This spectrum has had a large elastic tunneling background numerically subtracted out. There are several reasons to believe that no more than a monolayer coverage existed in these samples. 1) Tunneling resistances increase exponentially with barrier thickness.[10] The resistances of doped junctions were ~ 100 times larger than the resistances of undoped junctions on the same substrate: This corresponds to an additional effective tunneling thickness of a few tenths of nanometers, consistent with monolayer coverage. 2) Radioactively labeled molecules have been doped onto oxidized aluminum films in our and other laboratories[11] using the same techniques as for these experiments: the surface coverage as a function of solution concentration measured in this manner saturated at near monolayer coverage (e.g. 7x10(p14) p cm(p-2) p for benzoic acid). 3) The vibrational spectra lost the characteristic bulk mode at ~ 1700 cm^{-1} due to the C=O stretching vibration of the acid or aldehyde group. Instead the symmetric and antisymmetric COO$^-$ stretching vibration of the corresponding salt appeared[12]. The tunneling spectrum indicated that all of the molecules present were bonded to the alumina: no more than a monolayer was present.

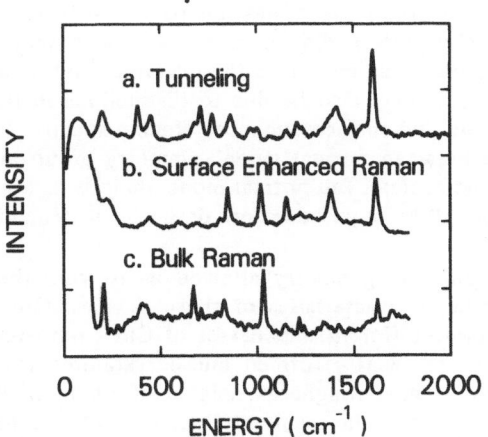

Figure 1 a. Inelastic electron tunneling spectrum of 4-pyridine-carboxylic acid absorbed on the oxide layer of an Al-Al$_2$O$_3$-Ag tunnel junction. A large elastic tunneling background has been subtracted out. b. Raman spectrum from 4-pyridine-carboxylic acid absorbed on the oxide layer in an Al-Al$_2$O$_3$-Ag tunnel junction fabricated on 80 nm CaF$_2$ (on glass) to provide surface roughness. c. Bulk Raman scattering spectrum from a solid pellet of 4-pyridine-carboxylic acid.

Our Raman spectra were taken at room temperature, with 10-50 mwatts incident power line focused, with a standard 3/4 meter double monochromator and photon counting. Care was required to avoid damaging the molecular layer with excessive laser power; some molecular layers were more susceptible to damage than others. A few Raman spectra were taken with the substrate at low temperatures (<2K): no qualitative differences from the room temperature spectra were observed.

The Raman scattering spectrum of 4-pyridine-carboxylic acid in an Al-Al$_2$O$_3$-Ag junction fabricated on a rough substrate is shown in Fig. 1b. The substrate was roughened by evaporation of $\sim$ 80 nm of CaF$_2$ on a glass slide prior to sample fabrication. Also included for comparison (Fig. 1c.), is the Raman scattering spectrum of a pressed pellet of 4-pyridine-carboxylic acid, which is a solid at room temperature. In this spectrum a large elastic scattering background has been subtracted out. The Raman scattering signal intensity per molecule of the 1598 cm^{-1} mode of 4-pyridine-COOH in the tunneling junction is $\sim$ 5x10^4 times larger than that for the molecule in bulk. Since the electromagnetic screening length in Ag is $\sim$ 20 nm, the effective cross section for this vibrational mode is $\sim$ 3x10^6 times larger in the junction than in bulk.

In addition to the truly remarkable absolute enhancements, there are also interesting differences in the intensity patterns for these three spectra. Tunneling spectroscopy is sensitive to both infrared and Raman active vibrational modes[13-15]. Although 4-pyridine-carboxylic acid has no inversion center of symmetry, and therefore has no separation of Raman active and infrared active vibrational modes, one might expect the tunneling spectrum to exhibit a larger number of strong vibrational modes than the Raman spectrum. This indeed was the case. All of the strong modes observed in the Raman spectrum of the tunneling junction had their counterparts in the tunneling spectrum, but the reverse was not true. Further, the surface enhanced Raman scattering intensity pattern did not agree well with that of the molecules in bulk. Part of this difference may be due to the different environment in the tunneling junction geometry, but part may also be due to distinctions in the scattering mechanisms responsible for bulk and surface enhanced Raman spectra. Initial analysis shows no obvious correlation between normal mode symmetry group and relative intensity enhancements for this molecule: a full normal mode analysis of a molecule with known orientation and position will be required before definitive statements can be made.

The tunneling junction geometry allowed us to vary the surface roughness without varying any other characteristics of these devices. One method of varying roughness was to evaporate different thicknesses of CaF$_2$ onto the glass slides before junction fabrication[16]. The CaF$_2$ formed microcrystallites with r.m.s. correlation lengths of $\sim$ 50 nm and r.m.s. roughness heights of up to 5 nm, as measured by reflectivity data[16]. Fig. 2 shows the resultant Raman spectra from a set of identically fabricated Al-Al$_2$O$_3$-Ag junctions, doped with 4-pyridine-COOH, on different mass thicknesses of CaF$_2$. Experiments with radioactively labeled benzoic acid molecules showed that the surface concentration only increased by a factor of 3 as the CaF$_2$ thickness was increased over this range: the Raman scattering intensity increased by more than a factor of 50.

We believe that junctions with optical scale roughness had larger Raman scattering signals because the incident radiation, as well as the Stokes scattered radiation, could couple more efficiently to electromagnetic resonances of the junction structure[17]. Electromagnetic resonances on planar metal surfaces cannot couple to external radiation because their speed of propagation is slower than that of light: momentum and energy cannot both be conserved in the transition. Momentum conservation can be satisfied if the surface is rough: The surface plasmon gains or loses momentum to surface roughness to radiate light. In the limit of large amplitude roughnesses, non-radiative electromagnetic resonances become localized radiative modes, which are the subject of several of the chapters in this volume. The view that electromagnetic resonances are involved in Raman scattering from tunnel junctions is supported by the dependence of the Raman scattering intensity on CaF_2 thickness. The Raman scattering intensity from a set of junctions evaporated on different thicknesses of CaF_2 saturated at a CaF_2 mass thickness of ~ 80 nm: this is the same thickness at which saturation occured for the reflectivity dip due to surface plasmons at 3.7 eV[16], and for the emission intensity from surface plasmons excited by tunneling electrons in light emitting tunnel junctions[18]. (More will be said about light emission from tunnel junctions below.) The saturation most likely occured because damping of the resonances by roughness compensated for any additional coupling to light[19].

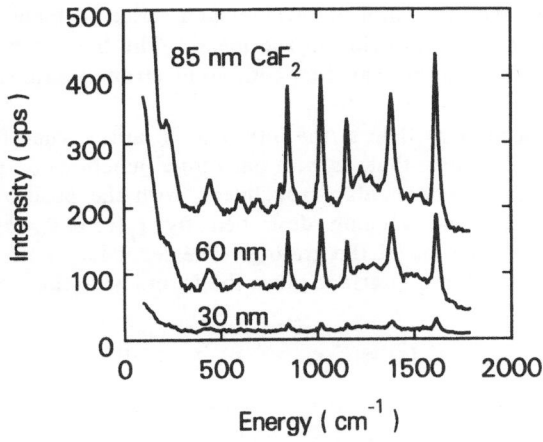

Figure 2. Raman spectrum of 4-pyridine-carboxylic acid absorbed on the oxide layers of Al-Al_2O_3-Ag junctions fabricated of 3 different thicknesses of CaF_2 on a glass substrate. As the CaF_2 underlayer was made thicker the tunnel junctions, which conformed to the topology of the substrate, became rougher, and larger Raman scattering signals were observed.

While the topology of randomly roughened surfaces can be characterized statistically, it is often difficult to make quantitative comparisons of theoretical models with experiment for these samples. We have therefore worked primarily with holographically produced, sinusoidal profile grating substrates[20]. These substrates allowed us to fabricate the tunnel junctions with only one Fourier component of known amplitude and periodicity at a time. They also allowed us to tune into the electromagnetic resonances of the tunneling structures, merely by changing the angle, energy, or polarization of the incident laser beam.

The gratings were fabricated on (100) orientation, polished Si wafers oxidized to a thickness of 80 nm. The wafers were coated with 100-200 nm of photoresist. The wafers were exposed to a hologram generated from the 325 nm line of a HeCd laser. The grating periodicities were varied by adjusting the beam geometry; the grating amplitudes were controlled by varying the exposure and development times. Grating periodicities ranging from 120-1250 nm were produced, with amplitudes ranging from 5-100 nm. The tunneling junctions were then fabricated on the photoresist films in the normal way. The grating periodicities of the completed junctions were obtained by measuring the first order diffraction angle. The grating amplitudes were obtained by fitting the first order diffraction efficiency in p-polarization to an expression by Heitman[21], or by electron microscopy.

Once the junction topology is known, it is relatively straightforward to characterize its electromagnetic resonances. The resonances are fairly complex for the multi-film geometry of a tunneling junction. Consider first the case of a plane electromagnetic wave incident on a metal-air interface. The metal (dielectric constant $\varepsilon(b1)$ $b(\omega)$) is assumed to take up the half plane $z > 0$. For a planar interface the solution to Maxwell's equations takes the form of incident and reflected plane waves in air, and an attenuated wave in the z direction in the metal. The boundary conditions can be matched and Maxwell's equations satisfied with no electromagnetic resonances.

However, a secondary field at the interface $\overline{E}_g$ arises when light is incident on a rough surface[22,23]. Assume that, as with our tunnel junctions on gratings, the metal surface has one Fourier component of roughness, with the position of the interface between the metal and the vacuum described by $\xi_g(\overline{r}) = \xi_g \exp(i\overline{g}\cdot\overline{r}_{||})$, where $|\overline{g}| = 2\pi/a$, if a is the period of the grating. The secondary field required to match boundary conditions across the interface takes the form inside the metal:

$$\overline{E}_g(\overline{r}) = \overline{E}_g e^{i\overline{K}_g \cdot \overline{r}_{||}} e^{-\gamma_g z} \quad z>0 \tag{1}$$

$$\gamma_g = (K_g^2 - \frac{\omega^2}{c^2}\varepsilon(\omega))^{1/2} \quad Re(\gamma_g > 0 \tag{2}$$

$\overline{K}_g$, the wavevector of the surface plasmon parallel to the interface is given by the momentum conservation condition: $\overline{K}_g = \overline{k}_t + \overline{g}$, where $\overline{k}_t$ is the wavevector of the incident light parallel to the surface. The secondary field takes the form outside the metal:

$$\overline{E}'_g(\overline{r}) = \overline{E}'_g e^{i\overline{K}_g \cdot \overline{r}_{||}} e^{\Gamma_g z} \quad z<0 \tag{3}$$

where

$$\Gamma_g = (K_g^2 - \frac{\omega^2}{c^2})^{1/2} \tag{4}$$

Matching boundary conditions across even this very simple non-planar interface is quite complex, but things can be simplified considerably by expanding to first order in $k_z\xi_g$, as has been described in detail by Jha et. al.[23] This expansion has been shown to give results in good agreement with numerical analysis of diffraction efficiencies for $K_g\xi_g < 1$[24]. Expansion to first order in $k_z\xi_g$ will not, of course, allow modeling of the the effects of grating structure on resonance lifetimes and dispersion curves. Therefore this first order theory should be applied only to gratings with relatively small amplitudes ($\sim$ 30 nm or less for an 800 nm grating). In the first order approximation the fields at the interface become resonantly large when the condition $\varepsilon_1 \Gamma_g + \gamma_g = 0$ is satisfied. This condition can be written as:

$$K_g = \frac{\omega}{c} \left(\frac{\varepsilon_1}{(\varepsilon_1 + 1)} \right)^{1/2}. \tag{5}$$

This is the familiar dispersion curve of the " fast " mode (shown in Fig. 3a). The mode is close to the light line ($K_g = \omega/c$)) and is referred to as a surface plasmon polariton if the frequencies are well below the maximum surface plasmon frequency ($\omega_p/\sqrt{2}$ in a free electron model). The dispersion curve bends away from the light line near the maximum surface plasmon frequency. Near the light line, the surface plasmon fields extend tens of nanometers into the vacuum. When the surface plasmons move away from the light line they become progressively more localized at the interface.

At resonance, the field enhancement factor, the ratio between the p-polarized secondary (E_{gz}^p) and incident (E_{gz}^p) fields, becomes:

$$\frac{E_{gz}^p}{E_{iz}^p} = \frac{4 C \xi_g K_g |\varepsilon_1|^{3/2}}{i \operatorname{Im}(\varepsilon_1)}, \tag{6}$$

where C is a factor of order 1 which depends on the scattering geometry. This is just as one would expect: at resonance the field strengths are proportional to the coupling factor $\xi_g K_g$ and inversely proportional to the loss in the metal.

The situation is somewhat more complex for the tunneling junction structure. Although complete calculations of the field strengths for junction structures on gratings under optical irradiation have not appeared, the similar problem of light emission from tunnel junctions has been worked out by Laks and Mills[25]. The resonance condition analogous to Eq. (5) becomes $D(\overline{K}_g, \omega) = 0$, where $D(\overline{K}_g, \omega)$ is a complex expression, given by Eq. 15 of Ref. 26.

The dispersion curve defined by $D(\overline{K}_g, \omega) = 0$ has two branches. One branch corresponds to the Ag-air interface mode and has a dispersion curve which is described by Eq. 5 very closely. The other branch corresponds to fields localized at the metal-oxide interfaces, and is called the junction or slow mode (Fig. 3b). This mode can only couple to radiation at infrared frequencies for the oxide thicknesses and grating periodicities we used. However, the junction mode can couple to light at optical frequencies if small wavelength random roughness (of order 10 nm or less) is present, or if the dielectric spacing between the metal electrodes is thick ($\sim$ 10 nm for roughness periodicities of $\sim$ 100 nm): field enhancements due to the slow modes should be considered as a possible contributor to the surface enhanced Raman effect in randomly roughened tunnel junction structures and in spacer experiments[27]. The incident laser beam couples into the fast surface plasmon polaritons in our controlled topology experiments.

The electromagnetic resonances observed in tunnel junction structures are somewhat broader and have correspondingly lower total field enhancements than for the idealized Ag-vacuum interface. There are several reasons for this broadening. The first is that Al[28] has a much larger imaginary part to its dielectric response than Ag[29].

The second reason is that radiative damping and scattering of the surface plasmons by residual random roughness, as well as the grating structure, tend to limit the surface plasmon lifetimes[19]. We can estimate the relative importance of these additional loss mechanisms by fitting the resonance widths predicted by Laks and Mills[25] to experiment, with the imaginary part of the Ag dielectric response as a fitting parameter.

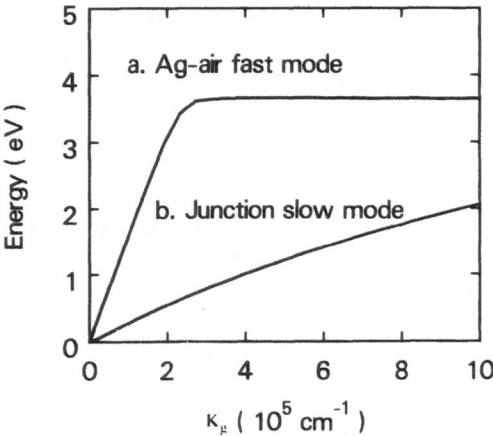

Figure 3. Theoretical dispersion curves for a. the Ag-air interface " fast " plasmon mode and b. the Al-Al$_2$O$_3$-Ag " slow " junction mode for our tunneling junction structures, with an oxide thickness of 3 nm and an Ag thickness of 20 nm.

The resonant energies and energy widths can be measured in a variety of ways. We have used three. The first and most direct method is to measure the specular reflectivity of the tunnel junction as a function of frequency, for a fixed incident angle[30]. The reflectivity undergoes a sharp dip at a frequency such that the momentum matching condition to surface plasmon polaritons can be satisfied.

A second method for measuring the resonant properties of these tunnel junctions is to observe the light emitted when they are biased to 1.5-4.0 volts[26,31]. Tunneling electrons excite the electromagnetic resonances of these structures, and the modes couple to light through the grating periodicity. Tunneling emission spectra are shown in Fig. 4 for a tunnel junction on a 5 nm amplitude, 820 nm periodicity grating. The tunnel junction has been biased to 2.5 volts at ~ 2 mA, direct current. These measurements have been made with very small collection angle apertures, to insure that the observed linewidths are intrinsic. The emission spectrum has a broad background with a linear cutoff at 2.5 eV: photons cannot be emitted with energies higher than the maximum tunneling electron energy. We attribute this background to residual roughness in the tunneling junction structure; possibly causing emission from the junction modes or localized modes through random short wavelength roughness, or from the fast modes through random long wavelength roughness. Superimposed on this background are sharp, angle tuneable emission peaks. The large intensity peak that moves to higher energies at higher angles corresponds to surface plasmon polaritons exchanging momentum with the grating periodicity (one " ruffion(p32) p "). The small intensity peak that moves to lower energies at higher angles corresponds to momentum exchange with two " ruffions ".

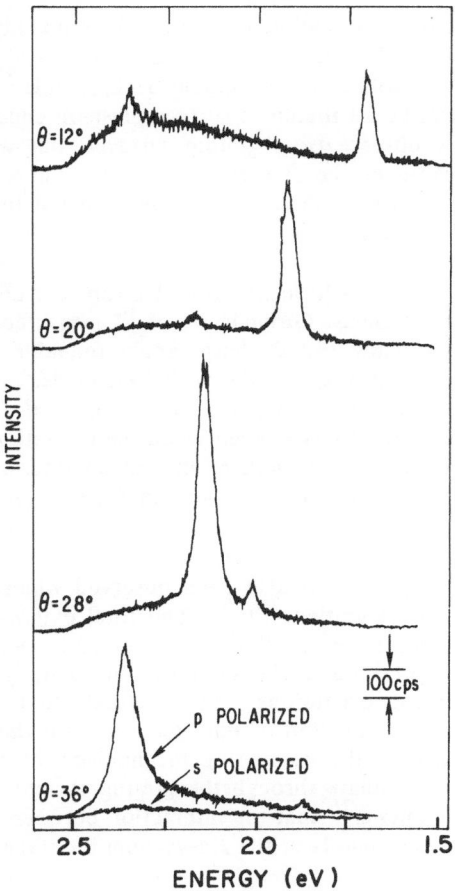

Figure 4. Light emission from an Al-Al$_2$O$_3$-Ag junction on a 5 nm amplitude, 820 nm periodicity photoresist grating. The junction was biased at 2.5 eV, at ~ 2 mA direct current. The light emission was primarily p-polarized, had an onset at the maximum tunneling energy (2.5 eV), and had sharp angle-tuneable emission peaks superimposed upon a broad continuum background. The sharp peaks resulted from coupling out of the Ag-air fast surface plasmon polariton through the grating periodicity.

A third method for observing the resonance energies and widths of electromagnetic resonances in our tunneling junction structures involves optical excitation[4].

The incident laser beam is set to the resonance angle. At resonance a large fraction of the incident power goes into exciting surface plasmons. The surface plasmons decay in energy and reradiate. The emission spectrum due to optical excitation of a metal film on a grating is very similar to that from tunneling excitation: there is a broad featureless background and sharp, angle tuneable emission peaks. In fact, the surface plasmon emission dispersion curves and linewidths are the same for optical excitation as for tunneling electron excitation.

Several investigators have pointed out that, in randomly roughened systems, those conditions for which the surface enhanced Raman scattering signal was the largest were also associated with large continuum backgrounds.[33] At least part of this continuum background can be attributed to surface plasmon emission analogous to the sharp emission peaks we observe from grating surfaces, but with a broad emission spectral distribution because of the random roughness. Surfaces which can couple light into and out from surface plasmons should be expected to give both large SER scattering and large surface plasmon emission signals.

There is good reason to believe that hot electrons are an intermediate state in the relaxation of surface plasmons. Girlando et al.[34] measured the optically excited emission from gold films. They found sharp, angle tuneable emission peaks. The emission peaks were strongly damped as the peak frequencies approached the surface plasmon frequency: no sharp emission peaks were observed above this frequency. These emission peaks could be observed even when the excitation laser was above the maximum surface plasmon frequency, indicating that an intermediate state that was not strongly damped by the interband transitions in Au was involved in the emission process.

Surface plasmon emission has also been observed when high energy electron beams were incident on metal gratings[30,35]. The total external quantum efficiency for light emission from thin films on gratings, if expressed in terms of the total power radiated, divided by the total power incident on the film, is $\sim 10^{-6}$ for optical, electron beam, and tunneling excitations. This suggests that all three experiments involve the same phenomena, injection of hot electrons into the films. The hot electrons relax in energy through the competing mechanisms of phonon and plasmon emission, and the plasmons radiate through the grating structure. There is evidence from the tunneling measurements[26] that the interaction between the hot electrons and the surface plasmons occurs primarily at the Ag-vacuum interface, as opposed to in the bulk of the Ag film.

The surface plasmon dispersion curves for tunneling junction structures are obtained by measuring the peak emission (or reflectivity dip) energies as a function of emission (or reflection) angle. The momentum matching condition for light in the plane defined by the grating periodicity wavevector and the normal to the junction surface is:

$$K_g = k_t + \sigma g = \frac{\omega}{c} \sin \theta + \sigma g \qquad (7)$$

where θ is angle between the light wavevector and the surface normal, and $\sigma = \pm 1, \pm 2,.$ is the scattering order: it describes how many " ruffions " are exchanged with the surface plasmon before it couples to external radiation. It is customary to display the data as a plot of ω vs. k_t. The possible resonance conditions in the laboratory frame are displayed by displacing the surface plasmon dispersion curve by σ. Such a plot is shown in Fig. 5 for light emission from a junction on a grating of 81.5 nm amplitude, 815 nm periodicity. The error bars represent the upper and lower half-maximum intensity points for this relatively large amplitude grating. Note that the $\sigma = -1$ peaks, which result from coupling to first order in the grating periodicity (one ruffion), were narrower than the two ruffion peaks. The two branches were strongly mixed at the zone center $g/2 = 3.95 \times 10^4$ cm^{-1}. The solid lines in Fig. 5 were calculated from Eq. 5. The dispersion curve calculated from the full expression[25] was indistinguishable from the curve shown in this energy range. Although the experimental points were lower than predicted theoretically because of interactions of the surface plasmons with the grating structure, we can assign the observed resonances to the Ag-air fast surface plasmon polariton.

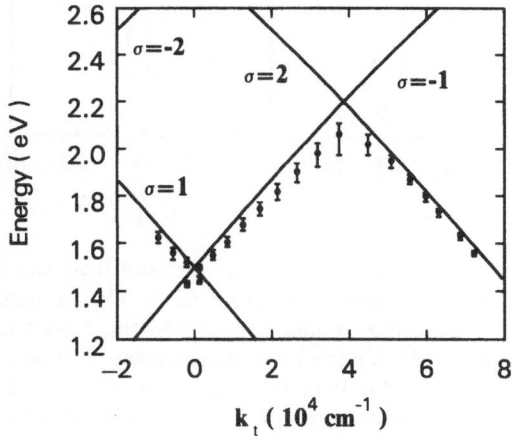

Figure 5. Measured Ag-air interface surface plasmon polariton dispersion curves (in the reduced zone scheme) for an Al-Al$_2$O$_3$-Ag tunnel junction on a 81.5 nm amplitude, 815 nm periodicity grating. The error bars represent the upper and lower half maximum intensity points for the emission peaks. The solid lines were generated from Eq. 5. Surface plasmon polaritons are scattered into external radiation to order σ in the grating periodicity.

Fig. 6 shows the dependence of the Raman scattering signal for the 1598 cm^{-1} line of a 4-nitrobenzoic acid monolayer in an Al-Al$_2$O$_3$-Ag tunnel junction on the incident and scattering angles. The angles were defined with respect to the normal in the plane set by the grating periodicity wavevector and the normal to the junction surface. The grating periodicity was 850 nm, the incident laser frequency was 514.8 nm. There was a sharp electromagnetic resonance (as evidenced by a sharp reflectivity minimum) at $\theta(\text{bi})$ b = 24° for the incident laser wavelength. If the Stokes scattering angle was held fixed and the Raman scattering intensity was plotted as a function of the incident angle, there was an increase of at least a factor of 50 between normal incidence and incidence at the resonance angle. There was a somewhat smaller resonance, of a factor of 5 between off- resonance and on-resonance intensities, when the excitation laser was held at resonance, and the Stokes scattered radiation was angularly analyzed. Combining these two factors, there was a total enhancement of ~ 200 when the electromagnetic resonance conditions for the junction structure were satisfied.

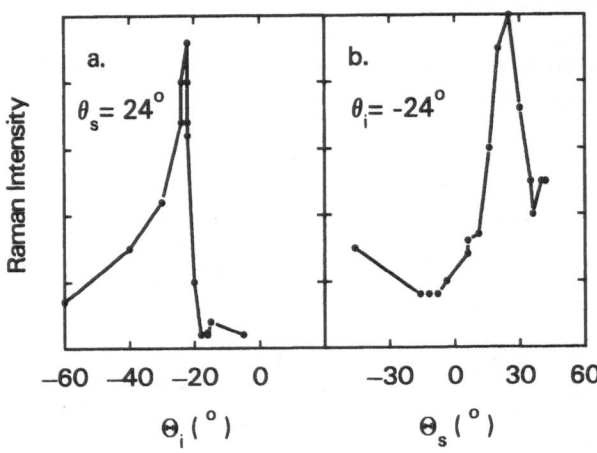

Figure 6. Plot of the relative Raman scattering intensity from the 1598 cm^{-1} line of 4-nitro-benzoic acid on the oxide layer of an Al-Al$_2$O$_3$-Ag junction on an 820 nm periodicity grating. There is resonant coupling to the Ag-air interface surface plasmon polaritons at θ_i = 24° for the laser energy used. a. If the collection angle relative to the junction normal is held fixed at resonance, the Stokes scattering intensity undergoes a strong resonance as a function of incident laser angle. b. A somewhat weaker resonance is observed as a function of scattered angle if the incident laser angle is held at resonance.

The Stokes scattered power per steradian from a molecule can be written as[23]:

$$\frac{dP_s}{d\Omega} = \frac{\omega^4}{8\pi c^3} \, | \, \hat{e}_s \cdot \overline{G}_s : \frac{\partial \overline{\alpha}}{\partial Q} : \delta Q \overline{E}_{loc} \, |^{\,2} \tag{8}$$

where $\hat{e}_s$ is the unit vector in the scattered direction, $\overline{G}_s$ is a tensor which describes the conversion of the near zone Stokes field to the radiation field outside the junctions, $\partial\overline{\alpha}/\partial Q$ is the derivative of the local polarizability tensor with respect to the normal coordinate displacement δQ, and $\overline{E}_{loc}$ is the electric field at the molecular site. The treatment of the Stokes scattering tensor is quite complex. Aravind, Hood, and Metiu[36] have shown that angular resonances should be observed in the emission from a dipole located near an Ag grating, just as was shown in Fig. 6. This resonance is apparently not particularly strong, since the signal on resonance in only a factor of 5 larger than that off resonance. We argue that the bulk of the enhancement in Raman cross sections in the tunneling junction geometry arises from 1) the enhancement of the local field strength at the molecular site over the incident field, and 2) an enhancement of the polarizability derivative with respect to the molecular vibrational normal coordinate for the molecular-Ag interface over that of the molecule alone.

We first wish to address the question: is the local field enhancement large enough to explain our large Raman cross sections for molecules in the tunneling junction geometry? The local field enhancement can be obtained from the expressions of Jha et al.[23] if they are corrected for the fact that the molecular layer is separated from the Ag-air interface by ~ 20 nm of Ag. To a good approximation, this correction factor is given by:

$$\frac{E_{gz}^{p}(z = -L)}{E_{gz}^{p}} = \frac{e^{-\gamma_g L}}{\varepsilon_2} \tag{9}$$

where ε_2 is the dielectric constant of the oxide (often taken to be 3) and γ_g^{-1} is the attenuation length in Ag (given by Eq. 2).

Direct comparison of experiment to theories that are first order in $K_g \xi_g$ should be made only for junctions on gratings with very small amplitudes. An Al-Al$_2$O$_3$-Ag junction, doped with 4-nitro-benzoic acid, on a 23 nm amplitude, 925 nm periodicity grating, had an enhancement of the Raman cross-section for the 1598 cm^{-1} line of ~ 10^3. For this measurement the incident laser wavelength was 568.2 nm, and the incident laser angle and polarization were set to excite the surface plasmon resonance. An overestimate for the local field enhancement at the molecular site can be obtained by inserting the smooth film dielectric constants of Ag, along with the measured values of the grating amplitude and periodicity into Eq. 6, and using the correction for the junction dielectric screening (Eq. 9). This estimate for the field enhancement squared is ~ 250. One might therefore be tempted to attribute the entire enhancement to electromagnetic resonances. However this is a large overestimate, because the bulk dielectric constants of Ag were used, while the junction structure includes 40 nm of Al, which is more lossy at the frequencies of interest than Ag.

To get an estimate of how much weaker the local electric fields should be at resonance in a junction structure than in pure Ag, we modeled the light emission peaks from the fast modes of tunnel junctions on gratings. Fig. 7 shows the experimentally determined linewidths (solid dots) from a junction on an extremely low amplitude (~ 5 nm) 820 nm periodicity grating. The solid curve is the linewidth predicted for the actual Al-Al$_2$O$_3$-Ag structure by the theory of Laks and Mills[25], using the smooth film dielectric constants of Johnson and Christy[29] for Ag, and of Powell[28] for Al. The dashed curve is the prediction for an all Ag structure. Ag has an imaginary part of its dielectric constant of Im(ε_1) = 0.4 at 2.18 eV. If we fit the predicted linewidths with an imaginary part of the dielectric response for the composite film, we find Im(ε_{eff})= 0.7. This value would lead to a predicted field enhancement squared of ~ 80 for our 223 nm amplitude grating. This is probably also an overestimate, since the actual observed linewidths for the 5 nm grating were 30% larger than predicted by theory: those for the 223 nm grating were about twice as large as predicted by theory.

The observed emission linewidths for even the 5 nm grating are probably broader than predicted because of scattering from residual surface roughness as well as imperfect film quality. The emission linewidths grew broader as the grating amplitudes increased because of radiation damping and additional scattering due to the grating structure. All of these processes would imply further damping of the surface plasmon polariton resonance. Recently Weber and Ford[37] have presented a calculation of the maximum field enhancement to be expected from coupling of external radiation to surface plasmon polaritons. They obtain a maximum field enhancement squared of about 300 using the dielectric constants of Johnson and Christy[29] for an Ag-air interface. This number is consistent with the largest change in observed Raman scattering intensities in our tunneling junction structures when the incident laser was on and off resonance. In any case, the local field enhancements due to excitation of electromagnetic resonances in the junction structure were not large enough to explain our observed total enhancements.

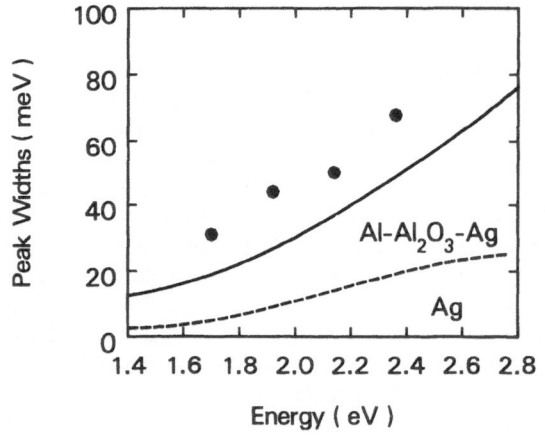

Figure 7. Tunneling emission peak widths as a function of emission energy for the junction of Fig. 5 (solid dots). The solid curve is the prediction of Laks and Mills (Ref. 20) for the full tunneling junction structure. The resonances would be much narrower for a simple Ag-air interface (dashed curve). We model the resonant field strengths by fitting the resonant energy widths.

We might argue that the extra enhancement could come from resonant outcoupling of the Stokes radiation through the surface plasmon polaritons. However, as can be seen in Fig. 6b, this resonance is not particularly strong, with only about a factor of 5 between on and off resonance. When we consider that the Stokes scattered radiation has to propagate through the Ag film, losing a factor of 7 or so in intensity, to reach the detector, it appears unlikely that this mechanism can increase the total scattering enhancement much over the classical field enhancement estimate of <80 for this sample.

The view that some enhancement mechanism besides electromagnetic resonances must be involved in Raman scattering from tunnel junctions is supported by experiments on different molecular species. We found that junctions doped with 4-pyridine-carboxylic acid, 4-nitrobenzoic acid, and 4-amino-benzoic acid gave large Raman scattering signals, but that junctions doped with formic acid and benzoic acid gave very small Raman signals, smaller than the differences in the molecular cross sections could account for. It might be argued that these differences could be due to some small scale roughness that depended on the identity of the molecular layer that the Ag was coating. We did two experiments to explore this possibility. First, light emitting tunnel junctions were fabricated with and without a molecular monolayer of 4-nitrobenzoic acid included. Both the sharp emission peaks and the broad emission background were the same for these two junctions when the tunneling current densities were normalized out. This meant that there was no difference in the coupling of surface plasmons to external radiation in the two junctions, and therefore no difference in the microstructural properties of the films on a size scale important for electromagnetic resonances. The second experiment involved mixing molecules known to have small enhancements with molecules with large surface Raman enhancements in the same junction. Fig. 8 compares the Raman spectra from three $Al-Al_2O_3-Ag$ tunnel junctions on the same glass slide, coated with 160 nm of CaF_2, but doped with: a. nitrobenzoic acid, b. a 1:3 mixture of nitrobenzoic acid and benzoic acid, and c. pure benzoic acid. The Raman intensities decreased monotonically with nitrobenzoic acid coverage. No signal could be observed for the pure benzoic acid sample. The relative count rates from a set of Al films on 160 nm CaF_2 on glass, for different solution ratios of radioactively labeled benzoic acid and unradioactive nitrobenzoic acid showed a linear relationship between count rate and benzoic acid concentration. This indicated that the molecules bonded to the Al_2O_3 surface in the same proportions as were present in solution. This is not surprising, since the binding energies should be very similar for these two molecules. The Raman spectra of benzoic acid and nitrobenzoic acid were measured in pellets under the same conditions as used for the tunneling junction measurements. The intensity under 647.1 nm irradiation of the 1598 cm^{-1} line of nitrobenzoic acid was comparable to the intensity of the 991 cm^{-1} line of benzoic acid. When the relative cross sections for these molecules are taken into account, the enhancement factor for a nitrobenzoic acid monolayer is at least 50 times larger than that for a benzoic acid monolayer in a tunneling junction geometry. The observation of different enhancement factors for different molecules within the same tunneling structure make it unlikely that the additional enhancement is due to localized electromagnetic resonances.

Our tunneling measurements have provided a clue to the additional enhancement mechanism. We have noted the correlation that the molecular monolayers that had large enhancements when included in the tunneling junction structure, also had large elastic tunneling backgrounds. This large background was due to a large change in the dynamic conductance with applied bias over the range required for tunneling spectroscopy: 0-0.5 volts. This large change in conductance has been associated with small effective barrier potentials between the molecular layer and the Ag top electrode in numerical fits of the current-voltage characteristics of doped tunnel junctions[38].

A small barrier potential between the Ag and the molecular layer can help to increase the local polarizability derivative at the Ag- molecular interface[23,39]. To understand how this can happen, consider the simple model of a step in the one-electron potential at z=0, of height ϕ_0 above the Fermi surface of the metal. (This subject will be treated in greater detail by S.S. Jha in this volume.) The surface electronic charge densities decay into the molecular layer with a z dependence given by:

$$n_0(z) = \exp\left(2\int_0^z K_0(z')dz'\right) \quad z<0 \tag{10}$$

where $K_0(z') = ((2m(\phi_0 + \phi_m)/\hbar^2)^{1/2}$, ϕ_0 is the time averaged one electron potential seen by the conduction electrons in the molecular layer, and ϕ_m is the oscillating potential due to the molecular vibration. The surface charge density decay into the bulk of the metal is given by a similar expression. The oscillating molecular potential modulates the decay rate of the surface charges into the molecular layer and into the bulk metal, and thus modulates the surface dipole moment. (We call this the modulated surface dipole moment (MSDM) mechanism). Since the polarizability of the Ag-molecular layer interface depends on both the magnitude and the spatial distribution of the interface charges, the system polarizability will oscillate at the molecular vibrational mode frequency.

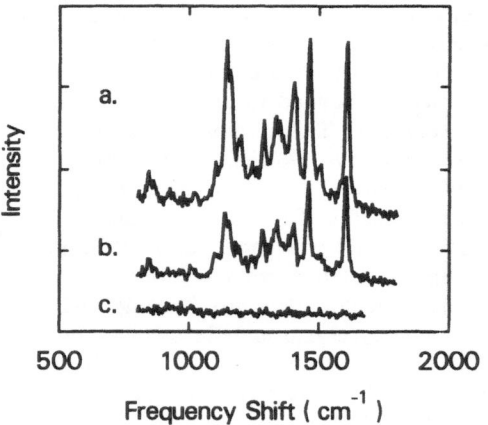

Figure 8. Raman scattering spectra from molecular monolayers absorbed on the oxide layer of Al-Al$_2$O$_3$-Ag tunnel junctions fabricated on 80 nm CaF$_2$ on glass. The molecular layers are: a. 4-nitrobenzoic acid, b. a 1:3 mixture of 4-nitrobenzoic acid and benzoic acid, and c. benzoic acid.

If we take the simple model of a single partial charge Ze located a distance a from the metal surface, with vibrational amplitude δR, allow it to interact with the electron gas through a Coulomb potential[15], and take the limit $K_g << K_0$, we find that the local polarizability derivative due to the component of surface charge density outside the surface z=0 is given by

$$\frac{\partial \beta}{\partial \overline{R}} = \frac{-Z_{eff}\, a_0^2}{2(K_0 a_0)^3}\left(\frac{K_i}{K_0 + K_i}\right)e^{-2K_0 a}(1 + 2K_0 a) \tag{11}$$

where k_f is the bulk metal Fermi wave vector, and $a_0 = \hbar^2/me^2$ is the Bohr radius.

If the charge is oscillating parallel to the surface, there is no large short range polarizability enhancement. The polarizability derivative from MSDM will be large relative to the intrinsic molecular polarizability only if $K_0 a_0$ and $K_0 a$ are both small: the effective barrier height ϕ_0 and the charge- image plane distance a must be small. The polarizability derivative then falls off rapidly on a scale of angstroms with molecule-metal spacing. Beyond about 1 nm, the long range interaction due to the modulation of the charge inside the metal surface dominates, but it is of the same order of magnitude as typical molecular polarizabilities. A different treatment of the long-range term has been presented by A. Otto.[40]

Recently J.C. Lau and R.V. Coleman[38] have fit the current-voltage character-istics of Al-Al_2O_3-Ag and Al-Al_2O_3-Pb junctions doped with a number of different molecular monolayers. They used a model barrier potential with 2 barriers- one corresponding to the oxide and one to the molecular layer. They found that: 1) the effective metal-molecular layer barriers varied from $\sim$ 8 eV to < 1 eV for different molecular layers, 2) junctions with large curvature in their current-voltage characteris-tics had small effective barriers between the top electrode and the molecular layer, 3) junctions with Ag had tunneling barriers typically a factor of 2 smaller than those with Pb barriers, and 4) lower effective tunneling barriers were associated with anomalously small tunneling intensities for the C-H stretch modes. All of the molecular layers that we have tried that had large Raman scattering signals also had highly curved current-voltage characteristics, as well as anomalously small C-H stretching mode intensities.

If we take the low but still reasonable effective barrier height of 0.5 eV and a charge-image plane distance of 0.1 nm, the polarizability derivative due to MSDM is $\sim$ 12 $Z_{eff} a_o^2$. Molecular polarizability derivatives are typically 10^{-16} cm^2 = 3.55 a_o^2 [41] The effective partial charges associated with molecular vibrations, as determined by fitting infrared[42] and tunneling[15] spectroscopies, are typically 0.1 e. Taking this value would yield small enhancements for reasonable values of ϕ_0. However, since the modulation of the surface dipole moment occurs over a relatively short length scale, it may be more appropriate to take the screened nuclear charge for Z_{eff}, in which case large polarizability derivative enhancements could be derived from the modulation of the surface dipole moment.

The MSDM model is purely phenomenological, and as such, has little pre-dictive value. It does, however, provide a simple explanation for the experimental correlation between low effective barrier heights and large Raman scattering cross sections in tunnel junction structures. Further, it may be useful to point out that this model, like all short range models, predicts that the symmetry of a molecule is reduced at the interface, such that previously symmetry forbidden vibrational modes should be observable[43]. While the simple model predicts that modes with vibrational motion normal to the interface should appear more intense in surface enhanced Raman scattering than modes with motion parallel to the interface, it is likely that a more detailed treatment of the molecule-electron interaction would modify this result to reduce the predicted asymmetry, as has been shown for tunneling spectroscopy[15].

In conclusion, Raman scattering was observed from molecular layers included at the oxide-metal interface of metal-insulator-metal tunnel junctions. The enhancements of the Raman scattering cross sections in tunnel junctions over those in bulk were up to $\sim 5 \times 10^4$. Part of this enhancement can be attributed to the excitation of electromagnetic resonances in the junction structure, but an additional enhancement mechanism is required to explain the large intensities observed. One possible mechanism for this enhancement is the modulation of the surface charge distribution by interaction with the vibrating molecular potential. These experiments show that tunnel junction structures provide an easily fabricated, readily characterizable, and durable system for the study of surface enhanced Raman scattering.

References

1. J.C. Tsang and J.R. Kirtley, Anomalous Surface Enhanced Molecular Raman Scattering from Inelastic Tunneling Spectroscopy Junctions, Solid State Comm. 30:617(1979).
2. J.C. Tsang and J.R. Kirtley, Raman Spectroscopy of Molecular Monolayers in Inelastic Electron Tunneling Spectroscopy Junctions, in " Light Scattering in Solids, " J.L. Birman, Herman Z. Cummins, and Karl K. Rebane, ed., Plenum, New York, (1979) p.499.
3. J.C. Tsang, J.R. Kirtley, and J.A. Bradley, Surface Enhanced Raman Scattering and Surface Plasmons, Phys. Rev. Lett. 43:772 (1979).
4. J.C. Tsang, J.R. Kirtley, and T.N. Theis, Surface Plasmon Polariton Contributions to Stokes Emission from Molecular Monolayers on Periodic Ag Surfaces, Solid State Comm. 35:667 (1980).
5. J.R. Kirtley, J.C. Tsang, T.N. Theis, and S.S. Jha, Surface Enhanced Raman Scattering from Tunnel Junction Structures, in " Proceedings of the VIIth International Conf. of Raman Spec. " , W.F. Murphy, ed., North Holland, New York (1980) p386.
6. J.C. Tsang, J.R. Kirtley, T.N. Theis, and S.S. Jha, Surface Enhanced Raman Scattering from Molecules in Tunneling Junctions, submitted to Phys. Rev. B.
7. A complete review of sample fabrication and analysis for inelastic electron tunneling spectroscopy appears in: P.K. Hansma, Inelastic Electron Tunneling, Phys. Lett. 30C:145 (1977).
8. John Kirtley and Paul K. Hansma, Vibrational Mode Shifts in Inelastic Electron Tunneling Spectroscopy: Effects Due to Superconductivity and Surface Interactions, Phys. Rev. B 13:2910 (1976).
9. John Kirtley, Inelastic Electron Tunneling Spectroscopy, in " Vibrational Spectroscopies for Absorbed Species ", (ACS Symposium Series, No. 137), A.T. Bell and M.L. Hair, eds., American Chemical Soc., New York (1980) p. 217.
10. A good review of tunneling concepts appears in: " Tunneling Phenomena in Solids ", E. Burstein and S. Lundqvist, eds., Plenum, New York (1969).
11. J.D. Langan and P.K. Hansma, Can the Concentration of Surface Species be Measured with Inelastic Tunneling?, Surface Science 52:211 (1975).
12. A.E.T. Kuiper, J. Medema, and J.J.G.M. Van Bokhoven, Infrared and Raman Spectra of Benzaldehyde Adsorbed on Alumina, J.Catalysis 29: 40 (1973).
13. J. Lambe and R.C. Jaklevic, Molecular Vibration Spectra by Inelastic Electron Tunneling, Phys. Rev. 165:821 (1968).

14. M.G. Simonsen, R.V. Coleman, and P.K. Hansma, High Resolution Inelastic Tunneling Spectroscopy of Macromolecules and Adsorbed Species with Liquid Phase Doping, J. Chem. Phys. 61:3789 (1974).

15. John Kirtley, D.J. Scalapino, and P.K. Hansma, Theory of Vibrational Mode Intensities in Inelastic Electron Tunneling Spectroscopy, Phys. Rev. B 14:3177 (1976); John Kirtley and James T. Hall, Theory of Intensities in Inelastic Electron Tunneling Spectroscopy: Orientation of Absorbed Molecules, Phys. Rev. B 22:848 (1980).

16. J.G. Endriz and W.E. Spicer, Study of Aluminum Films. I. Optical Studies of Reflectance Drops and Surface Oscillations on Controlled Roughness Films, Phys. Rev B 4:4144 (1971).

17. A good early review of these concepts appears in: E. Economou, Surface Plasmons in Thin Films, Phys. Rev. 182:539 (1969).

18. S.L. McCarthy and John Lambe, Enhancement of Light Emission from Metal-Insulator-Metal Tunnel Junctions, Appl. Phys. Lett. 30:427 (1980).

19. Bernardo Laks and D.L. Mills, Roughness and the Mean Free Path of Surface Polaritons in Tunnel Junction Structures, Phys. Rev. B 21:5175 (1980).

20. I. Pockrand, Reflection of Light from Periodically Corrugated Silver Films Near the Plasma Frequency, Phys. Lett. 49A:259 (1974).

21. D. Heitman, Refraction of Light from Weakly Modulated Silver Surfaces, Opt. Comm. 20:292 (1977).

22. A. Marvin, F. Toigo, and V. Celli, Light Scattering from Rough Surfaces: General Incidence Angle and Polarization, Phys. Rev. B 11:2777 (1975).

23. S.S. Jha, J.R. Kirtley, and J.C. Tsang, Raman Scattering from Molecules Adsorbed on a Metallic Grating, Phys. Rev. B 22:3973 (1980).

24. " Electromagnetic Theory of Gratings ", R. Petit, ed. Springer-Verlag, Berlin, (1980).

25. Bernardo Laks and D.L.Mills, Light Emission from Tunnel Junctions: the Role of the Fast Surface Polariton, Phys. Rev. B 22:5723 (1980).

26. John Kirtley, T.N. Theis, and J.C. Tsang, Light Emission from Tunnel Junctions on Gratings, submitted to Phys. Rev. B.

27. C.A. Murray, D.L. Allara, and M. Rhinewine, Silver-Molecule Separation Dependence of Surface Enhanced Raman Scattering, Phys. Rev. Lett. 46:571 (1980).

28. C.J. Powell, Analysis of Optical and Inelastic-Electron Scattering Data II. Application to Al, J. Opt. Soc. Am. 60:78 (1970).

29. P.B. Johnson and R.W. Christy, Optical Constants of the Noble Metals, Phys. Rev. B 6:4370 (1972).

30. Y.Y. Teng and E.A. Stern, Plasma Radiation from Metal Grating Surfaces, Phys. Rev. Lett. 19:511 (1967).

31. John Kirtley, T.N.Theis, and J.C. Tsang, Diffraction Grating Enhanced Light Emission from Tunnel Junctions, Appl. Phys. Lett. 37:435 (1980).

32. K.B. Kirtley, private communication.

33. A. Otto, Raman Spectra of (CN)⁻ Adsorbed at a Silver Surface, Surf. Sci. 75:392 (1978); E. Burstein, Y.J. Chen, C.Y. Chen, S. Lundqvist and E. Tossati, " Giant " Raman Scattering by Adsorbed Molecules on Metal Surfaces, Solid State Comm. 29:565 (1979); R.L. Birke, J.R. Lombardi, and J.I. Gersten, Observation of a Continuum in Enhanced Raman Scattering from a Metal-Solution Interface, Phys. Rev. Lett. 43:71 (1979); J.P. Heritage and J.G. Bergman, Picosecond Raman Gain Studies of Molecular Vibrations at a Surface,

in: " Proceedings of the 2nd USA-USSR Light Scattering Symposium ", J.L. Birman, H.Z. Cummins, and K.K. Rebane,ed. Plenum, New York (1979) p.167.

34. A. Girlando, W. Knoll, and M.R. Philpott, Plasmon Surface Polariton Luminescence from Periodic Metal Gratings, submitted to Solid State Comm.

35. D. Heitman, Radiative Decay of Surface Plasmons Excited by Fast Electrons on Periodically Modulated Surfaces, J. Phys. C 10:397 (1977).

36. P.K. Aravind, Eric Hood, and Horia Metiu, Angular Resonances in the Emission from a Dipole Located Near a Grating, submitted to Surface Science.

37. W.H. Weber and G.W. Ford, Enhancement of the Optical Field at a Metal Surface via Surface Plasmon Excitation, Bull. Am. Phys. Soc. 26:378 (1981).

38. M.F. Muldoon, R.A. Dragoset, and R.V. Coleman, Tunneling Asymmetries in Doped Al-AlO$_x$-Pb Junctions, Phys. Rev. B 20:416 (1979); C.S. Korman, J.C. Lau, A.M. Johnson, and R.V. Coleman, Studies of Aromatic-Ring Compounds Absorbed on Alumina and Magnesia Using Inelastic Electron Tunneling, Phys. Rev. B 19:994 (1979); J.C. Lau and R.V. Coleman, Ag vs Pb Electrodes in Inelastic Electron Tunneling Spectroscopy, submitted to Phys. Rev. B.

39. J.R. Kirtley, S.S. Jha, and J.C. Tsang, Surface Plasmon Model of Surface Enhanced Raman Scattering, Solid State Comm. 35:509 (1980).

40. A. Otto, Raman Scattering from Adsorbates on Silver, Surface Science 92:145 (1980).

41. G. Varsanyi, " Vibrational Spectra of Benzene Derivatives ", Academic Press, New York (1969).

42. G. Herzberg, " Infrared and Raman Spectra of Polyatomic Molecules ", Van Nostrand, New York (1945).

43. M. Moskovits and D.P. Dilella, Enhanced Raman Spectra of Ethylene and Propylene Adsorbed on Silver, Chem. Phys. Lett. 73:500 (1980); M. Moskowits and D.P. Dilella, Surface-Enhanced Raman Spectroscopy of Benzene and Benzene-d$_6$ Adsorbed on Silver, preprint.

VIBRATIONAL SPECTROSCOPY OF MOLECULES ADSORBED

ON VAPOR-DEPOSITED METALS

Martin Moskovits and Daniel P. DiLella

Department of Chemistry and Erindale College

University of Toronto, Toronto, M5S 1A1 CANADA

I.   INTRODUCTION

In the present chapter we will present the SERS spectra of a number of molecules adsorbed on silver and lithium metals vapor-deposited in vacuum onto a cooled substrate. The results will be used to illustrate a number of points. Among them:

(a) That microscopic surface roughness in which individual surface features, though smaller in dimensions than the wavelength of exciting light, are nevertheless large enough to be able to sustain conduction electrons, is necessary for large-scale SERS to be observed. In particular adatom type roughness, although undoubtedly present and perhaps useful as adsorption sites, does not contribute significantly to the enhancement. (b) That for molecules which are capable of making a chemical bond with the surface the Raman spectrum originating from the first monolayer is unusually strongly enhanced when compared with subsequent mono-layers. In those cases the "first layer" SERS spectrum is not simply an enhanced version of the gas-phase or solution-phase spectrum but shows remarkable differences, including differences in the relative intensities of the bands in the SERS spectrum from those in the spectrum of the bulk, unadsorbed molecule, and the presence of vibrations which are normally Raman-forbidden. (c) The possible role of the very strong field gradient which exists near a metal surface will be discussed as a possible cause for the appearance of forbidden modes. (d) We will illustrate the use of SERS in following surface chemical reactions including surface reorientation and decompositions.

II. EXPERIMENTAL

The apparatus and method have been described elsewhere.[1]
Briefly, silver was vaporized from a tantalum hairpin filament heated
with AC. The vapor was deposited onto a polished aluminum substrate
cooled to 11K by means of an Air Products Displex closed-cycle refri-
gerator. The adsorbate was immediately deposited on the cold surface. The aluminum substrate was so situated that only one-half of
it received the Ag vapor while all of it received the full dose of
the adsorbate. One-half of the cold substrate therefore served as
a reference. The Raman spectrum of bulk adsorbate was determined
by depositing a thick layer of adsorbate onto the aluminum substrate
in the absence of fresh silver. Spectra were excited with either a
Control model 554A or a Spectra Physics model 165 argon-ion laser,
dispersed with a Spex series 1400 double monochromator and recorded
using photon counting.

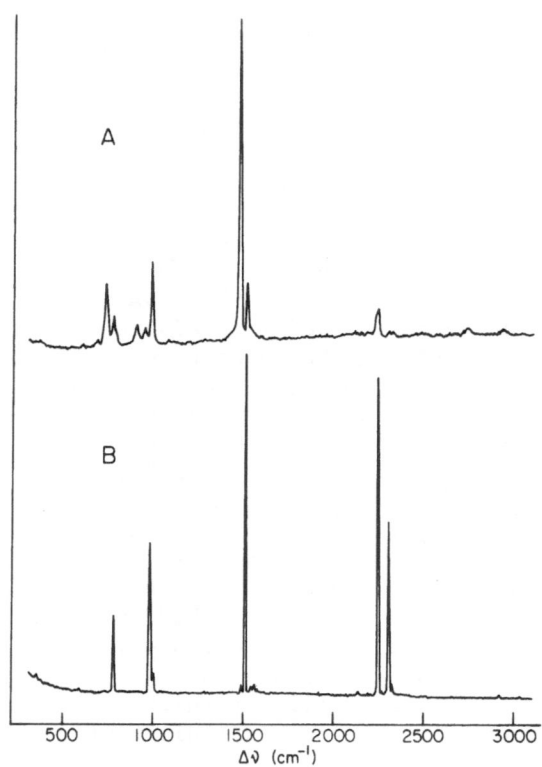

Fig. 1. (A) SERS spectrum of C_2D_4 adsorbed on silver. (b) Ordi-
        nary Raman spectrum of a thick, polycrystalline film of
        C_2D_4.

III. ALKANES, ALKENES AND ALKYNES

The SERS spectra of ethylene, propylene,[2] cis- and trans-2-butene, 1-butene and isobutene[3] adsorbed on silver have been previously reported, while those of ethylene-d_4, acetylene and acetylene-d_2 are shown in Figs. 1-3.

Several general observations can be made about the SERS spectra of these groups of molecules. For the alkenes and alkynes: (a) enhancements of the order of 10^6 are observed; (b) the CH or CD stretching vibrations are very weak in the SERS spectra, although strong in the spectra of the bulk materials; (c) the strongest bands in the SERS spectra are those which involve significant amounts of C=C or C≡C stretching motion regardless of the intensities of these modes in the corresponding spectra of the unadsorbed molecule; (d) for the alkenes, the band near 1650 cm^{-1} which belongs to the vibration with the greatest amount of C=C stretching character is shifted some 35-50 cm^{-1} to lower frequencies and broadened considerably relative to its counterpart in the spectrum of the polycrystalline material. In addition, when more than one monolayer of adsorbate is present, one notes the presence of another, weaker, unbroadened feature which comes precisely at the same frequency as in polycrystalline material; (e) no SERS spectra could be observed for the two alkanes tried, namely CH_4 and C_2H_6.[3]

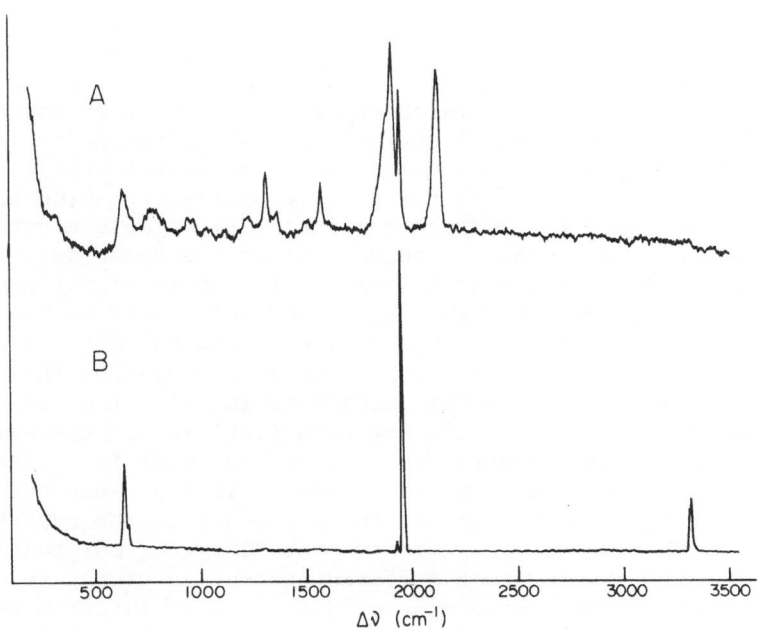

Fig. 2. (A) and (B) as in Fig. 1 but for acetylene.

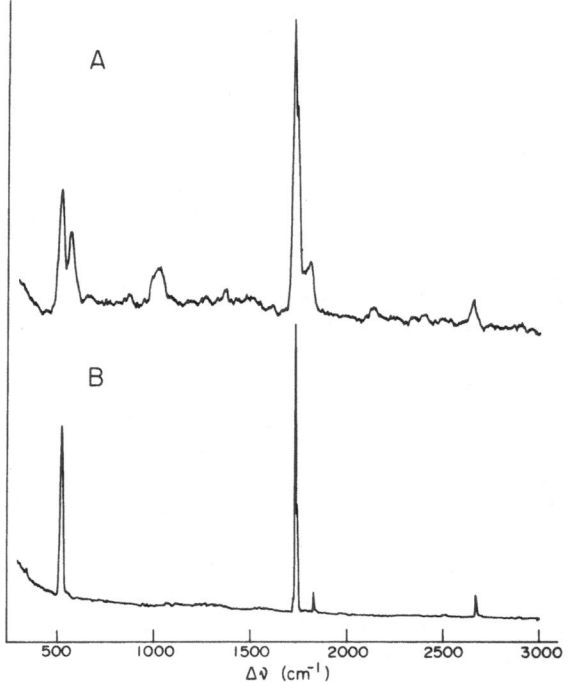

Fig. 3. (A) and (B) as in Fig. 1 but for C_2D_2.

The First Layer Effect

        Although many workers in the field now agree that excitation
of surface plasmons in small surface irregularities underlies a
major portion of the surface Raman enhancing effect, there remains
considerable controversy regarding the coupling mechanism between
the molecular vibration and the collective electronic excitation
and concerning the degree of enhancement of the Raman spectrum of
molecules which are not bonded directly to the metal surface.
Recently we proposed[4] that the first adsorbed layer, if bonded
"chemically" to the metal, could couple to the surface plasmons
through the periodic charge transfer which accompanies the adsorbate's vibration. Through this mechanism the vibration of the adsorbed molecule would modulate the polarizability of the surface
metal bump or colloidal metal particle to which it is bonded; an
inelastic component would therefore appear in the spectrum of the
light scattered by the system at the frequency of the adsorbate
vibration. This mechanism suggests that different molecular vibrations could be enhanced to a different extent according to the
degree of charge movement occurring during that vibration and that
the Raman spectrum of the first monolayer would be especially
enhanced over that of subsequent monolayers.

Gersten and Nitzan,[5] Kerker et al.,[6] and McCall et al.,[7] on the other hand, have all considered the enhancing process to be due to inordinately strong fields which exist near the small metal particle, or surface feature when the latter is illuminated by a radiation field of frequency close to the surface plasmon frequency of the metal particle. All three groups calculate a monotonic decrease in enhancement as a function of distance from the metal surface with no "special" enhancement for the first monolayer. The calculations of Gersten and Nitzan,[5] for example, indicate that on placing a molecule 25 Å above the metal surface one should see a Raman signal which is only some eight-fold less intense than that produced by the same molecule when it is 1 Å above the surface. Murray et al.[8] and Zwemer et al.[9] report such an enhancement-at-a-distance effect. If the Raman signal from the first layer could be distinguished somehow from that of subsequent layers as, for example, by means of the frequency shift resulting from the former's bonding to the surface, then one expects a stronger signal from, say, the outer 199 monolayers of a sample containing 200 monolayers of adsorbate as compared to that of the first monolayer, assuming that the spectra of all but the first monolayer overlap. Using the calculations of Gersten and Nitzan,[5] which go only to 25 Å, and assuming for ease of calculation that each monolayer is 2.5 Å thick, one finds that the sum of the Raman signal emanating from the outer nine monolayers is predicted to be roughly five-fold more intense than that scattered by the first monolayer. This ratio would be even higher if more monolayers were taken into account, but it would be smaller if metal features of very small radius of curvature were considered.[10] The above prediction may be tested in the following way.[3]

Figure 4A shows the SERS spectrum of approximately 200 monolayers of trans-2-butene adsorbed on silver, while Fig. 4C depicts the Raman spectrum obtained when the same quantity of adsorbate was condensed on the silver-free portion of our cold substrate.

Referring, for example to Fig. 4C, one notes that the excitation of a layer of trans-2-butene approximately 200 monolayers thick deposited on the air-oxidized, polished aluminum substrate produced no trace of the Raman spectrum of that molecule under the conditions of amplification used. When the silver-coated aluminum substrate with its load of about 200 monolayers of trans-2-butene is laser-excited, one observes that the frequency of the C=C stretch is sufficiently shifted from its value in bulk trans-2-butene to reveal the presence of another band some twenty times weaker than the strong band at 1632 cm^{-1}. The frequency of the weaker band falls exactly at the frequency of the C=C stretch in polycrystalline trans-2-butene. Since this band is not visible in Fig. 4C, one concludes that it is an enhanced Raman band originating from the roughly 199 monolayers of that molecule which are condensed on top of the first monolayer which, in turn, is producing the stronger, shifted peak

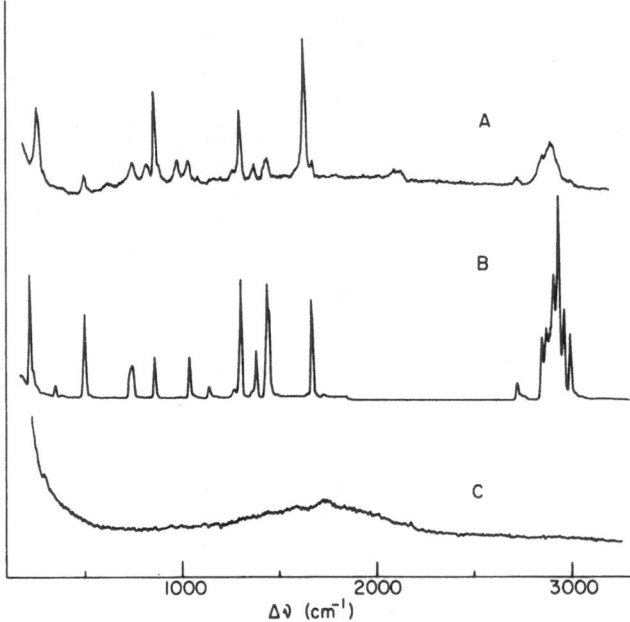

Fig. 4.   (A) SERS spectrum of approximately 200 monolayers of trans-
          2-butene condensed on silver (488 nm excitation).
          (B) Ordinary Raman spectrum of a thick, polycrystalline
          film of trans-2-butene. (C) Raman spectrum of about 200
          monolayers of trans-2-butene condensed on the silver-free
          part of the substrate.

at 1632 cm^{-1}. The former signal should have been five-fold
stronger than the latter according to the aforementioned calculation
rather than twenty times weaker. The Raman spectrum of the first
monolayer, which is bonded to the metal, is therefore at least one
hundred-fold (and, in some molecules, more than one thousand-fold)
more enhanced than the Raman signal proceeding from the second
monolayer. We contend that this comes about as a result of the
aforementioned charge coupling mechanism which grants the first
monolayer a special enhancement on top of that produced by the field
effect which is clearly also operating, as it should. That charge
coupling is present is also suggested by the observation that the
C=C stretch is the strongest band in the SERS spectra of all alkenes
studied to date, even in cases in which that mode is not the most
intense in the spectra of the respective bulk, polycrystalline
material (see, for example, the spectra of isobutylene and trans-
2-butene[3]). Moreover the next most intense one or two bands
invariably are those whose normal coordinate involves C=C stretching to a substantial degree. It stands to reason that a vibration
which causes the C=C to stretch will also result in the greatest

amount of charge injection into or withdrawal from the metal, causing in turn the greatest amount of coupling.

When no strong chemical bonding is possible, as between the metal and an alkane molecule, one expects a commensurately weaker SERS spectrum. In fact no SERS spectrum could be detected for methane and ethane on silver. Another example which illustrates the fact that not all bands are equally enhanced is provided by comparing the spectra of C_2H_4 and C_2D_4. Focusing our attention for the moment on vibrations ν_2 and ν_3, the former being mainly a C=C stretch with some methylene scissoring motion mixed in while the latter is mainly methylene scissoring with some C=C stretching motion, we see that in the SERS spectrum of C_2H_4 both vibrations (which come at 1585 and 1330 cm^{-1}) are about equally enhanced. In the SERS spectrum of C_2D_4 ν_3, which comes at about 972 cm^{-1}, is enhanced to a much lesser degree than ν_2 (1464 cm^{-1}) is. This makes sense in terms of the "first layer" effect since much less C=C motion is present in ν_3 of C_2D_4 than in the corresponding vibration of C_2H_4 owing to the greater separation in the frequencies of the two vibrations in the deuterated molecule; consequently ν_3 is expected to be less enhanced in C_2D_4 than in ordinary ethylene.

As a further example, the SERS spectrum of N_2 adsorbed on Ag is at least one hundred-fold weaker than that of CO even though the polarizabilities of the two molecules are virtually identical.[11] This can easily be gauged from Fig. 5 which shows that the ordinary Raman spectrum of an equimolar mixture of N_2 and CO consists of two lines of almost identical height and width although of differing frequencies. When this mixture is adsorbed on freshly deposited silver, the spectrum obtained consists of an intense, broad and shifted line corresponding to adsorbed CO and a weak, narrow, unshifted line originating with the N_2 component (Fig. 5). The shift or lack of it suggests that CO forms a bond, albeit a weak one, with the silver while N_2 doesn't. N_2 is known, of course, to be a much poorer π-acid than CO.[12]

Spectroscopy of Ethylene and Acetylene

In addition to providing insight into the SERS mechanism the molecules ethylene and acetylene afford some interesting surface spectroscopy. The SERS spectra of ethylene and its perdeuterio counterpart are summarized in Table 1. We see in it that several bands of these molecules have been seen clearly which have either never been seen previously or which have been seen only weakly. These include the A_u mode (ν_4) which is neither infrared nor Raman active and ν_6 (B_{1g}) and ν_{10} (B_{2u}), the former seen weakly in the Raman spectrum of the crystal while the latter has not been previously seen. In addition we see several modes in the SERS spectrum (for example ν_7 (B_{1u}) and ν_{12} (B_{3u})) which are known from the infrared spectra of these molecules but are normally Raman inactive.

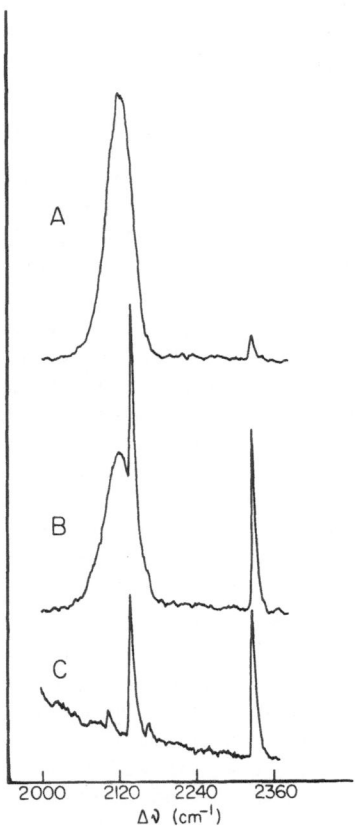

Fig. 5. (A) SERS spectrum of a layer of an equimolar mixture of CO
         and N_2 several monolayers thick, adsorbed on silver.
         (B) The Raman spectrum obtained when an additional layer
         several tens of thousands of monolayers thick was deposited
         on top of the above-mentioned sample. (C) As in (B) but
         recorded on the silver-free part of the spectrum.

        Similar phenomena are seen in the spectra of acetylene and its
deuterated form.[11] Table 2 shows that in addition to the interactions normally observed in the Raman spectrum of acetylene, v_5 (Π_u)
as infrared-active bending vibration is rendered active. Otherwise
the salient features of the spectra of the two alkynes are similar
to those of the alkenes: the C≡C vibration is the most greatly
shifted in frequency and the most greatly enhanced while the C-H or
C-D stretching vibrations are the least enhanced.

Table 1. Raman Spectra of C_2H_4 and C_2D_4

	s	C_2H_4		C_2D_4	
		Bulk	SERS	Bulk	SERS
1	a_g	2996vs	2996w	2244vs	2242wm
2	a_g	1625vs	1585vs	1507vs	1464vs
3	a_g	1325vs	1325vs	975s	972m
4	a_u	(1040)	1075m	(726)	b
5	b_{1g}	3073vs	3073w	2302s	a
6	b_{1g}	1220	a	(1011)	a
7	b_{1u}	(951)	977m	(720)	718m
8	b_{2g}	941m	955m	778m	761wm
9	b_{2u}	(3105)	b	(2345)	a
10	b_{2u}	(823)	825w	(584)	587vw
11	b_{3u}	(3021)	2975w	(2200)	2226wm
12	b_{3u}	(1440)	a	(1078)	1067vw

[a] Absent in SERS spectrum.
[b] Possible overlap with other modes prevents determination of the presence or absence of this mode.
[s] Symmetry species
Vibrational assignments from J.L. Duncan and E.
Hamilton, J. Mol. Struct. 76:65 (1981)

Table 2. Raman Spectra of C_2H_2 and C_2D_2

	s	C_2H_2		C_2D_2	
		Bulk	SERS	Bulk	SERS
1	Σ_g^+	3324 }m 3315	a	2670w	2645w
2	Σ_g^+	1961 } 1953 }s	1954 }m 1923 }s	1746m 1734vs 1725w	1778m 1736s 1716vs
3	Σ_u^+	(3226.3)	a		
4	Π_g	659 639 }m 632	634m	505 }s 533	508s
5	Π_u	(768.8) (760.6) } (747.5)	781m	(567.5) (561.8) } (551.4)	562s

[a] Does not appear in the surface spectrum.
The numbers in parentheses are infrared crystal fre-
quencies. G.L. Bottger and D.F. Eggers, J. Chem.
Phys. 40: 2010 (1964).

IV. ARENES AND THIA-ARENES

The spectra of benzene and benzene-d_6 have been reported previously.[13] Figs. 6 and 7 show the SERS spectra of benzene and benzene-1,3,5-d_3. A summary of the observed frequencies and their assignments is contained in Table 3 which also lists the SERS data obtained with benzene-d. A cursory look at the spectra or at Table 3 indicates clearly that several modes which are normally Raman inactive are present in the SERS spectrum. In the spectrum of normal benzene, for example, the lines which are observed at 397, 697, 1311 and 1473 cm^{-1} correlate well with ν_{11}, ν_{14}, ν_{16} and ν_{19}, all Raman-inactive modes of the free molecule. The most obvious explanation for the appearance of these modes in the SERS spectrum of benzene is that by bonding to the surface the molecule's symmetry has been lowered from D_{6h} to a point group of lower symmetry. With the aid of a correlation table[14] one determines that by reducing the symmetry of the molecule to C_{3v} (σ_d) all the frequencies

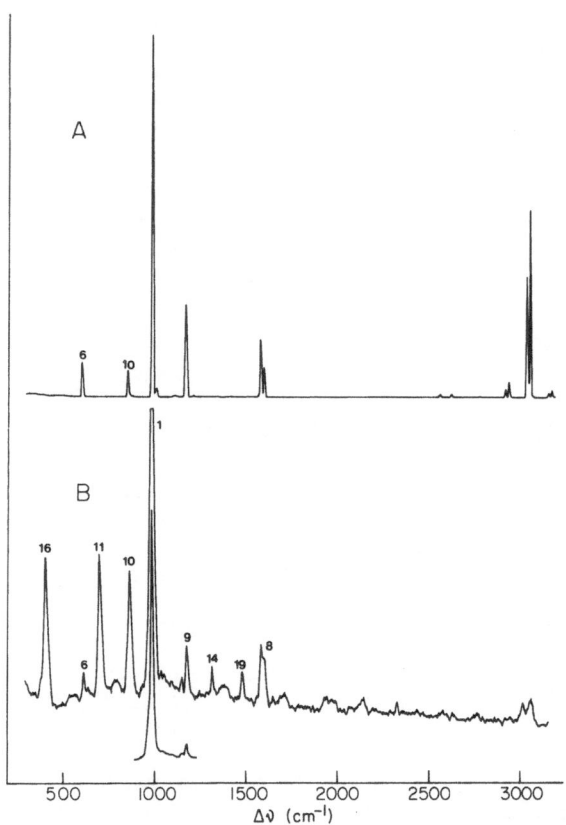

Fig. 6. (A) The Raman spectrum of a thick, polycrystalline film of
         benzene. (B) SERS spectrum of benzene adsorbed on silver.

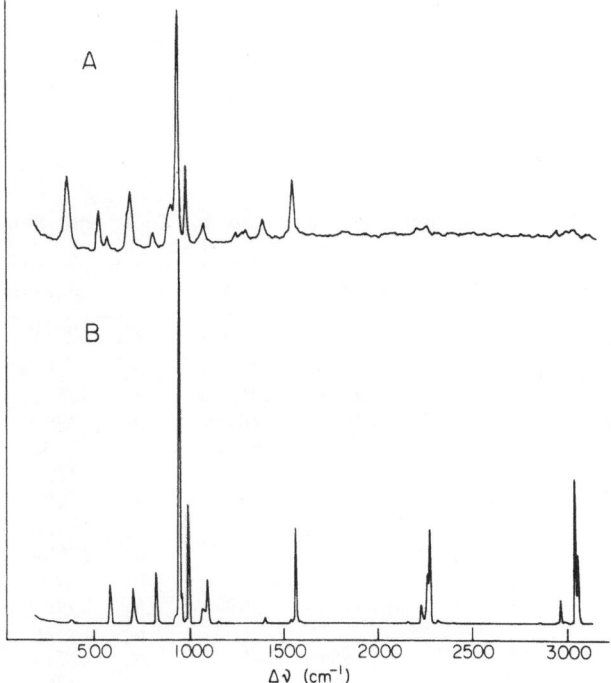

Fig. 7. As in Fig. 1 but for benzene-1,3,5-d$_3$.

observed are completely accounted for. This change in symmetry can only be effected by placing benzene atop three silver atoms in (111) patches of the metal surface oriented as in structure E of Fig. 8.

It is possible to conceive of an entirely different reason for the apparent reduction of the symmetry of benzene from D$_{6h}$ to C$_{3v}$ (σ_d), a reason independent of local metal geometry which would be equally applicable to the Raman spectrum of a molecule adsorbed on jellium.

The i-component of the dipole moment induced in a molecule is given by Ref. 15,

$$\mu_i = \alpha_{ij}E_j + \frac{1}{3}A_{ijk}(\partial E_j/\partial k) + G_{ij}B_j , \qquad (1)$$

where i, j, k may be x, y, z, E$_j$ and B$_j$ are the j-components of the electric and magnetic fields and α_{ij}, A_{ijk} and G_{ij} are, respectively, components of the normal polarizability, the polarizability which produces a quadrupole moment proportional to the field and the magnetic moment polarizability. Reference 16 should be consulted for more detail. [The tensor summation convention is observed in eq'n (1).] For fields associated with a light wave

Table 3. Raman Spectra of Benzene, Benzene-1,3,5-d_3 and Benzene-d_1

		Benzene			Benzene-1,3,5-d_3			Benzene-d_1	
	s	Bulk	SERS	s	Bulk	SERS	s	Bulk	SERS
1	a_{1g}	990vs	982vs	a_1'	956vs	952s	a_1	979vs	972s
2	a_{1g}	3059s	3060w	a_1'	3044s	3046vw	a_1	3062s	3057w
3	a_{2g}	(1346)	a	a_2'	1265vw	1267vw	b_1	1299w	1302w
4	b_{2g}	(703)	a	a_2''	(697)	696w,sh	b_2	703w	697m
5	b_{2g}	(989)	b	a_2''	911vw	b	b_2	(978)	b
6	e_{2g}	606m	605w	e'	594m	594w	a_1	600	603
							b_1	600	603
7	e_{2g}	3046s	b	e'	2275m	a	a_1	2265m	2273vw
							b_1	b	b
8	e_{2g}	1596m	1587m	e'	1574s	1569m	a_1	1589m	1581m
							b_1	1572m	1573m
9	e_{2g}	1178m	1174w	e'	1105m	1103w	a_1	1171m	1176m
							b_1	1076w	1068w
10	e_{1g}	849m	864m	e''	717m	711m	a_2	854m	859m
							b_2	787w	790m
11	a_{2u}	(670)	697m	a_2''	545vw	544m	b_2	622w	619m
12	b_{1u}	(1008)	a	a_1'	1004s	1002s	a_1	1005m	1005wm
13	b_{1u}	(3062)	b	a_1'	2284s	2285vw	a_1	b	b
14	b_{2u}	(1309)	1311w	a_2'	(1322)	1324vw	b_1	1320w	a
15	b_{2u}	(1149)	1149w	a_2'	(911)	b	b_1	1156m	1161w
16	e_{2u}	(404)	397m	e''	378w	374m	a_2	405vw	396
							b_2	382vw	384
17	e_{2u}	(966)	970vw	e''	932w	921w	a_2	970vw	b
							b_2	935w	929m
18	e_{1u}	(1036)	1032vw	e'	834m	835w	a_1	1032w	1028w
							b_1	(858)	b
19	e_{1u}	(1479)	1473w	e'	1413w	1410w	a_1	1475w	1475w
							b_1	1449w	1449w
20	e_{1u}	(3073)	b	e'	3061m	a	a_1	3048s	b
							b_1		b

a, b and s have the same meanings as in Table 1. Vibrational assignments from S. Brodersen and A. Langseth, Mat. fys. skr. Kong. Dansk. Vid. selsk. 1:1 (1956). Frequencies in parentheses do not appear in Raman spectra of the bulk.

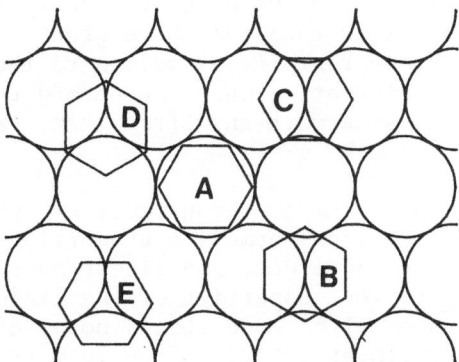

Fig. 8. The symmetries of five possible adsorption sites on a
         (111) face of an fcc metal crystal. (A) C_{6v}, (B) C_{2v}
         (X→X), (C) C_{2v} (X→Y), (D) C_{3v} (σ_v), (E) C_{3v} (σ_d).

travelling in the k-direction in vacuo,

$$\partial E_j/\partial k = i(2\pi/\lambda)E_j ,$$

while A_{ijk}/α_{ij} has the units and magnitude of a molecular dimension.
Consequently the ratio of the second term in eq'n (1) to the first
is of the order of a molecular dimension divided by λ or approximately 10^{-3}. Likewise for the third term. It is for this reason
that the last two terms in eq'n (1) are neglected in considering
ordinary Raman scattering.[17]

   Near a metal surface, on the other hand, the electric field
gradient may be substantial since classically the normal component
of the electric field drops to zero at the metal surface, while in
more realistic models[18] the field decreases rapidly near the surface, while executing damped oscillations within the metal. This
large decrease in the field is sustained over a region of molecular
dimensions so that the second term in eq'n (1) may now attain values
equal to those of the first term. The third term in eq'n (1) will
continue to be small, however. Naturally the field and field gradient near a raised metallic structure forming part of the surface
roughness will have a different spatial dependence from those near
a smooth surface, the latter being the case discussed in Ref. 18.
Since we postulated that the surface roughness generates the
enhancement,[19] it is clear that it is pertinent to inquire what the
spatial dependence of the field and field gradients near a surface
with non-zero curvature might be like. This might be undertaken
for a metal sphere as a first approximation, although there is evidence to suggest that metal films evaporated onto low-temperature
substrates form a columnar structure, growing outward towards the
direction of the metal vapor source.[20] We do not consider at this
point, as have others,[5,6] whether unusually large fields may be

created near a metal surface and hence unusually large field gradients which are even larger than the three orders of magnitude discussed above. Rather we restrict ourselves to the question of the
new selection rules which arise when the second term in eq'n (1)
becomes comparable in size with the first term, regardless of their
absolute magnitude.

The tensor $\hat{A}$ transforms as the product of three translations.
These are precisely the transformation properties of the tensor $\underset{\sim}{\beta}$
which gives rise to second-order, non-linear phenomena such as the
hyper-Raman effect.[21] The properties of the latter tensor have been
considered by Decius et al.[21] One should note, however, that $\hat{A}$ does
not give rise to non-linear effects; it simply transforms like a
tensor which does. The presence of very intense fields near the
surface will, of course, produce hyper-Raman scattering. This, however, is not the issue under discussion. Thus in a situation in
which the Raman scattering from a system proceeds with about equal
efficiency from the first two terms in eq'n (1), those modes which
are normally Raman-active and those modes which are normally hyper-
Raman-active will appear together in the Raman spectrum. For D_{6h}[21]
the modes which remain inactive in such a "combined" spectrum are
those of symmetries a_{2g}, b_{1g}, b_{2g} and a_{1u}. The last does not correspond to any vibration of benzene, while the first three are three
of the four representations which remain silent in the observed
spectrum of benzene adsorbed on silver.

Thus the Raman spectrum of benzene bonded so weakly to silver
that its geometry is not reduced from that of the unbound molecule
would, nevertheless, look like that of a molecule bonded to an
adsorption site which would reduce its symmetry to C_{3v} (σ_d) when the
first terms in eq'n (1) contribute to the Raman intensity. This may
lead to an erroneous identification of the local site symmetry of
the adsorption site. The appearance of formally forbidden lines in
the SERS spectra of ethylene and acetylene adsorbed on silver are
also interpretable in this fashion.

The azabenzenes are electronically and structurally very similar
to benzene and as a result have very similar vibrational spectra.
One of these, pyridine, was the first molecule for which SERS was
reported and remains the most studied molecule in the field. These
molecules differ from benzene in that they can bond easily to a metal
either as a π-bonded compound or as a σ complex through the lone
pairs of the ring nitrogens. The former form of bonding would cause
the molecule to "lie flat" on a metal surface while the other has the
molecule "standing up" on the surface. Benzene, on the other hand,
cannot form the σ complex without loss of a hydrogen. Both of these
geometries have been reported by Demuth[22] on the basis of an EELS
study of pyridine on silver.

Perhaps the most interesting of the azabenzenes is s-trazine

(1,3,5-triazabenzene) which retains a three-fold axis of symmetry and will therefore possess doubly degenerate vibrations and more restrictive selection rules than pyridine does.

A series of spectra of s-triazine adsorbed on a vapor-deposited silver film[11] is shown in Fig. 9. As with benzene a number of normally Raman-forbidden bands are seen. This is summarized in Table 4 along with the SERS data for pyridine and pyrazine (Fig. 10) on evaporated silver. Because of the aforementioned similarity to the spectrum of benzene the notation developed for benzene[23] is used to refer to the vibrational modes. (Note that for pyridine and pyrazine there are no doubly degenerate modes and hence two vibrations correlate with every doubly degenerate vibration of benzene.) Although similar

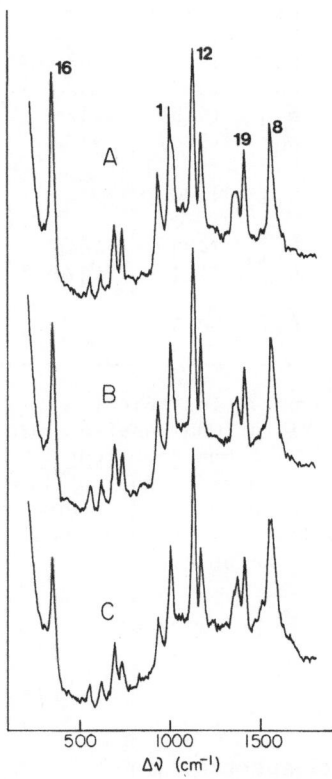

Fig. 9. Three SERS spectra of s-triazine adsorbed on silver taken after (A) 21 min, (B) 68 min and (C) 210 min of 100 mW, 488 nm cw-laser irradiation.

Table 4. Raman Spectra of Pyridine, Pyrazine and s-Triazine

	Pyridine			Pyrazine			s-Triazine		
	s	Bulk	SERS	s	Bulk	SERS	s	Bulk	SERS
1	a_1	989vs	1000s	ag	1015s	1015s	a_1'	990vs	999s
2	a_1	3054s	3055m	ag	3055s	3055w	a_1'	3040m	3040w
3	b_2	1230vw	a	b_{2g}	1346	1347w	a_2'	(?)	?
4	b_1	748vw	747vw	b_{3g}	756w	753w	a_2''	731vw	727m
5	b_1	(1007)	b	b_{3g}	983	b	a_2''	936vw	944m
6	a_1	602w	620m	ag	600vw	615m	e'	673m	682
	b_2	650m	656w	b_{2g}	699m	700w			693
7		c			c		e'	(3050)	b
	b_2	3035vw	b	b_{2g}	(3040)	b			
8	a_1	1581m	1589m	ag	1583m	1578s	e'	1545}m	1555}s
	b_2	1572m	1566w	b_{2g}	1524m	1522m		1551	1575
9	a_1	1218m	1215m	ag	1239w	1233m	e'	1172m	1164m
		c			c				
10	a_2	886vw	856	b_{1g}	(927)	922w	e''	1040w	1041vw
	b_1	945vw	950vw		c				
11	b_1	711vw	712w	b_{2u}	(785)	792m		c	
12	a_1	1029vs	1032m	b_{1u}	(1018)	1031w	a_1'	1122vs	1125s
13	a_1	3054s	3055m	b_{1u}	(3012)	a		c	
14	b_2	1357vw	1355vw	b_{3u}	(1149)	1154w	a_2'	(?)	1299vw
15	b_2	1146w	1147w	b_{3u}	(1063)	1088m		c	
16	a_2	380vw	382m	au	(350)	352m	e''	336m	344s
	b_1	408w	410m	b_{2u}	(418)	417m			
17	a_2	(980)	b	au	(960)	972w		c	
		c			c				
18	a_1	1068w	a	b_{1u}	1130	a		c	
	b_2	1056vw	1055w		c				
19	a_1	1481vw	1479w	b_{1u}	1483	1484m	e'	1407}m	1380w
	b_2	1438vw	1434wm	b_{3u}	1411	1407m		1413	1410m
20	a_1	3054s	3055m		c			c	
	b_2	(3079)	b	b_{3u}	3013	a			

a, b and s have same meaning as in Table 1. c: No corresponding mode for this molecule. Vibrational assignments: Pyridine--D.P. DiLella and H.D. Stidham, J. Raman Spectrosc. 9:1 (1980); Pyrazine --J. Zarembowitch and L. Bokobza-Sebagh, Spectrochim. Acta 32A:605 (1976); s-Triazine--J.E. Lancaster, R.F. Stamm and N.B. Colthup, Spectrochim. Acta 17:155 (1961). Frequencies in parentheses do not appear in Raman spectra of the bulk.

in many respects, the SERS spectra of s-triazine and benzene on silver differ considerably in the relative intensities of several bands. For example, the totally symmetric skeletal stretch, ν_1, is the most intense band in the SERS spectrum of benzene while ν_8, the doubly degenerate ring vibration, is relatively weak; ν_1 and ν_8 are of about equal intensity, however, in the SERS spectrum of s-triazine and both are somewhat less intense than ν_{12} which is almost undiscernible in the SERS spectrum of benzene. The last observation is not remarkable since ν_{12} is intense in the spectrum of the bulk, polycrystalline molecule. The former observation is intriguing, however, since ν_8 is roughly five-fold less intense than ν_1 in the spectra of both bulk benzene and s-triazine.

The most noteworthy difference between the SERS spectrum of s-triazine and that of benzene is that the former changes with time upon laser irradiation or sample heating to 40 K while the latter does not. This is shown in Fig. 9 in which are reproduced the SERS spectra of triazine adsorbed on Ag recorded after 21, 68 and 210 minutes of irradiation with 100 mW of 488 nm argon-ion laser light. Two observations bear mention: the first is the decline in intensity of all out-of-plane vibrations of s-triazine with respect to that of the in-plane vibrations (cf. ν_{16}, ν_4, ν_5 with respect to ν_1, ν_{12} and ν_8) and the second is the splitting of the doubly degenerate, in-plane vibrations into two components (cf. ν_8, ν_{19}) while out-of-plane, doubly degenerate modes do not seem to have their degeneracy lifted. Similar effects are observed when the surface is heated by conventional means.

We interpret these observations as follows: At high coverage we suggest that s-triazine favors the "standing up" adsorption geometry in which it interacts with the surface through the lone-pair electrons on one of its nitrogens. However, the majority of molecules approach the surface with an orientation which is favorable to the π-bonded, "flat" mode of bonding. Hence, most of the freshly deposited adsorbate will adopt this less favorable mode of bonding since there will not be sufficient thermal energy to allow the reorientation to take place on the cold (11-12 K) silver surface, the assumption being that the two geometries are separated by an activation barrier. Thermal or laser heating provides this energy.

Out-of-plane vibrations are expected to be enhanced more when the molecule is lying flat than when the molecule is standing up since in the former configuration an out-of-plane vibration causes the atoms of the molecule to recede from and approach the surface periodically while in the latter it will not. Moreover, the field gradient vector is parallel to the C_3 axis of the molecule while the latter is lying flat; consequently the degeneracy of doubly degenerate modes will not be lifted in that configuration. When the molecule stands up the C_3 axis is approximately normal to the field gradient and degeneracies may (therefore) be lifted. However, only the degeneracy of in-plane vibrations will be substantially removed because in the N-bonded molecule the two components of the degenerate mode will have, in general, different orientations with respect to the surface normal. One of these components, for example, may be

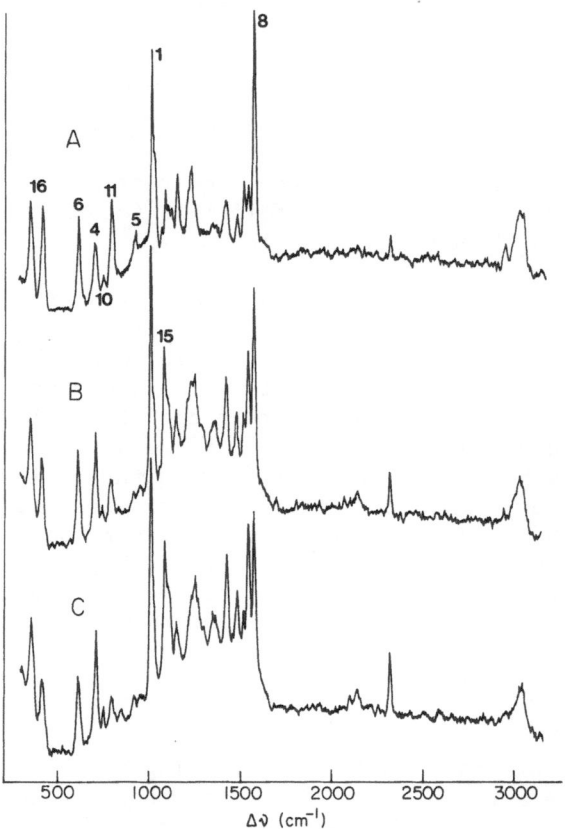

Fig. 10.   As in Fig. 9 but for pyrazine after (A) 13 min, (B) 92
           min and (C) 150 min.

along the surface normal while the other is tangential to the metal
surface. For out-of-plane, degenerate vibrations, on the other
hand, both components would be tangential to the metal surface when
the molecule is in the N-bonded form and will therefore remain
effectively degenerate.

    Although pyridine and pyrazine do not have degenerate vibra-
tions, they both show the same decline in the intensities of out-of-
plane vibrations upon laser irradiation or warmup to 40 K. Accordingly we interpret these changes in the same manner as with s-
triazine as arising from a surface unimolecular rearrangement. One
should state in passing that when laser irradiation was used the
reorientation effect was quite local, as evidenced by the fact that
when the laser beam was focussed on another part of the surface the
spectrum reverted to its original form.

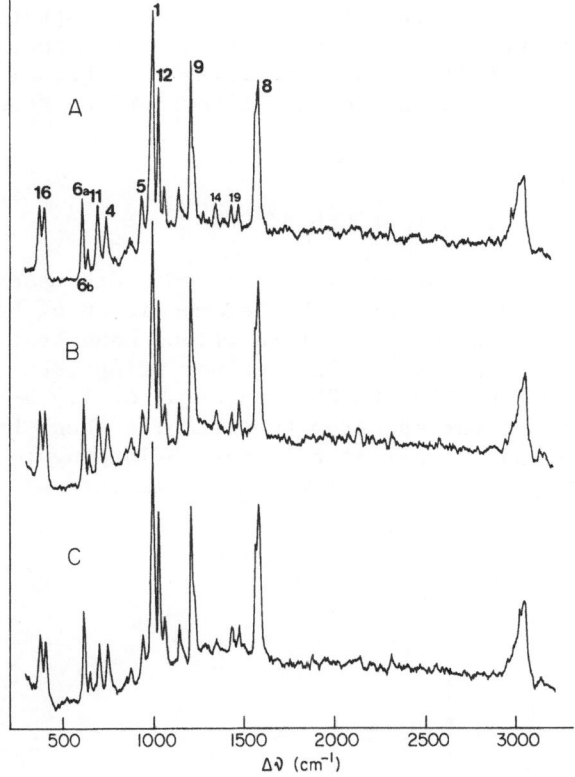

Fig. 11.   As in Fig. 9 but for pyridine after (A) 3 min, (B) 29 min
and (C) 90 min.

The relative intensities that we observe in the SERS spectrum
of pyrazine on Ag are quite different from those reported by Dornhaus et al.[24] for the same molecule adsorbed on Ag electrodes.

An even more striking difference exists between the SERS spectrum of pyridine adsorbed from the gas phase on the cold silver
surface and those reported for pyridine adsorbed from solution onto
silver electrodes. In particular, the out-of-plane modes are much
more intense in the former case (Fig. 11) as compared to the latter.
This difference is reduced somewhat, though not completely, upon
either laser irradiation or heating of the surface.

One should state in passing that the observed spectral changes
occasioned by heating or laser irradiation are clearly due to
changes in the molecule rather than in the surface since no significant change was observed in the SERS spectra of benzene or its
deuterio analogues when laser-irradiated or heated to moderate
temperature, although heating to larger temperatures (150-200 K)

caused a uniform decrease in the intensity of the spectrum and
eventually its disappearance to be replaced by the so-called cathe-
dral peaks resulting from carbon impurities[25] either diffusing out
of the silver or accumulating on the silver surface from the rather
poor vacuum.

V.   SURFACE DECOMPOSITION FOLLOWED BY SERS

In addition to the surface rearrangements discussed above, SERS
has been applied by us to follow the decomposition of haloginated
hydrocarbons on Ag. These types of reactions have been the subject
of several previous studies,[26] although not using SERS. As an
example we show in Fig. 12A the SERS spectrum of 1,3,5-trifluoroben-
zene. Fig. 12B shows the spectrum that results when the sample is
warmed briefly to 150 K. All of the lines belonging to 1,3,5-tri-
fluorobenzene have been drastically reduced in intensity while six

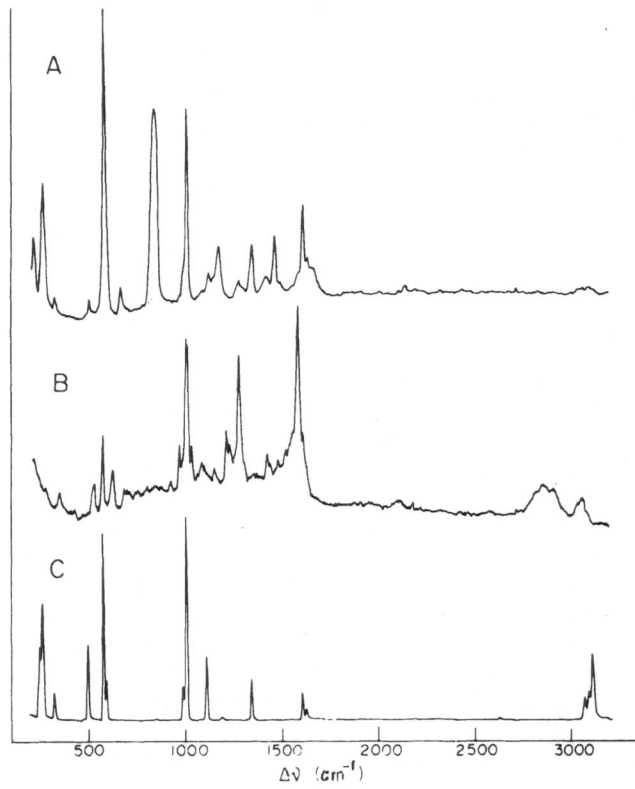

Fig. 12.   (A) SERS spectrum of 1,3,5-trifluorobenzene adsorbed on
           silver. (B) SERS spectrum obtained after several minutes
           warmup at 150 K. (C) Raman spectrum of the bulk, poly-
           crystalline material.

other lines have grown in, including a broad band in the CH stretching region characteristic of aliphatic or olefinic moieties.
Although a full analysis of the changes cannot yet be made it is
clear that the aromatic compound has decomposed on the surface to
give smaller, non-aromatic fragments. Further experiments are
underway to clarify the identity of the products. This result shows
clearly the possible utility of SERS in following surface reactions.

VI.   SERS FROM LITHIUM SURFACES

The majority of SERS studies have been carried out on silver
surfaces. SERS has also been reported for a few other metals
including copper, gold,[27] mercury,[28] platinum and nickel.[29] These
metals do not appear to be as good enhancers as silver, although
copper approaches silver under the proper conditions. Almost all of
the models based on the excitation of or coupling to surface plasmons
which have been presented for SERS have been able to predict the
especially good enhancement obtained with the group IB metals, while
most other transition metals are predicted to be rather poorer

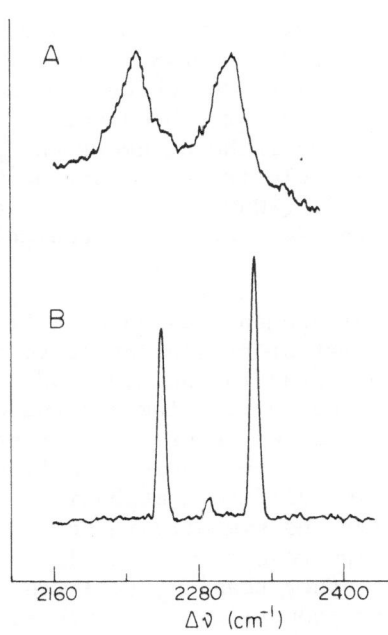

Fig. 13.   (A) SERS spectrum of an approximately equimolar mixture
           of $^{15}N_2$ and $^{14}N_2$ adsorbed on lithium.   (B) Raman spectrum
           of the bulk, polycrystalline material.

enhancers. On the other hand, alkali metals have been predicted to
be good enhancers.[19] This is borne out by our experiments with
lithium surfaces. (Yet unpublished results on colloidal potassium
obtained by Schulze et al.[30] predate these.) Figure 13A shows the
SERS spectrum obtained when a thin layer of a mixture of $^{14}N_2$ and
$^{15}N_2$ was condensed on a freshly vapor-deposited lithium mirror. The
Raman spectrum of a thick polycrystalline layer of the same mixture
is displayed in Fig. 13B. Lithium was vaporized from a stainless
steel knudsen cell; otherwise the experimental conditions are the
same as with silver.

Unlike the SERS for nitrogen on silver, the SERS band for nitrogen on lithium is both broadened and shifted relative to the corresponding polycrystalline band. As implied earlier, the SERS intensity for nitrogen on silver was largely due to the field effect and
probably included scattering from several monolayers. The SERS
intensity for nitrogen on lithium, on the other hand, shows a distinct
"first layer" effect involving chemical bonding. That nitrogen
interacts more strongly with a lithium surface than a silver surface
is not surprising. Under ordinary conditions of temperature and
pressure lithium and nitrogen readily react to form Li_3N whereas
silver is inert to nitrogen. At 12 K, however, Li and nitrogen do
not seem to form a very strong bond since the observed N_2 frequency
shift amounts to only 19 cm^{-1}.

Comparing the SERS spectra of N_2 on Li and Ag brings up an
interesting observation regarding SERS line widths. We have noted
that those bands in a SERS spectrum which are the most shifted in
frequency are also the most greatly broadened. This implies that at
least some of the broadening is homogeneous and probably caused by
coupling to the conduction electrons.[31] Strong SERS bands are
typically 20-40 cm^{-1} broad, (FWHM) implying a Raman lifetime of the
order of 1 ps. Bandwidths for the unadsorbed molecule range from
about 2 to 5 cm^{-1}.

The SERS spectrum of benzene adsorbed on lithium is shown in
Fig. 14 which includes that of benzene on silver for comparison.
All of the SERS bands observed for benzene and benzene-d_6 adsorbed
on silver are also observed for lithium and the observed frequencies
are nearly identical in the two cases. As for benzene adsorbed on
silver, the molecular symmetry of benzene on Li seems to be reduced
from D_{6h} to C_{3v} (σ_d). Two different mechanisms were presented as
possible explanations for the symmetry reduction on silver and it
was hoped that the lithium results would help to choose between them.
If an apparent symmetry other than C_{3v} (σ_d) was observed for benzene
adsorbed on lithium, the geometrical interpretation for symmetry
reduction would be supported. At room temperature lithium crystal-
lizes in a body-centered-cubic lattice which does not have a close-
packed plane equivalent to the fcc (111) plane used to explain the
silver results. There is no plane in a bcc lattice which could

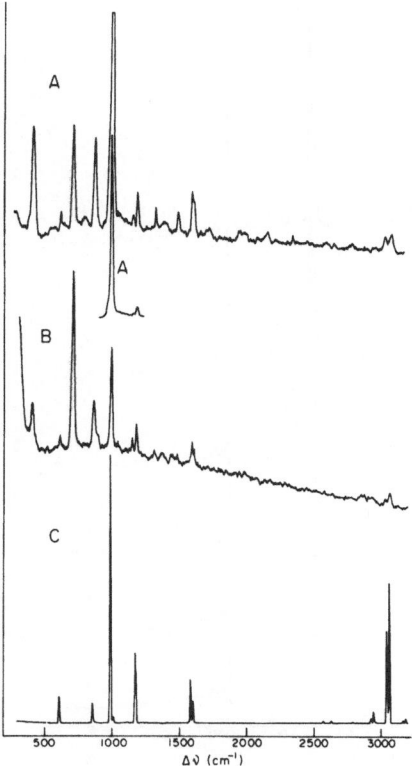

Fig. 14. A comparison of the SERS spectra of benzene adsorbed on
          (A) silver, (B) lithium, with bulk polycrystalline benzene
          (C).

provide C_{3v} adsorption sites. The geometrical interpretation cannot
be ruled out entirely, however, since it can transpire that the
molecular geometry of benzene is reduced on Li to a symmetry even
lower than C_{3v} but that the crucial bands discriminating between
the two groups are weak, or perhaps that lithium deposited on a
cold substrate at 12 K without annealing does not have bcc structure. The lithium atom is only 10 per cent smaller than the silver
atom; hence the geometries shown in Fig. 8 are equally pertinent to
lithium if a close-packed face were available.

    The relative intensities of the bands in the SERS spectrum of
benzene on lithium are quite different from those on silver (see
Fig. 14). The dissimilarities in the observed intensities support
the presence of a "first layer" effect assuming, of course, that
benzene adopts the same geometry relative to the metal surface in
the two cases. Advocates of the enhancement model based entirely
on intense electromagnetic surface fields predict relative intensities more or less proportional to those in the free molecule.[5,6]

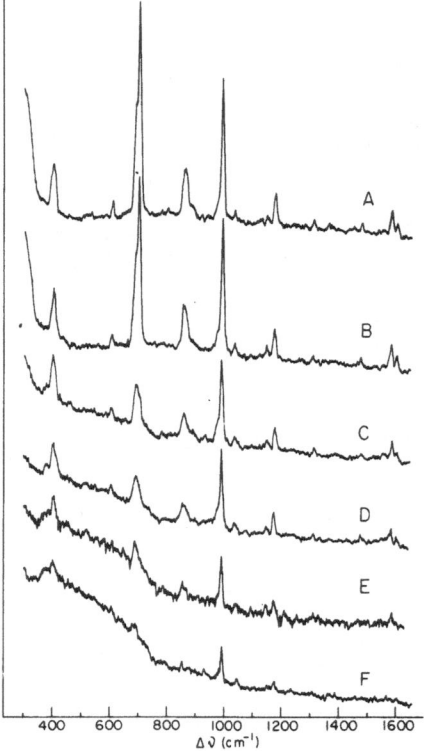

Fig. 15.   A series showing the effects of warmup on the SERS
           spectra of benzene adsorbed on lithium.   (A) As con-
           densed (12 K).   After warmup to (B) 28 K, (C) 40 K,
           (D) 72 K, (E) 114 K and (F) 137 K.

     One band in the SERS spectrum of benzene on lithium which
appears at 700 cm^{-1} and is nearly coincident with ν_{11} at 694 cm^{-1}
has no counterpart in the silver SERS spectrum.   This band behaves
quite differently from the other bands in the spectrum, having a
radically different excitation profile and being labile to warmup.
It disappears almost completely upon warmup to 40 K while the rest
of the spectrum remains almost unchanged.   The intensity of this
band also varies from sample to sample.   The other bands in the
spectrum gradually lose intensity upon warmup but don't disappear
completely until the temperature is raised above 150 K (Fig. 15).
The SERS spectrum of benzene-d$_6$ on lithium has an analogous band
at 517 cm^{-1} which also nearly coincides with ν_{11} (510 cm^{-1}).
Although we do not have, as yet, a clear understanding of the
source of this vibration, it appears from the isotope data that the
mode of ν_{11} perhaps belongs to a molecule adsorbed on a rather
unstable site.   Why only it among all the other modes of the molecule should be so greatly enhanced is still a mystery.

By way of pure speculation, however, we propose that this labile mode results from benzene molecules adsorbed on very sharp surface features at which the electric field gradient is very intense. v_{11} is particularly suited for activity through the field gradient mechanism since it causes the hydrogen atoms to follow a trajectory along which the field gradient varies most greatly, assuming, of course, that the molecule lies flat on the metal surface. The form of normal mode 11 is as follows:

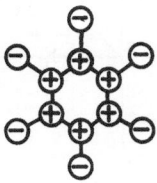

Careful excitation spectra for the lithium SERS are underway. Preliminary results indicate increasing enhancement towards the longer wavelength end of the range studied: 488.0 nm to 630 nm.

VII.   DIRECT OBSERVATION OF THE SPECTRUM OF SURFACE ROUGHNESS

In the preceding discussion the presence of microroughness was postulated to exist on the surface of the freshly deposited silver. It was also observed that as the silver film was annealed the enhanced Raman spectrum of the adsorbate began to disappear.

McBreen and Moskovits[32] have recently developed a technique which allows them to correlate the degree of microroughness with the observed surface Raman enhancement. This technique also afforded the first direct evidence for the presence of roughness on the scale proposed. Briefly the method works as follows: Light from a UV-visible source is passed through a polarizer, then through a stress modulator. The latter is excited with an electrical oscillator, causing periodic compression and dilation of a polished quartz block which acts as a retardation plate of periodically varying phase retardation. The light, whose polarization is modulated in this fashion, is reflected from the metal surface, passed through another polarizer, dispersed in a monochromator and detected by a photomultiplier tube (PMT).

A little mathematical analysis of this optical arrangement[32] indicates that when the first polarizer is perpendicular to the plane of incidence while the axes of the stress modulator and second polarizer are inclined at 45° to it, the output signal of the PMT consists of a DC term plus several sinusoidal components as follows:

$$I = K\{(DC\ terms) + 2(\rho_p^{\,2}-\rho_s^{\,2})J_2(\delta_0)\cos 2\omega t$$
$$+ 4\rho_p\rho_s \sin\Delta J_1(\delta_0)\ \sin\omega t + ...\} \quad (2)$$

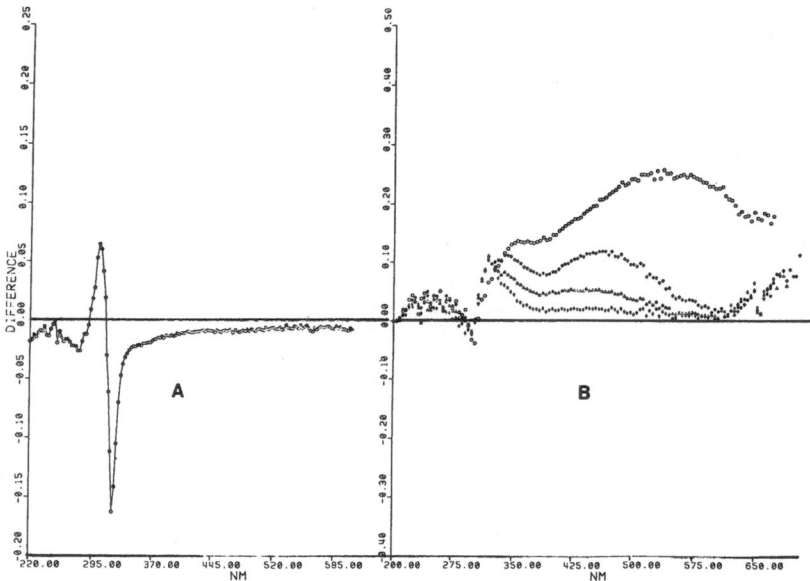

Fig. 16.   Portions of the UV-visible, stress modulator, ellipso-
           metric difference spectra of silver films (A) $B_{140}-B_{300}$
           for a film deposited at 300 K, then cooled to 140 K.
           (B) A series of $B_T-B_{300}^0$ where T is, top to bottom, 140,
           258, 268 and 300 K, for a film deposited at 140 K.

in which ρ_p, ρ_s and Δ are respectively the attenuation coefficients
following reflection of light polarized parallel to and perpendicular
to the plane of incidence and the phase difference between those two
components. J_1 and J_2 indicate Bessel functions of first and second
order, δ_0 is the maximum amplitude of retardation effected by the
stress modulator and ω the angular frequency of oscillation of the
latter. It follows from eq'n (2) that if one extracts the cosinus-
oidal component at frequency 2ω and the sinusoidal component at ω
and takes their ratio, all common contributions, indicated by K in
eq'n (2), cancel out and one is left with an expression which
depends only on the optical properties of the metal surface. This
quantity, which we call B, is given by

$$B = \frac{\tan^2\psi - 1}{2\tan\psi\sin\Delta} \cdot \frac{J_2(\delta_0)}{J_1(\delta_0)} \quad .$$

The quantities ψ and Δ are known as the ellipsometric parameters
and $\tan\psi = \rho_p/\rho_s$.

     Figure 16 shows a spectrum of the difference $B_{140}-B_{300}$ for a
silver film deposited on a substrate at 300 K at which temperature
the spectrum P_{300} was measured, then cooled to 140 K after which

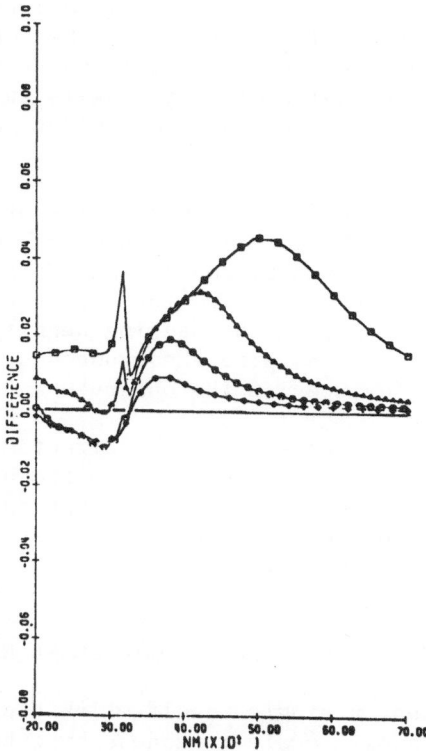

Fig. 17. A series of calculated ΔB curves attempting to reproduce
those of Fig. 16B by subtracting the B spectrum of a
smooth, silver film from those of silver films covered
with spherical silver particles with packing densities
(top to bottom) of 0.75, 0.55, 0.35 and 0.25.

the spectrum P_{140} was registered. The positive-negative couplet
observed results from the shift of an interband transition of
silver.[33] When the process was reversed and a silver film was
deposited at 140 K, then annealed to various temperatures (including
finally 300 K), one obtained the series of spectra shown in Fig. 16.
The spectra shown are $B_T - B^o_{300}$ where B^o_{300} refers to a well-annealed
silver film deposited at 300 K. The spectrum ($B_{140} - B^o_{300}$) shows, in
addition to the positive-negative couplet seen above and centered at
around 290 nm, an enormous absorption centered at 550 nm. On annealing, this feature decreases in both height and breadth and shifts to
higher photon energies, eventually almost disappearing when the film
is annealed to 300 K. This behavior is exactly that predicted by the
rough film model. When deposited on a cold substrate the film is
covered with very small, closely packed irregularities. Their small
size results in a broad band while their high density causes the
collective electron oscillation to occur in low photon energies.

On annealing, the small features coalesce into larger ones, causing
the bandwidth to decrease; at the same time the density of surface
bumps is reduced, producing the band center shift to lower wavelengths. Eventually the degree of roughness is sufficiently reduced
to make the conduction electron resonance feature almost impercep-
tible.

We have successfully modelled[32] the series of spectra shown in
Fig. 16B by calculating a B spectrum for a smooth silver film using
the optical constants reported by Johnson and Christy[34] and subtracting it from the B spectrum calculated for a silver surface covered
with silver spheres and varying the packing density. The results
are shown in Fig. 17 which, despite the crudeness of the model,
shows all the salient features of the observed spectrum.

The spectra of Fig. 16 are therefore direct manifestations of
the submicroscopic roughness features of the freshly deposited silver
film. At the very least they show that the optical properties of
silver films which produce enhanced Raman scattering are radically
different from those that do not.

VIII. LARGE-SCALE ROUGHNESS VERSUS ATOMIC-SCALE ROUGHNESS

Finally, we present a single result which tends to argue
strongly against the atomic scale roughness hypothesis. In Fig. 18
we show two excitation profiles associated with the CO stretching
frequency of CO adsorbed on silver. One of them was recorded for
CO adsorbed on a vapor-deposited silver surface,[11] the other for CO
adsorbed on well isolated, colloidal silver particles.[4] Although
the Raman spectra themselves obtained from the two samples are
almost identical, their excitation profiles are very different. In
the former case the profile shows a monotonic rise towards the red
while in the latter case the profile rises towards the blue. This
behavior is easily understandable in terms involving surface plasmons because the surface plasma frequency of spherical, well-
isolated, silver particles falls in the near ultraviolet while that
associated with surface bumps lies in the red as a result of dipolar
coupling among the closely-spaced surface irregularities, and
because of the non-spherical form of the surface features; both
effects are known to push the surface plasma frequency to lower
energies. The difference in the excitation profiles of the two
samples is not clearly explicable in terms of the adatom model,[35]
which ultimately implies that the adatom-adsorbate system forms a
colored complex whose absorption spectrum should not vary appreciably with the form of the underlying silver substrate.

We do not wish to conclude, however, with a statement regarding
the enhancement mechanism, since we believe that that issue has been
resolved, at least in broad terms. Instead we draw attention to the

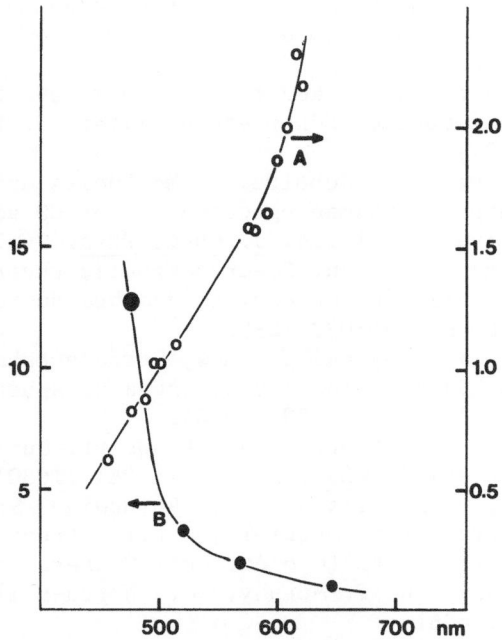

Fig. 18. Excitation profiles of the CO stretching SERS band of CO
          adsorbed on (A) a vapor-deposited, polycrystalline silver
          film and (B) on silver colloidal particles.[4]

utility of SERS in the study of surface reactions, the spectra of
adsorbed molecules with the added feature of the excitation of
normally forbidden modes, and subtle metal surface effects associated with the surface field gradient.

ACKNOWLEDGEMENTS

     We are grateful to Imperial Oil, the donors of the Petroleum
Research Fund and NSERC for financial assistance and to Dr. Peter
McBreen and Mr. Robert Lipson for many stimulating discussions. Our
thanks to almost all the authors in this book for sending to us
manuscripts before their publication.

REFERENCES

1. D.P. DiLella, A. Gohin, R.H. Lipson, P. McBreen and M. Moskovits,
      Enhanced Raman spectroscopy of CO adsorbed on vapor-deposited

silver, J. Chem. Phys. 73:4282 (1980).

2. M. Moskovits and D. P. DiLella, Enhanced Raman spectra of ethyl-
    ene and propylene adsorbed on silver, Chem. Phys. Lett.
    73:500 (1980).

3. D. P. DiLella and M. Moskovits, The surface-enhanced Raman spec-
    tra of some butenes adsorbed on silver, J. Phys. Chem. in
    press (1981).

4. H. Abe, K. Manzel, W. Schulze, M. Moskovits and D. P. DiLella,
    Surface-enhanced Raman spectroscopy of CO adsorbed on col-
    loidal silver particles, J. Chem. Phys. 74:792 (1981).

5. J. Gersten and A. Nitzan, Electromagnetic theory of enhanced
    Raman scattering by molecules adsorbed on rough surfaces,
    J. Chem. Phys. 73:3023 (1980).

6. M. Kerker, D.-S. Wang and H. Chew, Surface enhanced Raman scat-
    tering (SERS) by molecules adsorbed at spherical particles,
    Appl. Opt. 19:3373, 4159 (1980).

7. S. L. McCall, P. M. Platzman and P. Wolff, Surface enhanced
    Raman scattering, Phys. Lett. 77A:381 (1980).

8. C. A. Murray, D. L. Allara and M. Rhinewine, Surface enhanced
    Raman scattering in multi-layer film structures, in: "Pro-
    ceedings of the 7th International Conference on Raman
    Spectroscopy," W. F. Murphy, ed., North-Holland, Amsterdam
    (1980), p. 406.

9. D. A. Zwemer, C. V. Shank and J. E. Rowe, Surface-enhanced
    Raman scattering as a function of molecule-surface separa-
    tion, Chem. Phys. Lett. 73:201 (1980).

10. See, for example, Fig. 3 of reference 5.

11. D. P. DiLella and M. Moskovits, to be published.

12. K. G. Caulton, R. L. DeKock and R. F. Fenske, A comparison of
    carbon monoxide and nitrogen as ligands in transition metal
    complexes, J. Amer. Chem. Soc. 92:515 (1970).

13. M. Moskovits and D. P. DiLella, Surface-enhanced Raman spectros-
    copy of benzene and benzene-d_6 adsorbed on silver, J. Chem.
    Phys. 73:6068 (1980).

14. E. B. Wilson, Jr., J. C. Decius and P. C. Cross, "Molecular
    Vibrations," McGraw-Hill, New York (1955).

15. A. D. Buckingham, Permanent and induced molecular moments and
    long-range intermolecular forces, in: "Advances in Chemi-
    cal Physics," 12:107 (1967), J. O. Hirschfelder, ed.

16. J. K. Sass, H. Neff, M. Moskovits and S. Holloway, Electric
    field gradient effects on the spectroscopy of adsorbed mole-
    cules, J. Phys. Chem. 85:621 (1981).

17. A. D. Buckingham and L. D. Barron, Rayleigh and Raman scat-
    tering from optically active molecules, Mol. Phys. 20:1111
    (1971).

18. P. J. Feibelman, Microscopic calculations of electromagnetic
    fields in refraction at a jellium-vacuum interface, Phys.
    Rev. B 12:1319 (1975).

19. M. Moskovits, Surface roughness and the enhanced intensity of
    Raman scattering by molecules adsorbed on metals, J. Chem.

Phys. 69:4159 (1978); Enhanced Raman scattering by molecules adsorbed on electrodes--a theoretical model, Solid State Commun. 32:59 (1979).

20. J. R. Anderson, B. G. Baker and J. V. Sanders, Structure and properties of evaporated metal films, J. Catal. 1:443 (1962).

21. S. J. Cyvin, J. E. Rauch and J. C. Decius, Theory of hyper-Raman effects (nonlinear inelastic light scattering): Selection rules and depolarization ratios for the second-order polarizability, J. Chem. Phys. 43:4083 (1965).

22. P. N. Sanda, J. E. Demuth, J. C. Tsang and J. M. Warlaumont, this volume.

23. G. Varsányi, "Vibrational Spectra of Benzene Derivatives," Academic, New York (1969).

24. R. Dornhaus, M. B. Long, R. E. Benner and R. K. Chang, Time development of SERS from pyridine, pyrimidine, pyrazine and cyanide adsorbed on Ag electrodes during an oxidation-reduction cycle, Surf. Sci. 93:240 (1980).

25. M. R. Mahoney, M. W. Howard and R. P. Cooney, Carbon dioxide conversion to hydrocarbons at silver electrode surfaces. Raman spectroscopic evidence for surface carbon intermediates, Chem. Phys. Lett. 71:59 (1980).

26. R. G. Meisenheimer and J. N. Wilson, Interaction of oxygen and ethylene dichloride with silver surfaces, J. Catal. 1:151 (1962).

27. B. Pettinger and H. Wetzel, this volume.

28. R. Naaman, S. J. Buelow, O. Cheshovsky and D. R. Herschbach, Surface-enhanced Raman scattering from molecules adsorbed on mercury, J. Phys. Chem. 84:2692 (1980).

29. H. Yamada and Y. Yamamoto, Surface enhanced Raman spectra of pyridine adsorbed on silver, gold, nickel and platinum metals, Chem. Phys. Lett. 77:520 (1981).

30. W. Schulze and M. Moskovits, unpublished results.

31. H. Metiu and W. E. Palke, The infrared spectroscopy of chemi-sorbed molecules: A dynamical theory of the line shape, J. Chem. Phys. 69:2574 (1978).

32. P. H. McBreen, Ph.D. Thesis, "U.V. Visible Spectroscopy of Molecules Adsorbed on Metals," University Microfilms, Ann Arbor, Michigan (1981); D. P. DiLella, R. H. Lipson, P. McBreen and M. Moskovits, Metal molecules, metal clusters and metal bumps, J. Vac. Sci. Tech., in press (1981).

33. F. Wooten, "Optical Properties of Solids," Academic, New York (1976).

34. P. B. Johnson and R. W. Christy, Optical constants of noble metals, Phys. Rev. B 6:4370 (1972).

35. A. Otto, I. Pockrand, J. Billmann and C. Pettenkofer, this volume.

ELECTROCHEMICAL EFFECTS

M. Fleischmann and I.R. Hill

Chemistry Department, University of Southampton

Southampton, SO9 5NH, Great Britain

INTRODUCTION

The first *in situ* Raman spectroelectrochemical measurements on thin films of Hg_2Cl_2, Hg_2Br_2 and HgO formed on small droplets of mercury electrochemically deposited on inert substrates are illustrated in Fig. 1.[1]   These experiments already showed that as little as two monolayers of the solid phases could be detected on a liquid substrate using the spectrometer systems then available.   Since that time there has been only a limited number of measurements of the formation of solid phases on electrodes (such as of $Pb(OH)_xCl_y$ species formed on lead electrodes in aqueous chloride solutions of varying pH[3] as well as of non-electrochemical corrosion films on lead[4]).   Although measurements on systems of this type would undoubtedly be important both in the basic study of corrosion as well as in the investigation of the mechanisms of the surface enhancement of Raman spectra (especially experiments on liquid substrates, including droplets of defined size), the bulk of the work so far has been concerned with the investigation of enhanced spectra of molecules and ions adsorbed at silver, copper and gold electrodes (see Table 1).

In this article we shall concentrate mainly on the observations which have been made on the SERS of adsorbed pyridine, halide ions and the pseudo-halide cyanide ion on macroscopic silver electrodes since these systems have been most extensively investigated and we will incorporate some of our recent results.   Comments on the mechanism of enhancement will be restricted to those experiments which indicate the local structure of the scattering species.

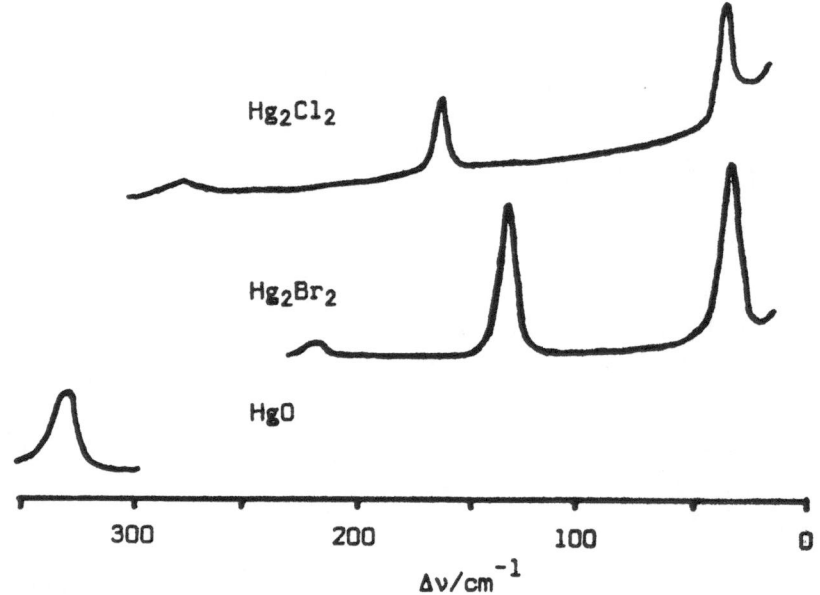

Fig. 1 *In situ* Raman scattering from thin films of electrochemically generated mercury compounds.

The Ag/pyridine/halide ion system

In spite of the considerable amount of work which has been
carried out on this system (especially using chloride solutions),
many of the results are still incompletely defined and the interpretation of the data thus remains uncertain. We therefore give a brief
(and to some extent chronological) review of work on this system.

The spectrum of pyridine adsorbed on silver electrodes from
0.05M pyridine/0.1M KCl was first recorded after the application of
about 150 linear potential sweeps between +200mV and -300mV[2] so as
to roughen the surface and increase the area (Fig. 2). At 0.0V (SEC)
the ring stretching mode region revealed bands at 1036, 1025, and
1008 cm^{-1} which all changed markedly in intensity as the potential
was increased to -1.0V. The 1025 cm^{-1} band was tentatively assigned
as pyridine chemisorbed to silver via Lewis acid coordination through
nitrogen, whereas the 1036 and 1008 cm^{-1} bands, being similar in frequency to those in aqueous solution were assigned as pyridine physisorbed at the electrode/electrolyte interface via water. The maximum
intensity of these two bands was at -0.6V, close to the point of zero
charge where maximum adsorption is expected for neutral molecules.
The two bands also shifted slightly with potential, and this shift
was thought to reflect reorientation of both water and pyridine in
the electrical double layer. Fig. 3 illustrates the model which was
proposed for adsorption at potentials positive and negative of the
point of zero charge.

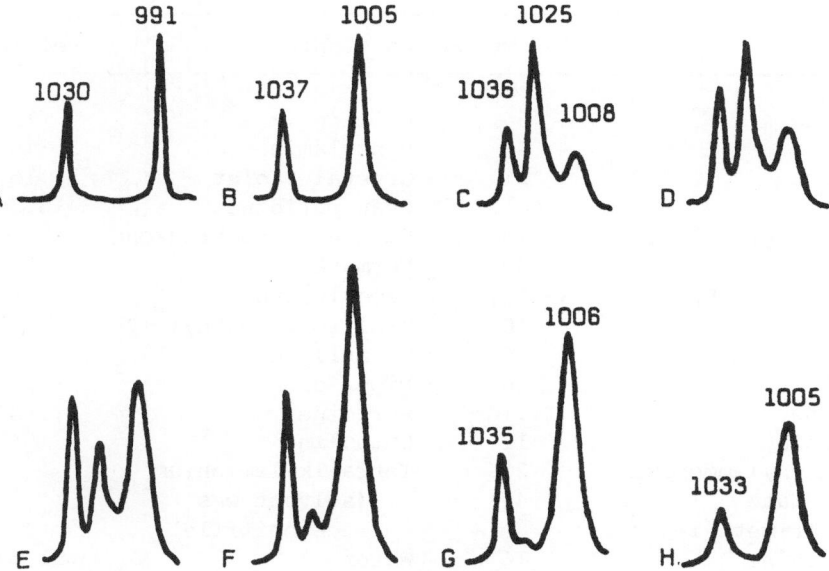

Fig. 2   Raman spectra of pyridine in solution and at the silver
         electrode. (A) liquid pyridine; (B) 0.05M aqueous
         pyridine; (C) silver electrode in 0.1M KCl/0.05M pyridine
         solution at 0.0V (SCE); (D) -0.2V; (E) -0.4V; (F) -0.6V;
         (G) -0.8V; (H) -1.0V.

Fig. 3   Possible models for pyridine adsorbed to a silver electrode
         at anodic and cathodic potentials[2].

Table 1. SERS spectra of species adsorbed at silver,
          copper and gold electrodes.

Silver		Silver	
Adsorbate	Reference	Adsorbate	Reference
Acridine	54	Aniline	16
Azide	40	Benzylamine	16
Carbonate	21	Crystal Violet	16
Cyanide	11,46-50,53	Cyanopyridines	16,36
NN-Dimethylaniline	16	Diphenylthiocarbazone	37
EDTA	39	Formate	21
Halides (Cl, Br, I)	33,51,52	Isoquinoline	54
Methyl Orange	16	p-Nitrosodimethyl-	
Oxalmethyline	16	aniline	38
Phenol	20	Piperidine	16
Pyrazine	11,16	Pyridine	2,5-33,53
Pyrimidine	11	Quinoline	54
Tetramethylammonium	20	Tetraalkylammonium	
Thiocyanate	42	(solvent was	
2,4,6 Trimethyl-		acetonitrile)	43
pyridine	16	Water	51,52
Glycine	41	Inosine	41
Glutamine	41	β-Alanine	41
Guanine	41		

Copper		Gold	
Pyridine	13,20,26, 44,45	Pyridine	20,26,44

Note. This table includes only in situ electrochemically generated
       SERS and does not include all investigations of the spectra
       of adsorbed pyridine.

Following these initial measurements Jeanmaire and Van Duyne[16]
and Albrecht and Creighton[5] showed that even more intense bands
could be obtained from electrodes which had been subjected to a
single roughening cycle only: prolonged roughening leads to a reduction in intensity. Enhancements 10^4 to 10^6 times compared to the
intensities predicted from the scattering cross sections for bulk
pyridine have been claimed. Electrodes roughened with a single

cycle and especially electrodes roughened using potential steps rather than ramps showed only a weak band due to Lewis acid coordinated pyridine at 1025 cm^{-1}.

Since these initial investigations, numerous further observations have been made which we summarise as follows:

(a) Using Ag/Cl$^-$/pyridine there is a large change in the relative intensity of the 1008 and 1036 cm^{-1} bands of pyridine as the applied potential is changed from 0.0V to -0.6V (already apparent in Fig.2). If F$^-$ ions, which do not specifically adsorb, are used then over the same potential range, the change in relative intensity is much smaller but is still present. Clearly there must be interactions between pyridine and the anions in the double layer. Using piperidine instead of pyridine no such changes are observed (see (c)(iv) below) in F$^-$ or Cl$^-$ so it seems that π electrons of the pyridine ring must be involved in these interactions.

(b) Marinyuk et al[18] have suggested that, because the pyridine bands at 617, 1005, 1215 and 1590 cm^{-1} all increase in intensity between -0.3V and -0.8V, the pyridine molecule probably reorientates from an end-on coordination to a flat coordination on the surface at -0.8V, a conclusion somewhat at variance with measurements of adsorption of aromatic systems by direct electrochemical techniques.[57,58] In the flat configuration the overlap of wave functions between electrons in the metal and the π electrons of the pyridine rings is at a maximum, leading to higher band intensities. Wetzel et al[33] have recently reported weak bands at 170 and 250 cm^{-1} which also reach a maximum near -0.8V. These bands were neither cation nor anion dependent and so could arise from pyridine in a nearly-flat configuration complexed to a silver adatom. Independently, we have also observed these bands using potential modulated Raman spectroscopy for pyridine in fluoride solutions.

(c) Measurements of the differential capacitance of electropolished (100), (110) and (111) silver single crystal faces and of polycrystalline silver are strongly dependent on the nature of the surface as well as the composition of the solution. Such measurements have shown that:

(i)   Electrochemical roughening cycles applied to polycrystalline electrodes as well as (100), (110) and (111) single crystal faces produce capacitance/voltage curves which closely resemble each other and which can be interpreted as a superposition of the curves for the individual single crystal faces[55]. It follows that the roughened electrode consists of an ensemble of faceted microcrystals.

(ii) The differential capacitance in the potential region -0.05 - -1.45V of a silver microelectrode roughened in the dark in

0.1M KCl/0.05M pyridine is identical to that measured with a
laser beam irradiating the whole of the surface. It follows
that the laser beam is sampling the species normally present
in the double layer[56].

(iii) Nonetheless roughening of the surface in solutions containing
      chloride ions and pyridine but with the laser beam incident
      during the single roughening cycle leads to the observation of
      an intense band due to chemisorbed pyridine (band at 1025 cm^{-1}).
      It follows that sites where Lewis acid coordinated adsorption
      can take place may be formed by photolysis of silver chloride
      and the adsorbed species are stabilised in the presence of
      chloride ions[34].

(iv) The capacitance curves indicate that adsorbed layers of
     chloride ions restructure with increasing positive potential.
     Pyridine is most strongly adsorbed on (110) faces less strongly
     on (100) and weakest on (111) faces in the presence of chlor-
     ide ions; the adsorption of chloride ions is in turn stabil-
     ised by the adsorption of pyridine (see (iii) above also (a) (d)).

     (d) Pettinger and Wenning[24] have reported the SERS of pyridine
adsorbed on (100), (111) and polycrystalline silver electrodes subjected to a minimal roughening cycle. Notwithstanding the results
in (c)(i) above, it was found that the most significant difference
between the surfaces was the observation of a band at 1025 cm^{-1} on
the (100) and polycrystalline surfaces which shifted to 1018 cm^{-1} on
the (111) surface. This shift may be related to the changes in the
strength of adsorption, see (c)(iv) above; we have also observed a
weak unresolved feature at 1017 cm^{-1} in the SERS spectrum of roughened polycrystalline silver and this may be due to weakened adsorption on (111) facets.

     (e) Dornhaus and Chang[12] report that, for the Ag/Cl^-/pyridine
system, the band seen at 243 cm^{-1} at +0.05V shifts to 216 cm^{-1} at
-0.65V and to 210 cm^{-1} at -1.0V. A similar shift is seen using
chloride solutions alone[33] although the chloride ion is then desorbed
by -0.6V, see (c). A lower frequency for the vibration of the
chloride ion in the double layer is expected at reduced coverage[59];
however, addition of pyridine does not lead to an increase in this
frequency which should accompany the increased specific adsorption.
Using Br^- and I^- corresponding bands are observed at 182 cm^{-1} (-0.8V)
and 116 cm^{-1} (-0.75V) respectively[12]. Atkinson et al[8] report similar
results for Ag/Cl^-/py and add that the intensity ratio of the 1038
and 1011 cm^{-1} bands of pyridine exhibits a similar potential dependence to the intensity of the $\sim$240 cm^{-1} band, indicating a surface
complex involving both pyridine and chloride.   Pettinger and
Wetzel[27] have accurately measured the low frequency region (down to
20 cm^{-1}) of the SERS of the Ag/Cl^-/pyridine system and, for an
applied potential of -0.3V, report bands at 231, 150, 121 and 53 cm^{-1}.

This compares with bands reported by Fleischmann *et al*[51] for silver electrodes in chloride solutions alone at -0.2V located at 240, 150 and 110 cm^{-1}. These bands are also observed on dry electrodes, thus excluding water (or water containing species) as a possible source of the spectra, (Fig. 4). Both groups talk about the possibility of surface complexes, rather than simple adsorption.

(f) The fact that the interactions of the species at the surface are even more complicated than those described above is shown by the data of Regis and Corset[29]. These authors have found that the pyridinium ion, rather than pyridine, is preferentially adsorbed at the Ag electrode surface in the presence of specifically adsorbed Cl$^-$ ions. The pyridinium ions are strongly bound because, when the pH is increased to 8 the SERS of the pyridinium species is still observed, which would not be the case if the electrode had been roughened at this pH. The pyridinium and chloride ions are adsorbed as an ion pair and are desorbed near -0.6V. The spectrum of the pyridinium ion does not reappear at more negative potentials, indicating the importance of the nitrogen lone pair in the adsorption, even when the π electrons of the ring are available. The 1025 cm^{-1} band of the pyridinium ion must not be confused with the 1025 cm^{-1} band of chemisorbed (Lewis acid) pyridine. The former species has a band of similar intensity near 1011 cm^{-1} which distinguishes the two. Marinyuk *et al*[19] have employed D_2O solutions to show that the 1025 cm^{-1} band of pyridine adsorbed in the presence of I$^-$ ions arises from Lewis acid rather than $C_5H_5ND^+$ which would be at 994 cm^{-1}. These authors have pointed out that the adsorbed halide ions and not the metal may be acting as Lewis acid sites. The spectrum of Lewis acid pyridine can be separated from that of physisorbed pyridine by diluting away the pyridine from the electrolyte[34,35] when the 1025 cm^{-1} band maintains its intensity.

(g) In addition to the interactions described in (f), measurements of the spectra of H_2O and D_2O both in solutions containing halide ions alone and in solutions containing halide ions and pyridine give further information about the nature of the adsorbed species. It is found that:

(i)   Using 1.0M KCl, electrochemical roughening leads to the observation, at -0.2V, of a band at 3498 cm^{-1} arising from adsorbed water[51,52] (Fig. 4). If pyridine is then added to the electrolyte the intensity of the 3498 cm^{-1} band is maintained. However, using 0.2M KCl, addition of pyridine results in a reduction in intensity of the band at 3498 cm^{-1}, while the bands due to pyridine appear.

(ii) Using 1.0M KCl, the spectrum of adsorbed water shows that rapid exchange between H_2O and D_2O takes place when D_2O is added, by the appearance of bands associated with HDO and D_2O.

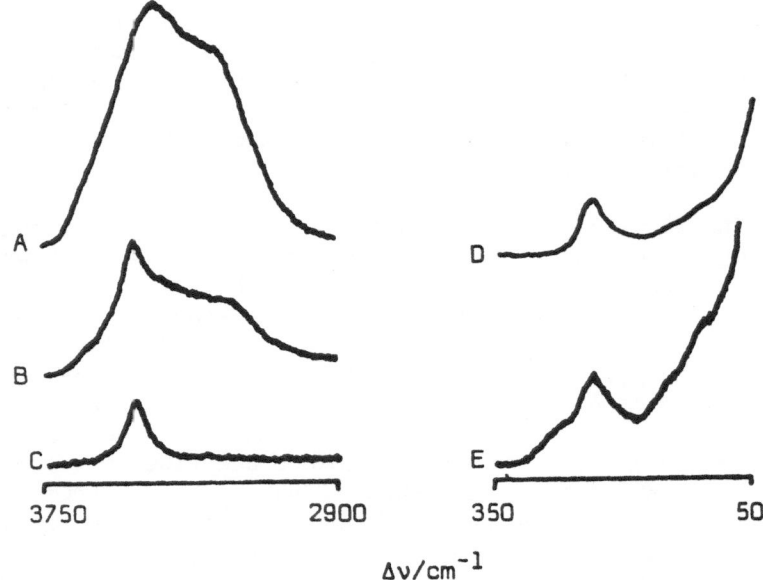

Fig. 4   Raman spectra of a silver electrode in 1M KCl/H$_2$O.
         (A) Polished electrode (bulk electrolyte only); (B), (D)
         roughened electrode at -0.2V; (C) electrode pushed against
         cell window, (E) dried, roughened electrode in air[51].

(iii) Using 1M KCl, addition of pyridine and HCl results in the
      observation of the SERS spectrum of the pyridinium ion which
      decreased in intensity at more negative potentials along with
      that of the chloride ion[29].   However, the pyridinium ion does
      not displace the 3498 cm^{-1} band.

      From (i) it appears that pyridine needs to be adsorbed at the
silver electrode for enhancement to occur whereas, from (iii) the
pyridinium ion can be coordinated via the chloride species.   From
(ii) water is not strongly bound in the double layer although it is
neither displaced by pyridine nor by pyridinium, and this may be due
to a size effect.   Fleischmann *et al*[2] originally suggested a model
of the double layer region in which pyridine was adsorbed to the
surface via water molecules, Fig. 3.   We have since failed to locate
a SERS spectrum of such water using potential modulation methods;
pyridine is therefore probably in direct contact with silver.   We
now believe that the chemisorbed form of pyridine is an essentially
Ag^I surface species whereas the physisorbed form is a Ag^0 surface
species[34] (compare results for the CN$^-$ system below).

The Adsorption of Cyanide Ions

      The adsorption of cyanide to silver electrodes has been studied
by several workers (see Table 1).   Furtak[48], using 0.1M Na$_2$SO$_4$/

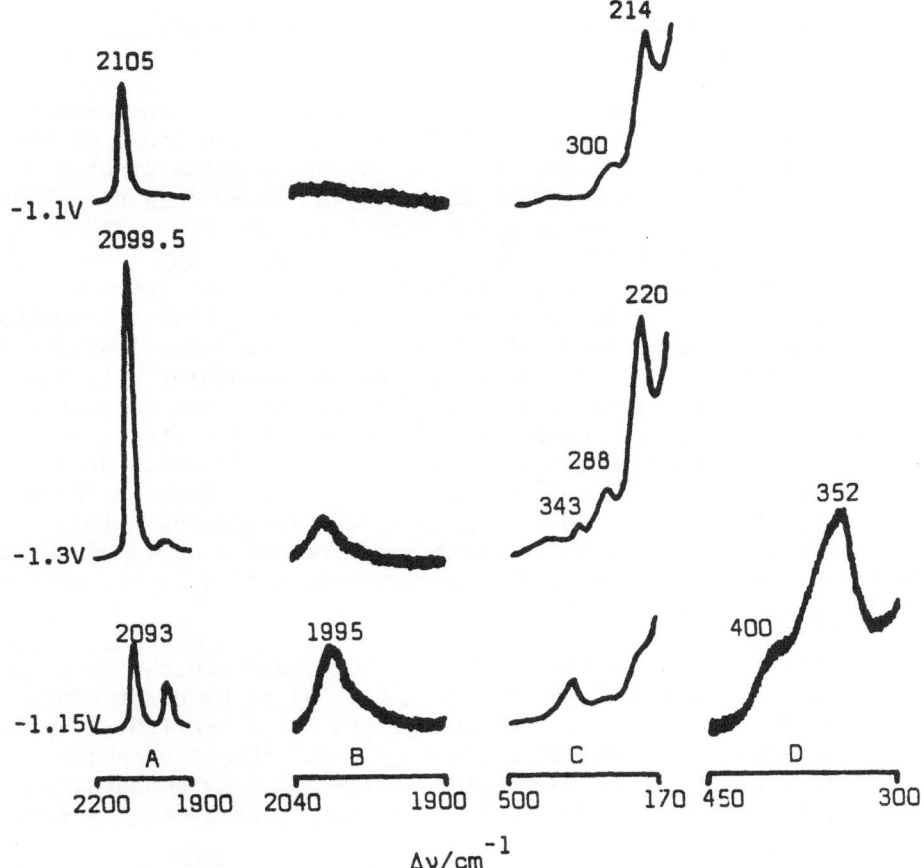

Fig. 5   SERS spectra from Ag/0.5M KCN showing the appearance of a
         new species at more negative potentials.   Spectra (B) and
         (D) are enlargements of bands in (A) and (C) respectively.

0.01 KCN, obtained a SERS spectrum of cyanide on silver at -0.8V
with strong bands at 2114 and 226 cm^{-1}, and weaker ones at 315 and
150 cm^{-1}.   From a comparison of the frequency of the CN stretching
mode in the SERS spectrum with those of di, tri and tetra-cyano-
argentate(I) ions in solution, the surface complex was assigned as
tri-cyanoargentate[47].   Dornhaus *et al*[11,46] used an OMA system to
relate the SERS of cyanide on Ag with the voltammogram.   A band was
seen at 2140 cm^{-1} and -0.3V which rapidly shifted to 2110 cm^{-1} by
-0.5V.   This shift was interpreted as a change in coordination from
2 to 3 or 4 cyano groups.   The non-Lorentzian 2110 cm^{-1} band was
fitted with 3 Lorentzian peaks located at 2095, 2110 and 2140 cm^{-1}
(for the 3 complexes).   Furtak *et al*[49] have shown that the formation
of a Ag(1) compound is important in the SERS activation process and
that the potential needs to be changed rapidly between 0.0V and

-0.8V otherwise the SERS is lost through dissolution of silver as
cyanide complexes.

From our own measurements it seems, however, that one cannot
simply compare the frequencies of the SERS bands with those of the
solution species[60].   Thus the frequency of the cyanide stretch
shifts from 2108 cm^{-1} at -1.0V to 2093 cm^{-1} at -1.5V (Fig.5).   These
frequencies may be compared with solution frequencies of 2108 cm^{-1}
for Ag(CN)$_3^{2-}$ and 2097 cm^{-1} for Ag(CN)$_4^{3-}$.   Hence, if the frequency
of the surface species reflects its coordination, then there should
be a change in coordination between -1.0V and -1.5V and a concomitant
change in band shape as this takes place.   The band at -1.0V is, in
fact, asymmetric to the low frequency side but maintains this same
asymmetry through to -1.5V, and this therefore does not reflect a
change in coordination.   Moreover we have detected a strong band
arising from coadsorbed water at 3521 cm^{-1} and -1.0V which progressively shifts to 3431 cm^{-1} at -1.5V (Fig. 6).   Both the 2108
and 3521 cm^{-1} bands shift in a roughly linear fashion over this
potential range, so the cyanide stretching frequency would appear to
reflect the strength of the metal-cyanide interaction rather than the
formation of different complexes.

Substitution of ^{13}CN for ^{12}CN leads to measurable shifts of the
bands at 2102.5, 2015, 342 and 221 cm^{-1} (-1.2V) to bands at 2058.5,
1969, 334, 289 and 213 cm^{-1}, confirming that all arise from cyanide
modes.   The weak bands at 400 and 154 cm^{-1} in ^{12}CN also appear
to shift to lower frequencies but the shifts are only comparable
with the error of the measurements.   For complexes involving more

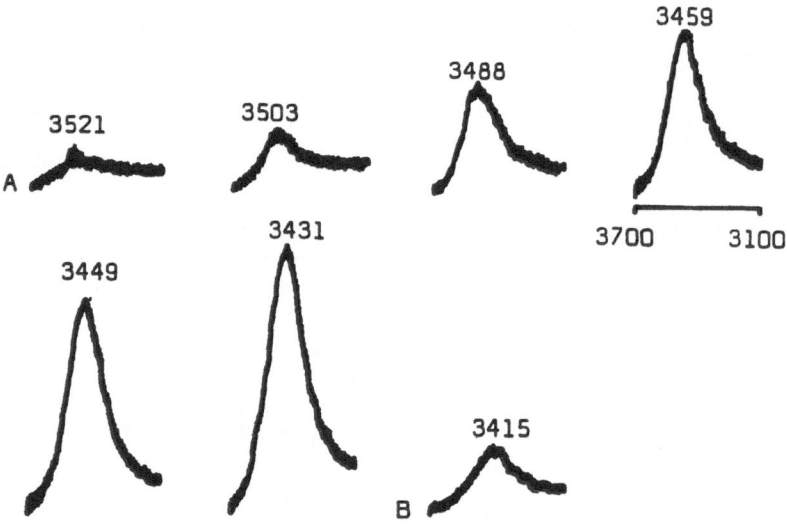

Fig. 6   SERS spectra of adsorbed water, obtained from Ag/0.5M KCN.
         The potential was increased in steps of 0.1V from -1.0V (A)
         to -1.6V (B).

than one ligand, isotopic mixtures can be used to determine the coordination number.   For example, a 50:50 mixture of ^{12}CN and ^{13}CN
in a solution of silver ions of the appropriate concentration will
form the di-cyanoargentate ions $[^{12}CN \ Ag \ ^{12}CN]^-$, $[^{12}CN \ Ag \ ^{13}CN]^-$and
$[^{13}CN \ Ag \ ^{13}CN]^-$ in the ratio 1:2:1.   Considering vibrations involving Ag-C bonds we would expect bands of 1:2:1 intensity ratios at
221, $\sim(221+213)/2$ and 213 cm^{-1} respectively.   Unfortunately, in
practise, this band is too broad to distinguish simple CN^- adsorption
from di-, tri- and tetra-cyanoargentate complexes.   However, substitution of ^{12}CN to give $[^{12}CN \ Ag \ ^{13}CN]^-$ also leads to a loss of
the centre of symmetry and does, in fact, lead to a 3.3 cm^{-1} frequency shift of the Raman active cyanide stretching mode[61].   For
unresolved bands in our 50:50 mixture the observed frequency shift
would only be one half of 3.3 cm^{-1}.   The shift we have observed is
1.3 cm^{-1}, in excellent agreement with the predicted 1.6 cm^{-1}.   Hence
the use of the isotopic mixture proves the presence of a complex.
Naturally detailed force constant analysis would need to be carried
out to confirm that the complex involves 2 rather than 3 or 4
cyanides.   Isotopic mixtures therefore will have a useful rôle to
play in the investigation of surface structures.

From a general comparison of the cyanide surface spectrum with
the Raman spectra of the di-, tri- and tetra-cyanoargentate complexes
in solution, one sees that the relative intensities of di-cyanoargentate compare favourably with the surface spectrum.   Hence it is
reasonable to assign the surface complex to di-cyanoargentate(1),
with all bands shifted to lower frequency under the influence of the
applied potential[62].

An interesting new finding[60] has been the spectrum of what
appears to be the reduced form of the complex at very negative
potentials (Fig. 5).   The bands at 2097, 286 and 218 cm^{-1} at -1.4V
are accompanied by new bands at 2006, 400 and 348 cm^{-1}.   The positions of these bands are consistent with the reduction of $Ag^{I}(CN^-)_x \rightarrow$
$Ag^0(CN^-)_x$.   The observed bands are all broad with the 2006 cm^{-1}
band wider on the low frequency side and the 348 cm^{-1} band wider on
the high frequency side.   This complex is probably only weakly bound
to the surface.   The importance of this finding is that the previously observed spectra can be seen to be due to essentially Ag^+
species present at the interface even at fairly negative potentials.

A further assessment of the SERS of adsorbed halide ions

The nature of the cyanide complexes raises the issue of whether
Cl^- is present as simple adsorbed ions or, for example, in a complex
$(AgCl_2)^-$.   The low frequency region of the SERS spectrum of Cl^- on
Ag contains several bands[27,51].   Regis and Corset[29] have pointed out
that the Cl_2^- radical, which has been observed in doped KCl crystals,
has an optical absorption band at 750 nm and gives a strong resonance

Raman effect with the internal stretching vibration between 225 and 264 cm^{-1} depending on the alkali metal in the $M^+Cl_2^-$ ion pair. Such resonance enhancement could account for the increase in SERS intensity going from blue to red excitation. However our own measurements on pyridine adsorbed to silver from KF solutions also show a large increase in the SERS of pyridine going from blue to red laser excitation. Cl_2^- ions alone therefore do not account for the enhancement effects. The 240 cm^{-1} band is too broad for mixed isotope analysis to be of use in identifying coordination but indirect evidence for a complex may be obtained in the very low frequency region of the SERS spectrum. Weitz *et al*[32] reported an extremely intense band at 8 cm^{-1} in the SERS of Ag/Cl$^-$/pyridine. The band was said to be absent without pyridine but has since been observed using Cl$^-$ alone[51] and has also been seen to shift to even lower frequency using Br$^-$ and I$^-$ ions[51]. Weitz *et al*[32,66] also observed this band in the SERS of Ag/CN$^-$ and have assigned it to a rocking or libration of the adsorbed species about the point at which it is bonded to the surface. We have failed to locate this band using Ag/F$^-$/pyridine and so we believe that the band must be associated with a complex species such as $[AgX_2]^-$ in both cyanide and, by analogy, chloride.

We have also observed that, for the Ag/F$^-$/pyridine system, the SERS of the background due to carboxy species[13,63] (1580 and 1360 cm^{-1}) does not increase in intensity going from blue to red excitation, although the spectrum of pyridine does[10]. Hence it is possible that resonance Raman effects have a rôle to play in the further enhancement of the SERS of particular species[20].

As mentioned earlier, SERS spectra have been obtained of a chloride species in Ag/Cl$^-$ [33] along with a spectrum of coadsorbed water[51,52]. Further, the frequencies of both the chloride[33] and adsorbed water[62] bands shift with potential. Here we note that this adsorbed water band is not observed in 1M HCl but that it appears as soon as small quantities of KCl are added to the electrolyte[61]. Hence the adsorbed water appears to be in a bridging position between the chloride and potassium ions. Current work shows large changes in the SERS of water with the nature of the metal ion. For example, using 1M KI two bands arising from coadsorbed water are seen at 3491 and 3549 cm^{-1} (-0.9V), whose relative intensities invert between -0.9V and -1.1V. However, changing to NaI results in the observation of only one band at 3594 cm^{-1}; addition of a little KI to the NaI electrolyte again results in the observation of the bands associated with KI[62].

General Summary

It can be seen that for a number of systems it is now possible to obtain spectroscopic information about adsorbed anions, cations, solvent and neutral adsorbates.   This information gives a much more detailed picture of the species present in the double layer than can be derived from electrochemical measurements alone.

The experiments described here indicate that surface complexes involving adatoms of silver[24,47] are produced during the oxidation-reduction cycles.   The number of bands found at low frequencies for pyridine, chloride and cyanide imply the presence of such complexes which must contain more than one ligand attached to a silver atom; adatoms seem necessary from spatial requirements.   These complexes may be bound more by electrostatic forces than by chemical bonding, so a large part of the spectroscopic changes with applied potential may arise from changes in these electrostatic forces as well as from variations of the structure of the complex and its composition.   The quenching of SERS at high negative potentials[33] also indicates that adatoms are important.   In support of this, addition of silver and chloride complexes to the electrolyte after quenching has led to the recovery of SERS[33].   Furtak et al[49] believe that electrostatic forces are not sufficient in order to obtain SERS, otherwise bands arising from free cyanide near a silver surface would be observed. Hence a particular type of chemical bond between the ligand and silver is required.   However, the SERS of coadsorbed water[51,52] is observed at a frequency representative of electrostatic interaction rather than in the region of strong hydrogen bond formation.   The results we have briefly reported here on the cation dependence of the SERS of adsorbed water imply that the water molecule is adsorbed between cation and halide ions rather than directly to silver.

Although considerable progress has been made, it is now clearly essential to combine Raman (and indeed infra red[64]) spectroscopic information with conventional measurements of surface excesses in the compact double layer.   Such combined measurements should show $inter$ $alia$ whether the species being detected are the majority species present at the interface.   We note here that the structural information which has been obtained hitherto from SERS is necessarily somewhat limited because, in general, all normally highly polarised bonds are effectively depolarised[10,16].   We offer for further consideration some possible models (Fig. 7) for the structure of the species in the double layer for the three systems we have reviewed (the complexes are probably of C_{2v} symmetry rather than linear $D_{\infty h}$.

Finally, the conclusions which can be drawn at this time are also restricted by the small number of systems which have been studied.   Very weak spectra have already been obtained on substrates other than silver, copper and gold (e.g. of CO on Pt[65]) and it is possible that these spectra are not enhanced or only weakly enhanced.

The development of multiplexing spectrographs should allow the extension of the range of systems which can be studied by Raman

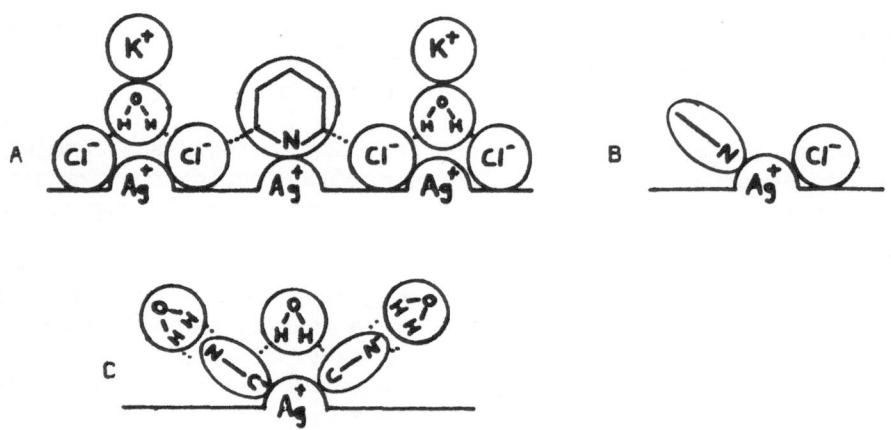

Fig. 7   These diagrams are simple representations of some possible
         surface complexes.   (A) Chemisorbed pyridine, stabilised
         by the presence of chloride ions.   Coadsorbed water is also
         shown.   A similar diagram could represent physisorbed
         pyridine, with the pyridine molecule in electrostatic inter-
         action with Ag⁰; (B) An alternative model for chemisorbed
         pyridine; (C) A diacyanoargentate complex.   Only one
         "sharp" band is seen in the SERS spectrum of water, so only
         one of these water molecules may be observed in the spectrum.

spectroscopy and thereby provide a more generally useful surface
analytic probe and a wider base for the testing of theories of
scattering phenomena at interfaces.

REFERENCES

1. M. Fleischmann, P. J. Hendra, and A. J. McQuillan, Raman spectra
        from electrode surfaces, J. Chem. Soc. Chem. Commun. (1973),
        p. 80.
2. M. Fleischmann, P. J. Hendra, and A. J. McQuillan, Raman spectra
        of pyridine adsorbed at a silver electrode, Chem. Phys. Lett.
        26:163 (1974).
3. E. S. Reid, R. P. Cooney, P. J. Hendra, and M. Fleischmann, A
        Raman spectroscopic study of corrosion of lead electrodes in
        aqueous chloride media, J. Electroanal. Chem. 80:405 (1977).
4. R. J. Thibeau, C. W. Brown, A. Z. Goldfarb, and R. H. Heiders-
        bach, Infrared and Raman-spectroscopy of aqueous corrosive
        films on lead in 0.1 M solution, J. Electrochem. Soc. 127:37
        (1980).
5. M. G. Albrecht and J. A. Creighton, Anomalously intense Raman

spectra of pyridine at a silver electrode, J. Am. Chem. Soc. 99:5215 (1977).

6. M. G. Albrecht, J. F. Evans, and J. A. Creighton, The nature of an electrochemically roughened silver surface and its role in promoting anomalous Raman scattering intensity, Surf. Sci. 75:L777 (1978).

7. M. G. Albrecht and J. A. Creighton, Intense Raman spectra at a roughened silver electrode, Electrochim. Acta 23:1103 (1978).

8. G. F. Atkinson, D. A. Guzonas, and D. E. Irish, Raman spectral studies at the silver surface of the Ag|KCl, pyridine electrode, Chem. Phys. Lett. 75:557 (1980).

9. R. L. Birke, J. R. Lombardi, and J. Gersten, Observation of a continuum in enhanced Raman scattering from a metal-solution interface, Phys. Rev. Lett. 43:71 (1979).

10. J. A. Creighton, M. G. Albrecht, R. E. Hester, and J. A. D. Matthew, The dependence of the intensity of Raman bands of pyridine at a silver electrode on the wavelength of excitation, Chem. Phys. Lett. 55:55 (1978).

11. R. Dornhaus, M. B. Long, R. E. Benner, and R. K. Chang, Time development of SERS from pyridine, pyrimidine, pyrazine, and cyanide adsorbed on Ag electrodes during an oxidation-reduction cycle, Surf. Sci. 93:240 (1980).

12. R. Dornhaus and R. K. Chang, Comments on the 210-243 cm^{-1} mode in surface enhanced Raman scattering from the pyridine-Ag system, Solid State Commun. 34:811 (1980).

13. M. Fleischmann, P. J. Hendra, A. J. McQuillan, R. L. Paul, and E. S. Reid, Raman spectroscopy at electrode-electrolyte interfaces, J. Raman Spectrosc. 4:269 (1976).

14. R. M. Hexter, Enhanced Raman intensity of molecules adsorbed on metal surfaces. Experiments and theory, Solid State Commun. 32:55 (1979).

15. M. W. Howard, R. P. Cooney, and A. J. McQuillan, The origin of intense Raman spectra from pyridine at silver electrode surfaces: The role of surface carbon, J. Raman Spectrosc. 9:273 (1980).

16. D. L. Jeanmaire and R. P. Van Duyne, Surface Raman spectroelectrochemistry. Part I. Heterocyclic, aromatic and aliphatic amines adsorbed on the anodized silver electrode, J. Electroanal. Chem. 84:1 (1977).

17. B. H. Loo and T. E. Furtak, The giant Raman effect from pyridine on a chemically modified gold substrate, Chem. Phys. Lett. 71:68 (1980).

18. V. V. Marinyuk, R. M. Lazarenko-Manevich, and Ya. M. Kolotyrkin, Raman resonance scattering of pyridine adsorbed on silver, Elektrokhimiya 14:1019 (1978).

19. V. V. Marinyuk, R. M. Lazarenko-Manevich, and Ya. M. Kolotyrkin, The effect of halide-ions on the adsorption properties of silver, Elektrokhimiya 14:1747 (1978).

20. V. V. Marinyuk, R. M. Lazarenko-Manevich, and Ya. M. Kolotyr-

kin, Nature of the interaction of adsorbate molecules with metal adatoms, J. Electroanal. Chem. 110:111 (1980).

21. A. J. McQuillan, P. J. Hendra, and M. Fleischmann, Raman spectroscopic investigation of silver electrodes, J. Electroanal. Chem. 65:933 (1975).

22. V. V. Marinyuk and R. M. Lazarenko-Manevich, Raman scattering cross section of pyridine adsorbed on silver, Elektrokhimiya 14:452 (1978).

23. R. L. Paul and P. J. Hendra, Minerals Sci. Engng. 8:171 (1976).

24. B. Pettinger and U. Wenning, Raman spectra of pyridine adsorbed on silver (100) and (111) electrode surfaces, Chem. Phys. Lett. 56:253 (1978).

25. B. Pettinger, U. Wenning, and H. Wetzel, Angular resolved Raman spectra from pyridine adsorbed on silver electrodes, Chem. Phys. Lett. 67:192 (1979).

26. B. Pettinger, U. Wenning, and H. Wetzel, Surface plasmon enhanced Raman scattering frequency and angular resonance of Raman scattered light from pyridine on Au, Ag and Cu electrodes, Surf. Sci. 101:409 (1980).

27. B. Pettinger and H. Wetzel, Surface enhanced Raman spectroscopy of pyridine on Ag electrodes. Surface complex formation, Chem. Phys. Lett. 78:398 (1981).

28. B. Pettinger, Surface enhanced Raman spectroscopy of pyridine on Ag electrodes. Evidence for overtones, Chem. Phys. Lett. 78:404 (1981).

29. A. Regis and J. Corset, A chemical interpretation of the intense Raman spectra observed at a silver electrode in the presence of chloride ion and pyridine: Formation of radicals, Chem. Phys. Lett. 70:305 (1980).

30. W. Suetaka and M. Ohsawa, Potential modulation Raman spectrum of species on metal-electrode surface, Appl. Surf. Sci. 3:118 (1979).

31. R. P. Van Duyne, Laser excitation of Raman scattering from adsorbed molecules on electrode surfaces, in: "Chemical and Biochemical Applications of Lasers," Vol. 4, C. B. Moore, ed., Academic Press, New York (1978), p. 101.

32. A. Z. Genack, D. A. Weitz, and T. J. Gramila, Very low frequency surface enhanced Raman scattering, Surf. Sci. 101:381 (1980).

33. H. Wetzel, H. Gerischer, and B. Pettinger, Surface enhanced Raman scattering from silver-halide and silver-pyridine vibrations and the role of silver ad-atoms, Chem. Phys. Lett. 78:392 (1981).

34. M. Fleischmann, I. R. Hill, and M. E. Pemble, submitted to J. Electroanal. Chem.

35. A. Kuhn, private communication.

36. C. S. Allen and R. P. Van Duyne, Orientational specificity of Raman scattering from molecules adsorbed on silver electrodes, Chem. Phys. Lett. 63:455 (1979).

37. J. E. Pemberton and R. P. Buck, Dithizone adsorption at metal

electrodes. 2. Raman spectroelectrochemical investigation of effect of applied potential at a silver electrode, J. Phys. Chem. 85:248 (1981).

38. G. Hagen, B. S. Glavaski, and E. Yeager, The Raman spectrum of an adsorbed species on electrode surface, J. Electroanal. Chem. 88:269 (1978).

39. H. Wetzel, B. Pettinger, and U. Wenning, Surface enhanced Raman scattering from ethylenediaminetetraacetic-disodium salt and nitrate ions on silver electrodes, Chem. Phys. Lett. 75:173 (1980).

40. R. Kunz, J. G. Gordon II, M. R. Philpott, and A. Girlando, Surface enhanced Raman spectra from silver electrodes in azide solution, J. Electroanal. Chem. 112:391 (1980).

41. S. Venkatesan, G. Erdheim, J. R. Lombardi, and R. L. Birke, Voltage dependence of the surface-molecule line in the enhanced Raman spectrum of several nitrogen containing compounds, Surf. Sci. 101:387 (1980).

42. R. P. Cooney, E. S. Reid, M. Fleischmann, and P. J. Hendra, Thiocyanate adsorption and corrosion at silver electrodes: A Raman spectroscopic study, J. Chem. Soc. Faraday I 73:1691 (1977).

43. V. V. Marinyuk, R. M. Lazarenko-Manevich, and Ya. M. Kolotyrkin, Resonance Raman scattering of organic cations adsorbed on silver, Dokl. Acad. Sci. U.S.S.R. 242:1382 (1978).

44. U. Wenning, B. Pettinger, and H. Wetzel, Angular-resolved Raman spectroscopy of pyridine on copper and gold electrodes, Chem. Phys. Lett. 70:49 (1980).

45. R. L. Paul, A. J. McQuillan, P. J. Hendra, and M. Fleischmann, Laser Raman spectroscopy at the surface of a copper electrode, J. Electroanal. Chem. 66:248 (1975).

46. R. E. Benner, R. Dornhaus, R. K. Chang, and B. L. Laube, Correlations in the Raman spectra of cyanide complexes adsorbed on Ag electrodes with voltammograms, Surf. Sci. 101:341 (1980).

47. J. Billmann, G. Kovacs, and A. Otto, Enhanced Raman effect from cyanide adsorbed on a silver electrode, Surf. Sci. 92:153 (1980).

48. T. E. Furtak, Anomalously intense Raman scattering at the solid-electrolyte interface, Solid State Commun. 28:903 (1978).

49. T. E. Furtak, G. Trott, and B. H. Loo, Enhanced light scattering from the metal/solution interface: Chemical origins, Surf. Sci. 101:374 (1980).

50. J. Timper, J. Billman, A. Otto, and I. Pockrand, Surface enhanced light scattering from silver electrodes: Background and CN stretch vibration, Surf. Sci. 101:348 (1980).

51. M. Fleischmann, P. J. Hendra, I. R. Hill, and M. E. Pemble, Enhanced Raman spectra from species formed by the coadsorption of halide ions and water molecules on silver electrodes, J. Electroanal. Chem. 117:243 (1981).

52. B. Pettinger, M. R. Philpott, and J. G. Gordon II, Contribution of specifically adsorbed ions, water, and impurities to the surface enhanced Raman spectroscopy (SERS) of Ag electrodes, J. Chem. Phys. 74:934 (1981).

53. C. Y. Chen, E. Burstein, and S. Lundquist, Giant Raman scattering by pyridine and CN^- adsorbed on silver, Solid State Commun. 32:63 (1979).

54. A. Girlando, J. G. Gordon II, D. Heitmann, M. R. Philpott, H. Seki, and J. D. Swalen, Raman spectra of molecules on metal surfaces, Surf. Sci. 101:417 (1980).

55. M. Fleischmann, J. Robinson, and R. Waser, An electrochemical study of the adsorption of pyridine and chloride ions on smooth and roughened silver surfaces, J. Electroanal. Chem. 117:257 (1981).

56. M. Fleischmann and P. Graves, in preparation.

57. B. E. Conway, R. G. Barradas, P. G. Hamilton, and J. M. Parry, Electrochemical adsorption of neutral and ionic components in solutions of pyridine and derived ions, J. Electroanal. Chem. 10:485 (1965).

58. R. G. Barradas and B. E. Conway, Some applications of ultraviolet spectrophotometry to studies of adsorption at copper, nickel and silver electrodes, J. Electroanal. Chem. 6:314 (1963).

59. H. Nichols and R. M. Hexter, in press.

60. M. Fleischmann and I. R. Hill, submitted to J. Electroanal. Chem.

61. L. H. Jones, Vibrational spectrum and structure of metalcyanide complexes in the solid state. I. $KAg(CN)_2$, J. Chem. Phys. 26:1578 (1957).

62. M. Fleischmann and I. R. Hill, unpublished material.

63. M. R. Mahoney, M. W. Howard, and R. P. Cooney, Carbon dioxide conversion to hydrocarbons at silver electrode surfaces. Raman spectroscopic evidence for surface carbon intermediates, Chem. Phys. Lett. 71:59 (1980).

64. A. Bewick, K. Kunimatsu, and B. S. Pons, Infra-red spectroscopy of the electrode-electrolyte interphase, Electrochim. Acta 25:465 (1980).

65. R. P. Cooney, M. Fleischmann, and P. J. Hendra, Raman spectrum of carbon monoxide on a platinum electrode surface, J. Chem. Soc. Chem. Commun. (1977), p. 235.

66. Editors' note: This structure has now been identified with inelastic Mie scattering from the silver itself. See D. A. Weitz, T. J. Gramila, and A. Z. Genack in this volume.

ORGANIC AND INORGANIC SPECIES AT Ag, Cu, AND Au ELECTRODES

Bruno Pettinger and Herbert Wetzel

Fritz-Haber-Institut der Max-Planck-Gesellschaft

1000 Berlin (West) 33

INTRODUCTION

Surface enhanced Raman scattering (SERS) allows the characterisation of a variety of organic and inorganic adsorbates on several metal electrodes. Hitherto, most investigations have been performed in order to obtain insights into the enhancement mechanisms as the basis for future spectroscopic employments. Since the detection of SERS by Fleischmann and coworkers[1,2] and Van Duyne et al.,[3,4] SERS effects have been observed only for specifically prepared metal surfaces.[5-10] Intense SERS can be generated, for instance, at Ag, Cu, and Au substrates (in situ and under UHV conditions) by roughening or use of particular surface geometries such as small spheres, ellipsoids, islands, gratings, or thin films on hemicylinders.[11-20] In addition, experimental evidence has been given for the relevance of chemisorption on SERS. In particular, there is a growing amount of data which point to the importance of adatom-adsorbate structures for the enhancement.[21-29]

Based on the experimental results, distinct theoretical approaches have been made to explain the tremendous increase of the Raman cross section found for many species physi- or chemisorbed at specifically prepared metal interfaces.[30-36] The need of (sub)-microscopic surface structures provides evidence for the importance of electromagnetic surface modes in SERS. Only at that type of interface can surface polariton plasmons (SPP) or related modes be excited, resulting in a high density of states near the zone boundary. According to McCall et al.,[37] this intensity enhancement multiplies with the gain obtained by new radiation channels for the induced dipole, because the particular surface geometries operate as radiating antennas for the dipoles' near-field zone (which otherwise would not contribute to the scattered intensity).

In contrast to models of .this kind, Otto et al. proposed that adatoms may contribute to the enhancement via the excitation and scattering of electron-hole pairs.[23,36] Accordingly, one would expect the appearance of SERS signals when adatom-adsorbate complexes are created, even at smooth surfaces. Besides e-h excitation, other types of enhancement mechanisms may be induced by adatom-adsorbate complexes. Among them ranks the surface-induced resonance Raman process suggested by Efrima and Metiu,[38] or molecular resonance Raman effects.[39]

The question that has to be solved is whether non-local, electromagnetic or the more local, adsorbate-induced enhancement mechanisms are operating. The total enhancement ($10^5 - 10^6$) was suggested to result from combinations of local and non-local processes. An answer to this should be obtained in electrochemical investigations because electrochemistry provides the means to create and to alter (sub)microscopic structures and adatom-adsorbate complexes at the interface independently from each other. In this contribution, therefore, we confine ourselves to the presentation and discussion of important SERS parameters and properties found at electrodes, such as pretreatment, excitation profile, complex formation, quenching, and regeneration processes.

EXPERIMENTAL

The experimental arrangement used to record intensity as a function of exciting frequency, Raman shift, electrode potential, and polarization of incident and scattered light has been shown elsewhere.[40]

Massive metal plates or 2000 Å thick gold, silver, and copper films evaporated on glass substrates of optical quality were placed into the center of a cylindrical cell with a semicircular quartz window.[40] A Krypton or Argon ion laser (Coherent Radiation Model 3000 K or CR 4) served as the monochromatic light source focussed on the electrode. Standard electrochemical equipment and aqueous electrolytes prepared with triply distilled water and suprapure chemicals were used. All potentials are cited against the saturated calomel electrode (SCE).

RESULTS

Pretreatment

Intense SERS at electrodes occurs only after exposing the metals to an oxidation-reduction cycle (ORC). The understanding of the ORC influence on interfaces and the consequences for SERS is a continuing challenge.

The polarization of an electrode changes the potential drop at the metal-electrolyte interface occurring in the Helmholtz and diffuse Gouy-Chapman layer, which directly influences the rate of anodic or cathodic charge transfer at the phase boundary. If the applied potential is positive of the dissolution potential, surface atoms will be oxidized and may diffuse as ions into solution. In the opposite case, any dissolution is prohibited and metal ions can be (re)deposited at the electrode. Nucleation, e.g., three dimensional growth of metal clusters, is usually favored during metal deposition due to diffusion of ions and atoms within the electrolyte and along the surface. Therefore, surface morphology will be changed by strong anodizations. In addition, formation of new compounds in solution and at the surface may occur too.

The effect of a strong ORC on light scattering from the surface is shown in Fig. 1 with Cu, Ag, and Au electrodes as substrates. Note that the Raman intensities are drawn in logarithmic scale in order to demonstrate their different heights and also to resolve weak bands. Note also the upward shift of the logarithmic scale in the case of the Cu spectrum for the sake of clarity. In the experiments shown on both sides of the figure, the incident and reflected laser beam is guided through the electrolyte. Along its path, inelastic light scattering occurs. Therefore, some light scattered from the electrolyte in front of the electrode will be focussed on the spectrometer entrance slit in addition to photons possibly scattered from the surface.

On the left-hand side, a spectrum recorded immediately after placing a smooth electrode in the light beam (under -0.6 V vs SCE) is shown. Under this condition, one obtains a typical solution spectrum: No SERS is detectable at smooth, unanodized electrodes in spite of the fact that Cl^- ions, pyridine, or water molecules are adsorbed. This contrasts, however, to the reported SERS for unanodized, mechanically polished Ag electrodes by Schultz et al.[41]

The solution spectrum exhibits characteristic features for water and pyridine molecules such as the librational and bending vibrations of the solvent around 400 and 1642 cm^{-1}, and two breathing modes of the pyridine ring at 1004 and 1036 cm^{-1}. Other relatively weak Raman bands of pyridine cannot be resolved due to its low solution concentration.

After an ORC in the form of a triangular potential sweep from -.6 V to +0.05, +0.2, +1.3 V and back to -0.6 V vs SCE, the Raman intensity increases dramatically and by far exceeds the Raman signals from the solution as shown in the right-hand side of Fig. 1.

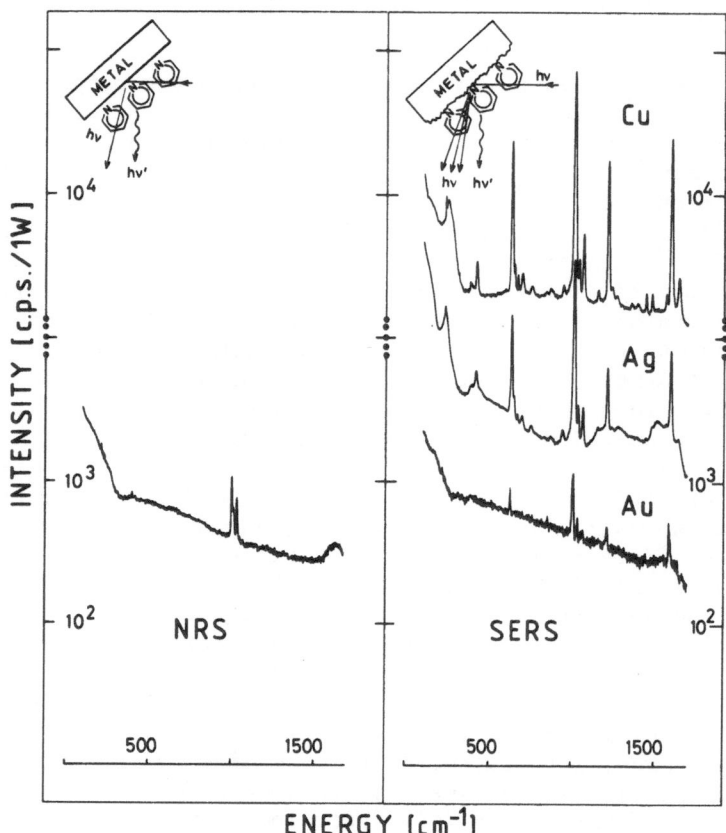

Fig. 1. Raman spectra at Cu, Ag, and Au electrodes before (left side) and after (right side) a strong ORC. Electrolyte: 0.1 M KCl + 0.05 M pyridine. ORC's: -.6 V to +0.05, +0.2, or +1.3 V to -.6 V vs SCE for the three substrates, respectively; scan rate: 20 mV/s. Excitation with 647.1 nm. Spectrum for Au represents the numerical difference of spectra recorded at -.6 V and +.5 V, respectively. Inserts indicate scattering geometry and topographical properties of the electrodes.

It is very significant that all pyridine fundamental vibrations appear in the SERS spectra of silver, many of them with larger relative intensities than in the normal Raman scattering (NRS) spectrum.[26] In addition, new bands appear, e.g., the AgCl vibrations at 240 cm^{-1}.[42] Analogous features are found for Cu and Au electrodes in this spectral region; however, their assignment has not been performed yet. Furthermore, a considerable background signal is observed which appears to be partly composed of weak SERS

lines.[26,43] Besides SERS bands of the adsorbate, impurity lines
appear in addition to broad spectral features which are indicative
for graphitized carbon at the interface.[44,45]

The enhancement of pyridine on Au is apparently lower than on
the other electrodes. The continuous decay in intensity with increasing Raman shift is partly the result of the time dependence
of SERS at these substrates.[11] In spite of the low intensity,
several SERS bands are clearly detectable, because all Raman scattering from the solution has been numerically eliminated in the
potential difference spectrum obtained by the numerical difference
of spectra recorded subsequently at U = -0.6 V and U = +0.5 V.

In summary, it may be said that SERS spectra for pyridine at
these metals are similar, differing solely in absolute intensity
and background.

A typical ORC as used in our experiments is a triangular potential sweep applied to the electrode. Its scan rate and limiting
potentials determine the amount of charge passed. This allows
control of the strength of anodization and, hence, of the degree of
surface morphology changes.[7,40]

In order to study the influence of different strengths of
anodizations on SERS, single crystalline Ag(111) films evaporated
on mica were used. These very smooth surfaces have been exposed
subsequently to ORC's with increasing positive limiting potentials
resulting in a rise of the amount of silver oxidized and redeposited.[40] It has been found that surface orientation is not changed
noticeably as long as the strength of anodization was not too
high.[40,46] As Fig. 2 shows, the SERS intensity rises approximately
linearly up to a charge transfer equivalent to three monolayers of
Ag. However, there is no indication of an increase of surface area.
Further, in electrochemical investigations (by determining the
relative surface area from the amount of charge passed in under-
potential deposition of lead on Ag after various ORC's; see curve
c in Fig. 2) and in electroreflectance (ER),[40] and in RHEED[46]
studies, no indication for roughening of the silver surface has
been observed employing weak ORC's. On the other hand, as shown
with curve c, roughening becomes more and more important the
stronger the anodization. Therefore, the SERS intensity normalized
to the total surface area is given as curve b. Obviously, roughening of the surface is noticeable above a charge transfer equivalent
of 3 monolayers of Ag, but the normalized relative SERS intensity
increases only slightly with stronger anodizations.

Excitation Frequencies

The excitation frequency determines the enhancement factor as
shown in Fig. 3. For the readers' convenience, excitation profiles

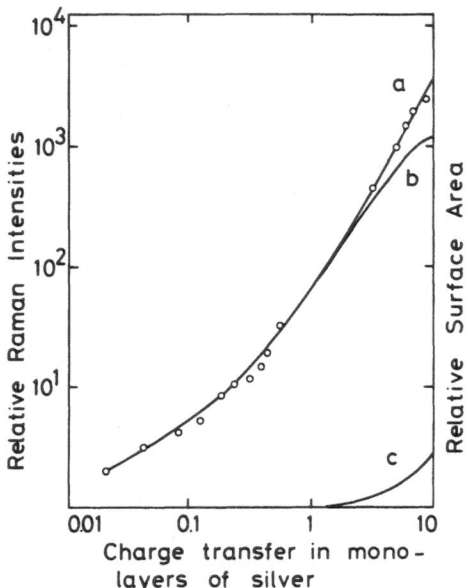

Fig. 2. Raman intensities of pyridine adsorbed at a silver(111) electrode (curve a), relative change of surface area (curve b), and Raman intensities normalized to unit surface area (curve c) as a function of monolayers of Ag anodized.

published by Creighton et al.[6] and Girlando et al.[47] were included in addition to a $\Delta R/R$ spectrum for pyridine on Ag showing the absorption of this system.[40] For Ag, the enhancement increases from the blue spectral region (457.1 nm) up to the red (676.4 nm) by a factor of 30. According to Girlando et al.[47] the enhancement reaches a maximum at 770 nm (see dotted curve). It follows approximately the absorption of the Ag/pyridine system as shown by the dotted line which reaches a maximum around 780 nm. Note that both SERS and absorption appear only after an ORC. The excitation profiles shown here rise the longer the excitation wavelength, a trend which we also observed for other adsorbates at Ag electrodes, such as halide or CN^- ions.[28,29] In contrast to Ag, which exhibits SERS in the whole visible region, Cu and Au electrodes show significant SERS only with red excitation frequencies.

Recent ER-investigations of Taddjedine and Kolb[48] performed on Ag and Au electrodes yielded further indications for remarkable absorption by pyridine at these interfaces if they have been exposed to an ORC. In these investigations, dispersion curves of SPP showed significant deviations from the usual course at 780 and 900 nm. These deviations may result from the interaction of electronic states of adsorbates with surface polariton plasmons.

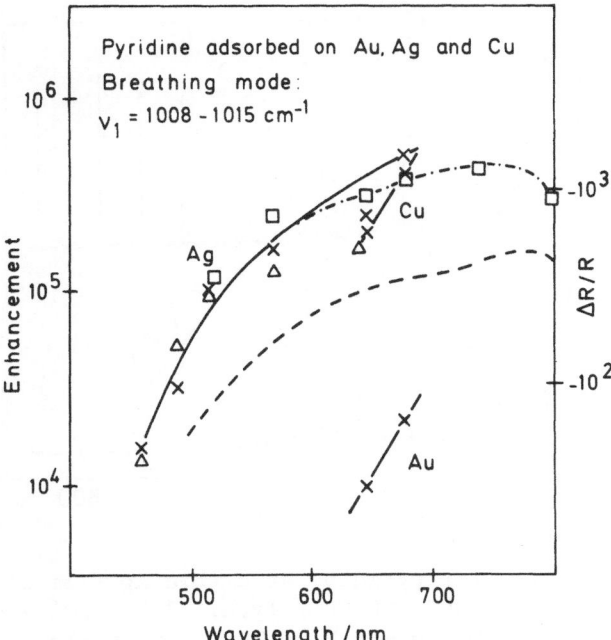

Fig. 3. Excitation profiles for pyridine on Ag, Cu and Au, and $\Delta R/R$ spectrum of pyridine on Ag. Electrolyte: 0.1 M KCl + 0.05 M pyridine, U = -.6 V. Data: □,[47] ∧,[6] X[40,49].

Direct SPP Excitation

It has been proposed that excitation of surface EM-modes and coupling of vibrational states of the adsorbate to collective oscillations of the electron gas are responsible for the enhancement.[15,35,33] In this model, the ORC creates surface roughness which provides the necessary momentum for SPP excitation. Hence, an equivalent experiment using an ATR configuration in order to allow direct excitation of SPP should be feasible.[20] A thin Ag film (500 Å) evaporated on a hemicylinder represents one possible configuration for a direct SPP excitation, because the momentum components, k_x, of photons and SPP match here at a certain, well defined angle of incidence. Figure 4 shows the comparison of two spectra obtained at direct illumination of the metal-electrolyte interface (upper curve) and under ATR conditions with the angle of incidence in SPP resonance (lower curve). The corresponding scattering geometries are shown in the inserts. Without an ORC, in both experimental configurations, no SERS is detectable. Obviously, excitation of SPP is a necessity, but not a sufficient condition for SERS. However, after applying an ORC (-.6 to +0.09 back to -0.6 V) which anodizes only a submonolayer of the silver surface,

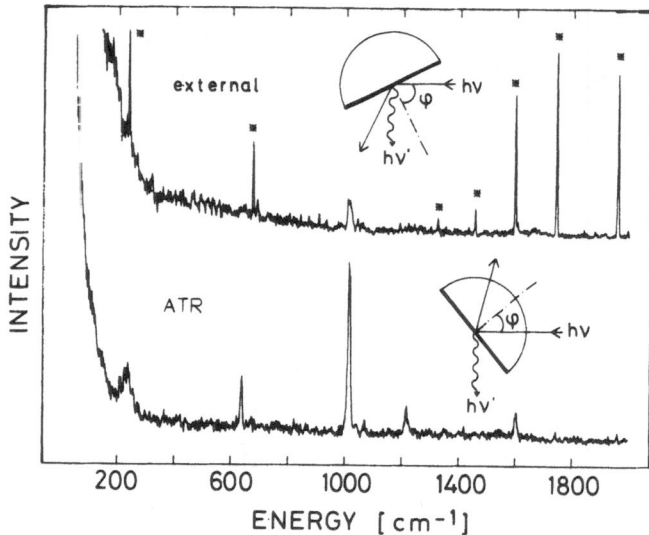

Fig. 4. Raman spectra for pyridine at a thin Ag film electrode on
a hemicylinder after an ORC. Electrolyte: 0.1 M KCl + 0.05 M pyridine. ORC: -.6 to +0.09 to -.6 V vs SCE. Excitation: 514.5 nm.

SERS signals peak in SPP resonance. As illustrated by the comparison of both spectra, the intensity is approximately five times
larger than for direct illumination. Analogous observations, in
particular the need of an ORC, have been obtained in this laboratory using gratings evaporated with silver or employing smooth
silver surfaces coated with Ag particles which have a narrow distribution of diameters around 100 Å.

Adatom-Adsorbate Complexes

Specific conditions for maximum SERS have been reported by
Jeanmaire and Van Duyne.[4] Among them rank a 1:2 ratio of pyridine and Cl^- ion concentrations and the fact that the presence
of halide ions in the electrolyte increases the SERS intensity
of pyridine vibrations, whereas pyridine in solution strengthens
that of Ag-Cl.[4,42,26] Coadsorption of organic adsorbates and ions
obviously leads to large scattering intensities. An example for
this is presented in Fig. 5 by a comparison of spectra for normal
Raman scattering (NRS) of pyridine in solution, and SERS from this
molecule at Ag electrodes activated in $NaNO_3$ or KCl electrolytes,
respectively. Again, the Raman intensities are drawn in logarithmic scale. Using a $NaNO_3$ electrolyte, there is a considerable
enhancement (ca. 10^4) which is, however, much larger if these

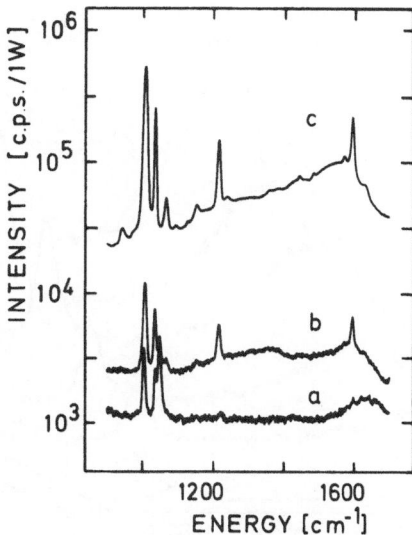

Fig. 5. NRS and SERS spectra for pyridine in NaNO$_3$ solution (a), pyridine at Ag anodized in NaNO$_3$ (b) or in KCl electrolyte. Concentrations: 0.1 M salt + 0.05 M pyridine. ORC: -0.6 to 0.2 (KCl) or 0.7 (NaNO$_3$) V to -0.6 V vs SCE. Excitation: 514.5 nm.

anions have been replaced by halide ions. According to Jeanmaire and Van Duyne, the enhancement decreases in the sequence of order:[4]

$$I^- \gg Br^- \cong Cl^- > SCN^- > HPO_4^= > SO_4^= > ClO_4^- \ .$$

The importance of anions (in particular of halides) on SERS has been verified in this laboratory for several distinct species at Ag electrodes such as EDTA (see next paragraph), CN⁻, water, and for pyridine on Cu electrodes.

A particularly interesting example of coadsorption is shown in Fig. 6 which gives a set of SERS spectra for the Ag/NO$_3$-Ethylendiaminetetraacetic disodium salt (EDTA-Na$_2$) recorded after various ORC's.[50] To our knowledge, this is the only example which shows an enhancement for NO$_3^-$ ions. These anions are not specifically adsorbed at Ag electrodes and, in contrast to halide ions, have considerably lower surface concentration; consequently, only weak SERS can be expected. Without an ORC, one obtains a solution spectrum displaying only one band, the total symmetric NO$_3^-$ vibration at 1050 cm^{-1}. The EDTA bands at 912 and 935 cm^{-1} are not resolved in this curve. The following spectra are recorded after

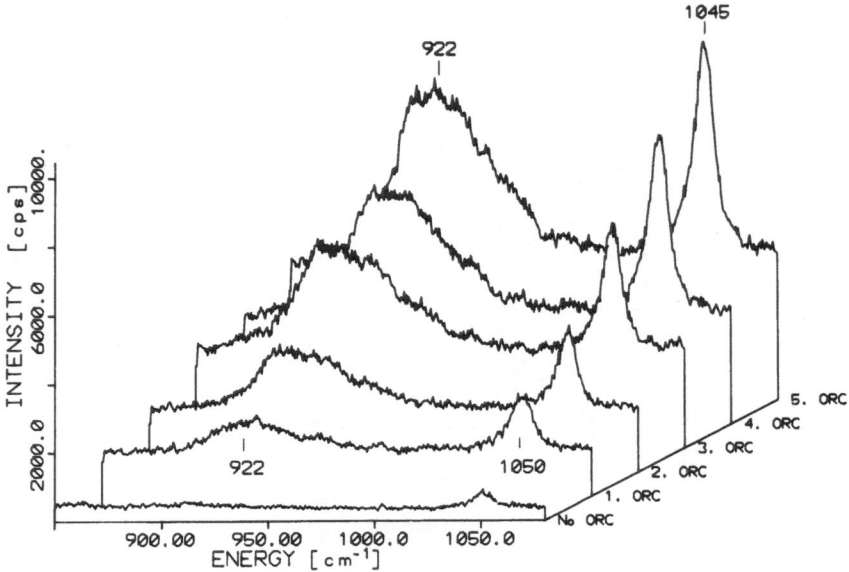

Fig. 6. Raman spectra for NO_3^- and Ethylendiamineteraacetic disodium salt at Ag electrodes after several ORC's. Electrolyte: 0.1 M $NaNO_3$ + 0.05 M EDTA-Na_2 ORC's: -.1 V to 400, 450, 500, 550, 600, 650, 700 mV to -0.1 V. Excitation: 514.5 nm, power 100 mW.

subsequent ORC's with increasing positive endpoints. In this situation, SERS appears for both species, for EDTA and NO_3^- ions, and the signal increases with the number of cycles. The nitrate line shifts by five wave numbers to lower frequency, which is indicative for a SERS band. Obviously, nitrate ions are coadsorbed with EDTA at the Ag electrode, and therefore a strong enhancement occurs for both species.[50]

Quenching of SERS

        SERS, which is sensitive to species adsorbed at the interface, does not occur for other layers in the metal electrolyte system, which are located beyond the first adsorbed layer. The adsorption and desorption processes can be controlled by the applied electrode-potential. Thus, the SERS intensity should follow roughly the amount of adsorbed species, once SERS has been generated by an ORC.

        Surprisingly enough, SERS for metal-chloride vibrations has been detected only under specific conditions, such as coadsorption (with organic molecules)[42,26] high salt concentration (pure electrolyte)[51,27] or excitation with the 676.2 nm Krypton laser line in more diluted solutions (ca. 0.1 M).[28] An interesting example is presented in Fig. 7, showing the influence of the potential on SERS of Ag-Cl (upper part) and Cu-Cl (lower part) vibrations. Note the enormous SERS intensity which is contrary to that expected, because bulk AgCl does not exhibit significant first-order Raman scattering (it is completely forbidden by symmetry).[52] Roughly, in accordance with the desorption of chloride ions, the SERS intensity decreases the more negative the electrode potential.

Surprisingly, the SERS effects do not reappear after applying the starting potential to the electrodes (U = -0.3 V), in spite of the readsorption of halide ions.[27] SERS remains quenched (only 5% of the original signal reappear). In order to obtain this quenching effect, it is sufficient to apply negative potential pulses of one sec duration.[28] This observation has been made for EDTA, NO_3^-, halide CN⁻ and SCN⁻ ions, as well as for pyridine and water molecules on Ag electrodes.[50,28,29]

Regeneration of SERS by Metal Deposition

The quenching of SERS after applying a sufficient negative potential pulse takes place within seconds. It has been inferred that during such a short time and at substantial negative potentials, no significant changes of submicroscopic and microscopic roughness

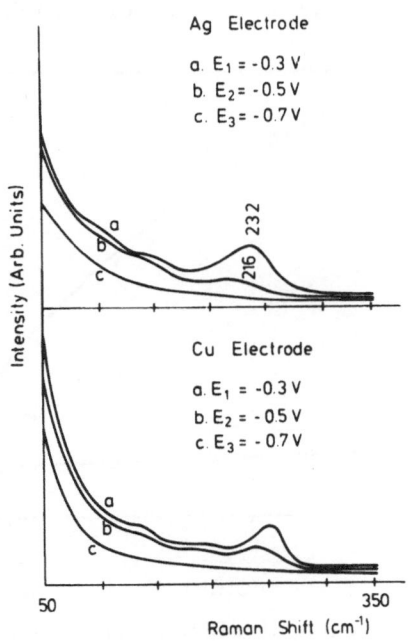

Fig. 7. SERS spectra for Ag-Cl and Cu-Cl vibrations at various electrode potentials. Electrolyte: 1 M NaCl or sat. NaCl, respectively. Excitation with 647.1 nm.

can occur[28],[29] particularly if one notes that the metal adsorbate
system showing SERS at more positive potentials is relatively stable
for hours.[26] Since the adsorption of chloride ions at rough metal
interfaces, for instance, is not a sufficient condition for SERS,
the entities causing SERS must differ in structure and composition
from the normal Ag-Cl units at the interface.[53] It is likely that
silver-adatom adsorbate complexes are formed at the surface. In
order to test this idea, such complexes have been added to the
electrolyte and the influence of metal deposition on the SERS
effects have been examined.[28],[29] An example for this is given in
Fig. 8, in which the SERS intensity for the Ag-Cl vibration at
240 cm^{-1} is drawn as a function of potential. Along the curve a
a → b the quenching occurs, as above described for silver complex
free solution. The quenching is nearly complete as curve b → c
shows; only 5% of the original intensity reappears. When a solution which contains silver complexes is added to the electrolyte,
a remarkably different behaviour must be noted, as indicated by
curve b → c

For the regeneration of SERS it is sufficient to redeposit
about a monolayer of Ag. Very conclusive is the fact that intense SERS can be regenerated only at rough surfaces. Even long
time deposition of Ag at smooth electrodes does not cause observable
SERS effects, which means that this deposition does not form
(sub)microscopic rough interfaces, a result which is in accordance

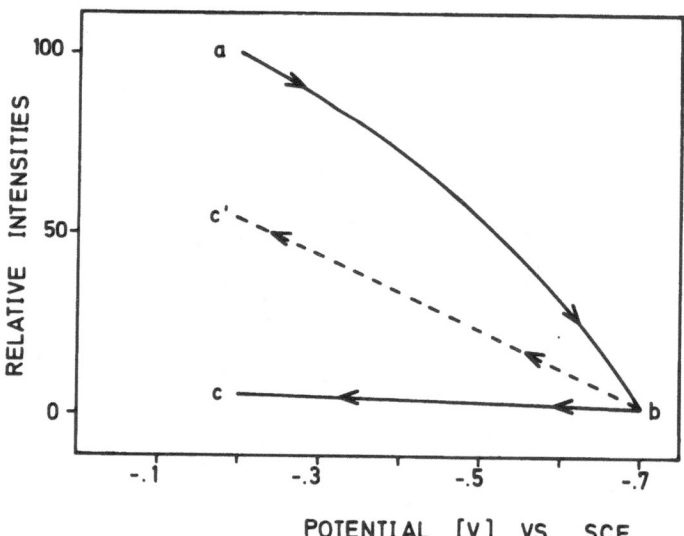

Fig. 8. Potential dependence of SERS intensity of Ag-Cl vibration
at 240 cm^{-1} without (a - b - c) and with AgCl complexes (b -c′) in
the electrolyte. Excitation: 514.5 nm.

with the observation of Billmann et al. in their new surface formation studies.[54]

Frozen Adatoms at Low Temperature

It has been argued that adatom-adsorbate complexes and their
stabilisation are relevant for SERS. In UHV experiments, the
stabilisation is basically caused by adatom immobility at low temperature,[36] whereas in in-situ experiments adatoms seem to be
trapped and isolated by complex formation with ligands such as
halides, CN⁻, and pyridine. The performance of low temperature
in-situ experiments reveals results analogous to UHV investigations.[55] At sufficiently low temperature, no quenching of SERS
takes place because of the immobility of adatoms.

Figure 9 shows a series of Raman spectra recorded in the
water stretch vibration region and their numerical difference
spectra. Whereas at room temperature SERS quenching for water
occurs at a potential negative of -0.6 V and is correlated with the
desorption of Cl⁻ ions, it takes place at low temperature only at
potentials negative of -1.1 V. Still, complete quenching can be

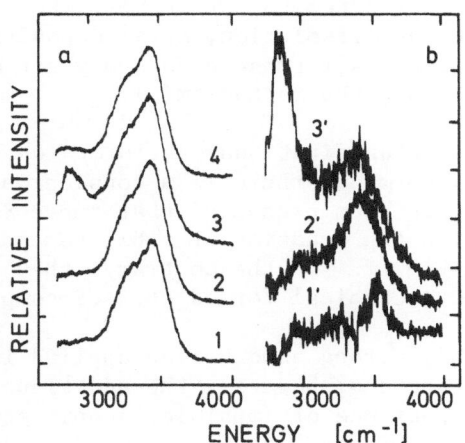

Fig. 9. Raman spectra of water at Ag electrodes using low temperature. T = 250°K. 1,2,3: $U_{1,2,3}$= -.3, -.6, -.9 V vs SCE. U_4
recorded at -1.1 V after applying -1.8 V. 1', 2',3' potential difference spectra for U_4-U_1, U_4-U_2, U_4-U_3, respectively. Excitation
with 514.5 nm.

achieved by applying -1.8 V to the electrode; however, this results
in a strong hydrogen evolution with a possible warming up of the
surface. Note that the curve for U_4 is recorded after applying
-1.8 V for several seconds to the electrode. It represents, therefore, basically a solution spectrum of water still containing very
weak SERS contributions.

It is significant that at low temperature SERS for water does
not require adsorption of chloride ions. It is more intense in the
potential region where these ions are, at least partly, desorbed.
Apparently, there is a competition between water and chloride adsorption at low temperature.

DISCUSSION

Intense SERS at the three metal electrodes, Ag, Cu, and Au, has
been observed only after applying an oxidation-reduction cycle (ORC).
Mechanical polishing is by far a less efficient pretreatment. At
smooth, annealed interfaces employed in UHV or electrochemical en-
vironments, however, no enhanced Raman processes have been detected.[14]
Using such substrates, an ORC is definitely a prerequisite for in-
situ SERS effects.[40]

What type of change does an ORC cause at metal interfaces which
act on the light scattering process?

Very different electrochemical processes take place in the
course of an ORC, such as dissolution, metal deposition, complex
formation and nucleation. All these events may act on SERS and,
hence, will influence the SERS mechanism(s).

In the case of weak anodizations, performed on evaporated metal
films such as Ag(111) on mica, there is a considerable enhancement
(ca. 10^4). Evidence has been presented that these SERS processes
are not correlated with the creation of submicroscopic roughness
(250 Å), as shown in Fig. 2. On the contrary, they indicate their
association with electrochemical formed surface complexes.[40,48]

On the other hand, strong anodization applied in an ORC leads
to roughening of the surface; however, the additional gain in SERS
intensity with the appearance of (sub)microscopic structures is
surprisingly low (only a factor of 30). The additional increase in
SERS intensity with roughening points to the importance of SPP in
SERS. This is further supported by the threshold frequencies found
in the excitation profiles (Fig. 3), which correspond to the surface
plasmon resonance frequencies:

$$(\hbar\omega_{SPP,Ag} < 3.6 \text{ eV}, \quad \hbar\omega_{SPP,Au} < 2.5 \text{ eV}, \quad \hbar\omega_{SPP,Cu} < 2.1 \text{ eV}).$$

Above these energies, surface plasmon excitation is less probable
Therefore, only at lower energies can significant SERS scattering
be observed.

The SERS results obtained by excitation of SPP at thin films
on hemicylinders or on gratings (Fig. 4) point in the same direction. Only after an ORC are SERS processes detectable, and only
then do SERS intensities peak at the SPP resonance angle. Interest-
ingly enough, anodizing a submonolayer of Ag is sufficient for large
SERS intensities.

At first sight, these results and those on island films and
spheres obtained by other groups indicate an electromagnetic enhancement mechanism.[13,18] However, the absorption spectrum of the
Ag-pyridine system is one of the experimental findings which speak
against models of this type (see Fig. 3). This absorption may be
considered as an indication of resonance Raman processes induced by
new electronic transitions in the surface complex.[40] However, the
absorption band is rather broad and, hence, molecular resonance
Raman enhancement processes caused by these transitions will be
weak. Moskovits suggested that the excited state is related to
conduction electron resonances occurring in small adsorbate covered
metal particles on top of the surface.[30]

Since SERS is much more intense when the adsorbate and ions
are coadsorbed (see Figs. 5 and 6), chemisorption and complex for-
mations, rather than physisorption, seem to play a crucial role in
the enhancement. Similar ideas have been developed independently by
other authors for CN^- on Ag,[56] or Au[25] electrodes, or benzoic acid
on Ag island films.[24] After cathodic desorption of the adsorbate
the SERS intensity is, of course, nearly zero. Unexpectedly, however,
readsorption without an ORC does not restore the previous intensity.
The SERS is irreversibly quenched (see Figs. 7 and 8). This is a
strong indication that specific molecular entities such as adatom-
adsorbate complexes are needed for large enhancements.[27,53,28,29]

It is possible to create adatoms electrochemically by a deposition process. However, the (re)generation of intense SERS signals
occurs only with metal deposition on previously roughened surfaces
(Fig. 8), not at smooth surfaces.[28,29] Herein, we believe, lies
the function of an ORC: It creates adatom-adsorbate complexes and
(sub)microscopic structures. This is illustrated in Fig. 10,
which shows a sequence of adsorption and desorption processes.
During the redeposition of metals, adatoms (black circles) are
created as a precursor of the nucleation process. The adsorption
of ligands, (A), for instance, pyridine and halide ions, stabilizes
the adatoms forming various complexes between these entities (10a).
Their complete incorporation in the crystal structure becomes less
probable when more kink sites, steps, etc. are blocked by the adsorbate. Due to the large number of complexes, intense SERS appears.

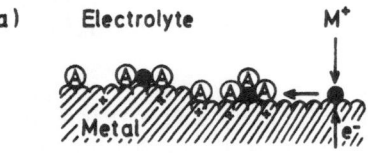

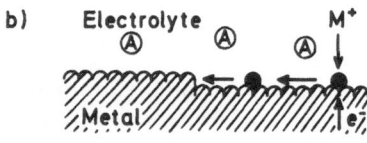

Fig. 10. Surface complex formation at various potentials. a) metal deposition at adsorbate covered interface. b) desorption of adsorbate by negative potential. Lower adatom concentration. c) readsorption at adatom free interface.

The amount of adsorption of the ligands can be controlled by the electrode potential. At sufficiently negative potentials, the adsorbate is desorbed and the adatoms diffuse along the surface and will be included in the metal structure (Fig. 10b). With desorption of the adsorbate, the SERS signals decrease. Applying a less negative potential, organic molecules and ions are readsorbed. However, the number of adatoms remains at a very low level (Fig. 10c). Thus, the SERS is irreversibly quenched.

Deposition of silver from complex containing solutions results in a high density of adatom-adsorbate complexes at the surface and, hence, in regeneration of the SERS effect.

Figure 11 elucidates the possible underlying enhancement mechanisms. Incident photons excite surface plasmon polaritons, if the law of conservation of momentum is satisfied. Rough surfaces as well as gratings or particularly shaped metals supply the necessary momentum parallel to the surface. Therefore, surface plasmons, polaritons, or other types of collective electron gas oscillations are created which represent the quanta of the electromagnetic wave

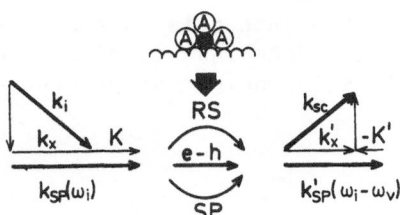

Fig. 11.   Combination of surface plasmon (SP) and adsorbate induced enhancement processes.

propagating along the metal surface.   This leads to a local field enhancement near the interface.

Analogous to photons, SPP can be inelastically scattered, possibly along various channels as illustrated in the figure.   Raman scattering (RS) in the usual sense may occur, e.g., by radiation of induced dipoles directly via their far-field zone or indirectly by their near-field zone via the rough surface.

Another scattering possibility results from the interaction of SPP with the molecular vibrations of the chemisorbed species, because the motions of the bonded atoms modulate the electron density at the surface which is already collectively oscillating with the frequency of the SPP.[57]

The excitation of electron-hole pairs may be an additional channel for interaction.   Scattered e-h pairs either radiate directly or decay back to SPP.   Adatoms irregularly located at the surface provide the necessary momentum for the excitation or radiation processes.[8] Because the regeneration of SERS occurs only at rough surfaces, a participation of e-h pairs in the enhancement is imaginable only if their rate of creation via SPP is much larger than via incident photons directly.

According to this scheme, the inelastic scattering leads to scattered SPP, which may radiate if the wavevector components parallel to the surface, k_{SP} and $k_{sc}-K'$, match.

CONCLUSION

Evidence has been presented for Ag electrodes that intense SERS results from a combination of electromagnetic and adatom-adsorbate induced enhancement processes.[28,58] The similarity of SERS effects for pyridine on Ag, Cu and Au and for halide ions and water on Ag and Cu electrodes indicates that this is also valid for the other substrates. To what extent the individual enhancement processes are participating in SERS is unclear at present and requires further investigations.

ACKNOWLEDGEMENTS

We thank Professor H. Gerischer for his continuing support and stimulating interest, Drs. D. M. Kolb and H. J. Lewerenz for helpful discussions, and Mrs. E. Seiler and R. Putzke for valuable technical assistance. We also gratefully acknowledge the preparation of cluster coated interfaces by Dr. Schultze.

REFERENCES

1. M. Fleischmann, P. J. Hendra, and A. J. McQuillan, Raman spectra of pyridine adsorbed at a silver electrode, Chem. Phys. Lett. 26:163 (1974).
2. M. Fleischmann, P. J. Hendra, A. J. McQuillan, R. L. Paul, and E. S. Reid, Raman spectroscopy at electrode-electrolyte interfaces, J. Raman Spectrosc. 4:269 (1976).
3. R. P. Van Duyne, Applications of Raman spectroscopy in electrochemistry, J. Physique 38:239 (1976).
4. D. L. Jeanmaire and R. P. Van Duyne, Surface Raman spectroelectrochemistry. Part I. Heterocyclic, aromatic and aliphatic amines adsorbed on the anodized silver electrode, J. Electroanal. Chem. 84:1 (1977).
5. R. P. Cooney, M. Fleischmann, and P. J. Hendra, Raman spectrum of carbon monoxide on a platinum electrode surface, J.C.S. Chem. Commun. 7:235 (1977).
6. J. A. Creighton, M. G. Albrecht, R. E. Hester, and J. A. D. Matthew, The dependence of the intensity of Raman bands of pyridine at a silver electrode on the wavelength of excitation, Chem. Phys. Lett. 55:55 (1978).
7. B. Pettinger and U. Wenning, Raman spectra of pyridine adsorbed on silver (100) and (111) electrode surfaces, Chem. Phys. Lett. 56:253 (1978).
8. A. Otto, Raman spectra of (CN)⁻ adsorbed at a silver surface, Surf. Sci. 75:L392 (1978).
9. E. Burstein, Y. J. Chen, S. Lundquist, and E. Tosatti, "Giant" Raman scattering by adsorbed molecules on metal surfaces, Solid State Commun. 29:567 (1979).

10. T. E. Furtak and J. Reyes, A critical analysis of theoretical
     models for the giant Raman effect from adsorbed molecules,
     Surf. Sci. 93:351 (1980).

11. B. Pettinger, U. Wenning, and H. Wetzel, Surface plasmon en-
     hanced Raman scattering. Frequency and angular resonances
     of Raman scattered light from pyridine on Au, Ag and Cu
     electrodes, Surf. Sci. 101:409 (1980).

12. H. Abe, W. Schulze, and B. Tesche, Surface-enhanced Raman
     spectroscopy of CO adsorbed on colloidal silver particles,
     Chem. Phys. 47:95 (1980).

13. J. E. Rowe, C. V. Shank, D. A. Zwemer, and C. A. Murray,
     Ultrahigh-vacuum studies of enhanced Raman scattering from
     pyridine on Ag surfaces, Phys. Rev. Lett. 44:1770 (1980).

14. T. H. Wood and M. V. Klein, Studies of the mechanism of
     enhanced Raman scattering in ultrahigh vacuum, Solid State
     Commun. 35:263 (1980).

15. C. Y. Chen and E. Burstein, Giant Raman scattering by mole-
     cules at metal-island films, Phys. Rev. Lett. 45:1287 (1980).

16. H. Seki and M. R. Philpott, Surface enhanced Raman scattering
     by pyridine on silver island films, Surf. Sci., in press
     (1981).

17. J. C. Tsang, J. R. Kirtley, and J. A. Bradley, Surface enhanced
     Raman spectroscopy and surface plasmons, Phys. Rev. Lett.
     43:772 (1979).

18. J. A. Creighton, C. G. Blatchford, and M. G. Albrecht, Plasma
     resonance enhancement of Raman scattering by pyridine
     adsorbed on silver or gold sol particles of size comparable
     to the excitation wavelength, J. Chem. Soc. Faraday II
     75:790 (1979).

19. H. Wetzel and H. Gerischer, Surface enhanced Raman scattering
     from pyridine and halide ions adsorbed on silver and gold
     sol particles, Chem. Phys. Lett. 76:460 (1980).

20. B. Pettinger, A. Tadjeddine, and D. M. Kolb, Enhancement in
     Raman intensity by use of surface plasmons, Chem. Phys.
     Lett. 66:544 (1979).

21. A. Otto, J. Timper, J. Billmann, G. Kovacs, and I. Pockrand,
     Surface roughness induced electronic Raman scattering, Surf.
     Sci. 92:L55 (1980).

22. A. Otto, Raman scattering from adsorbates on silver, Surf.
     Sci. 92:145 (1980).

23. A. Otto, J. Timper, J. Billmann, and I. Pockrand, Enhanced
     inelastic light scattering from metal electrodes caused by
     adatoms, Phys. Rev. Lett. 45:46 (1980).

24. E. Burstein, C. Y. Chen, and S. Lundquist, in: "Proceedings of
     the US-USRR Symposium on Inelastic Light Scattering in
     Solids, J. L. Birman, H. Z. Cummins, and K. K. Rebane, eds.,
     Plenum Publishing Corp., New York (1979), p. 479.

25. R. E. Benner, K. U. von Raben, R. Dornhaus, R. K. Chang, B. L.
     Laube, and F. A. Otter, Correlation of SERS with cyclic
     voltammetry for cyanide complexes adsorbed on Cu electrodes,

Surf. Sci. 102:7 (1981).

26. B. Pettinger and H. Wetzel, Surface enhanced Raman spec-
     troscopy of pyridine on Ag electrodes. Complex formation,
     Chem. Phys. Lett. 78:398 (1981).

27. B. Pettinger, M. R. Philpott, and J. G. Gordon II, Contribu-
     tion of specifically adsorbed ions, water and impurities
     to the surface enhanced Raman spectroscopy (SERS) of Ag
     electrodes, J. Chem. Phys. 74:934 (1981).

28. H. Wetzel, H. Gerischer, and B. Pettinger, Surface enhanced
     Raman scattering from silver-halide and silver-pyridine
     vibrations and the role of silver-adatoms, Chem. Phys.
     Lett. 78:392 (1981).

29. H. Wetzel, H. Gerischer, and B. Pettinger, Surface enhanced
     Raman scattering from silver-cyanide and silver-thiocyanate
     and the importance of ad-atoms, Chem. Phys. Lett. 80:159
     (1981).

30. M. Moskovits, Enhanced Raman scattering by molecules adsorbed
     on electrodes. A theoretical model, Solid State Commun.
     32:59 (1979).

31. R. P. Van Duyne, Laser excitation of Raman scattering from
     adsorbed molecules on electrode surfaces, in: "Chemical and
     Biochemical Applications of Lasers," Vol. IV, C. B. Moore,
     ed., Academic Press, New York (1979), p. 101.

32. R. M. Hexter and M. G. Albrecht, Metal surface Raman spec-
     troscopy: Theory, Spectrochim. Acta 35A:233 (1979).

33. M. Kerker, D.-S. Wang, and H. Chen, Surface enhanced Raman
     scattering (SERS) by molecules adsorbed at spherical
     particles: Errata, Appl. Opt. 19:4159 (1980).

34. F. W. King, R. P. Van Duyne, and G. C. Schatz, Theory of
     Raman scattering by molecules adsorbed on electrode sur-
     faces, J. Chem. Phys. 69:4472 (1978).

35. S. S. Jha, J. R. Kirtley, and J. Tsang, Intensity of Raman
     scattering from molecules adsorbed on a metallic grating,
     Phys. Rev. B 22:3973 (1980).

36. A. Otto, Surface enhanced Raman scattering (SERS), what do
     we know?, Appl. Surf. Sci. 6:309 (1980).

37. S. L. McCall, P. M. Platzman, and P. A. Wolff, Surface enhanced
     Raman scattering, Phys. Lett. 77A:381 (1980).

38. S. Efrima and H. Metiu, Light scattering by a molecule near a
     solid surface. II. Model calculations, J. Chem. Phys.
     70:2297 (1979).

39. A. Regis and J. Corset, A chemical interpretation of the
     intense Raman spectra observed at a silver electrode in the
     presence of chloride ion and pyridine: Formation of
     radicals, Chem. Phys. Lett. 70:305 (1980).

40. B. Pettinger, U. Wenning, and D. M. Kolb, Raman and reflectance
     spectroscopy of pyridine adsorbed on single crystalline
     silver electrodes, Ber. Bunsenges. Physik. Chem. 82:1326
     (1978).

41. S. G. Schultz, M. Janik-Czachor, and R. P. Van Duyne, Surface

enhanced Raman spectroscopy: A re-examination of the role of surface roughness and electrochemical anodization, <u>Surf</u>. <u>Sci</u>. 104:419 (1981).

42. R. Dornhaus and R. K. Chang, Comments on the 210-243 cm^{-1} mode in surface enhanced Raman scattering from the pyridine-Ag system, <u>Solid</u> <u>State</u> <u>Commun</u>. 34:811 (1980).

43. B. Pettinger, Surface enhanced Raman spectroscopy (SERS) of pyridine on Ag electrodes. Evidence for overtones, <u>Chem</u>. <u>Phys</u>. <u>Lett</u>. 78:404 (1981).

44. R. P. Cooney, M. R. Mahoney, and M. W. Howard, Intense Raman spectra of surface carbon and hydrocarbons on silver electrodes, <u>Chem</u>. <u>Phys</u>. <u>Lett</u>. 76:488 (1980).

45. M. R. Mahoney, M. W. Howard, and R. P. Cooney, Carbon dioxide conversion to hydrocarbons at silver electrode surfaces. Raman spectroscopic evidence for surface carbon intermediates, <u>Chem</u>. <u>Phys</u>. <u>Lett</u>. 71:59 (1980).

46. G. Lehmphuhl, private communication.

47. A. Girlando, M. R. Philpott, D. Heitmann, J. D. Swalen, and R. Santo, Raman spectra of thin organic films enhanced by plasmon surface polaritons on holographic metal gratings, <u>J</u>. <u>Chem</u>. <u>Phys</u>. 72:5137 (1980).

48. A. Tadjeddine and D. M. Kolb, The optical properties of the silver-pyridine surface complex, <u>J</u>. <u>Electroanal</u>. <u>Chem</u>. 111:119 (1980).

49. B. Pettinger, Evidence of overtones and combination modes in the surface enhanced Raman scattering, in: "Proceedings of the VIIth International Conference on Raman Spectroscopy," Ottawa, Canada, August 1980, W. F. Murphy, ed., North-Holland Publishing Co., Amsterdam, New York (1980), p. 412.

50. H. Wetzel, B. Pettinger, and U. Wenning, Surface enhanced Raman scattering from ethylendiaminetetraacetic-disodium salt and nitrate ions on silver electrodes, <u>Chem</u>. <u>Phys</u>. <u>Lett</u>. 75:173 (1980).

51. M. Fleischmann, P. J. Hendra, I. R. Hill, and M. E. Pemble, Enhanced Raman spectra from species formed by the co-adsorption of halide ions and water molecules on silver electrodes, <u>J</u>. <u>Electroanal</u>. <u>Chem</u>. 117:233 (1980).

52. G. Burns, F. Dacol, and R. Alben, Lattice dynamics of simple superionic conductors, <u>Solid</u> <u>State</u> <u>Commun</u>. 32:71 (1979).

53. B. Pettinger, M. R. Philpott, and J. G. Gordon II, Surface enhanced Raman spectroscopy (SERS) of metal halide vibrations on Ag and Cu electrodes, <u>J</u>. <u>Chem</u>. <u>Phys</u>., in press (1981).

54. J. Billmann, G. Kovacs, and A. Otto, Enhanced Raman effect from cyanide adsorbed on a silver electrode, <u>Surf</u>. <u>Sci</u>. 92:153 (1980).

55. I. Pockrand, A. Otto, Surface enhanced and disorder induced Raman scattering from silver films, <u>Solid</u> <u>State</u> <u>Commun</u>. 37:109 (1981).

56. I. Pockrand and A. Otto, Raman scattering from silver/vacuum

interfaces, Appl. Surf. Sci. 6:362 (1980).

57. P. M. Platzman, S. L. McCall, and P. A. Wolff, Surface en-
     hanced Raman scattering, in: "Proceedings of the VIIth
     International Conference on Raman Spectroscopy," Ottawa,
     Canada, August 1980, W. F. Murphy, ed., North-Holland
     Publishing Co., Amsterdam, New York (1980), p. 390.

58. B. Pettinger and H. Wetzel, Surface enhanced Raman scatter-
     ing from pyridine, water and halide ions on Au, Ag and Cu
     electrodes, Ber. Bunsenges. Physik. Chem., in press (1981).

METAL COLLOIDS

J. Alan Creighton

Chemical Laboratories
University of Kent
Canterbury, CT2 7NH, U.K.

INTRODUCTION

Since the first observations of surface-enhanced Raman scattering (SERS) by molecules adsorbed on silver surfaces it has been recognized that the intensity of Raman scattering is strongly dependent on the state of division of the metal surface. The early experiments and much of the subsequent work on SERS has been carried out on silver electrodes randomly roughened by an electrochemical oxidation-reduction cycle in aqueous electrolyte. Such optimally roughened surfaces give particularly large SERS signals, and are therefore ideal for investigations into the adsorbate chemistry of silver electrodes. However, they are less favorable for probing the physics of enhanced Raman scattering, since it is difficult to measure some of their other optical properties, particularly their absorption spectra, and to account for these properties in terms of the roughness in a precise way. Much attention has therefore also been directed to more regular finely divided metal surfaces, viz. colloidal dispersions or evaporated island films, or the surfaces of diffraction gratings, all of which also exhibit SER scattering. This chapter discusses the experimental features of SER scattering by molecules adsorbed on aqueous silver and gold colloids. Such measurements have been significant in providing the first clear demonstration that the enhancement in SER scattering is associated with the resonant excitation of electron density oscillations in the metal surface. They are also of considerable potential chemical interest, however, in view of the importance of supported metal particles as heterogeneous catalysts and the need for a spectroscopic method for studying reactions at the surface of these particles.

PREPARATION OF SOLS

It has long been known that stable dispersions of silver or gold particles of ca. 10-100 nm diam. may be prepared by reduction of dilute solutions of simple silver or gold salts. These sols have one or more extinction maxima in the visible range, the wavelength of which depends on the particle size and shape. The pleasing colors of the sols therefore vary, depending upon the preparation procedure, and may change with time due to aggregation or particle growth. It is thus important for optical studies that the particles are uniform in size and shape, and that there is control of their aggregation.

The usual source of silver or gold for the preparation of these colloids are salts of the Ag^+ or $[AuCl_4]^-$ ions, but a wide range of reducing agents have been used, including citrate and oxalate ions, hydroxylamine, $[BH_4]^-$, and ethylenediaminetetraacetate ions.[1-4] A comprehensive study of the size and shape of colloidal gold particles prepared with some of these reagents has been made by Turkevich et al.[1] using transmission electron microscopy. This study showed that citrate ions are particularly favorable as a reducing agent for giving particles of good sphericity and uniform particle size. Typically a solution of $H[AuCl_4]$ ($3 \times 10^{-4}M$, 95 ml) is heated to 90°C and tri-sodium citrate solution (9×10^{-4} M, 5 ml) is rapidly added with vigorous stirring. After continued gentle stirring at 90°C for 30 min a deep red colloid consisting of gold spheres of ca. 20 nm diam. results. It is important for the reproducibility of this procedure, and for the stability of the sols, that care is taken with the purity of the water used and with the cleanliness of glassware, but with attention to these matters the sols are stable for several weeks. By decreasing the concentration of citrate in order to decrease the rate of nucleation, larger particles may be prepared, and Frens[2] has given conditions for the preparation by the citrate method of colloidal gold particles of uniform controlled size within the range of diameters of 12-150 nm. Alternatively relatively large particles of controlled size may be grown by adding a dispersion of small gold particles prepared by the citrate method to a solution of $[AuCl_4]^-$ and hydroxylamine hydrochloride, where they nucleate further growth to give particles whose final size depends on the ratio of the concentration of nucleating particles to the total gold concentration in the growth medium.[1] Methods for the preparation of silver particles of good sphericity and uniform controlled size are less well developed. Transmission electron microscopy shows[5] that silver sols prepared by reduction of dilute silver nitrate solution with sodium borohydride, like gold sols prepared with this reducing agent,[6] contain particles of rather irregular sizes and shapes, and there is evidence from extinction spectra (see below) of partial aggregation resulting from some of the methods of preparation.[6,7]

Metal particles in aqueous colloidal dispersions usually bear a negative charge due to adsorbed anions, and provided the charge is sufficiently great the colloids are stable to aggregation on account of the electrostatic repulsion between the particles. Aggregation is induced however by the addition of neutral adsorbate molecules such as pyridine, which displace the adsorbed ions, thus reducing the charge on the particles to the point where collisions occur as a result of diffusional motion. On contact the particles are then held together by short range attractive forces. At high concentrations of neutral adsorbates, aggregation may be quite rapid and lead to complete precipitation, unless the colloid is stabilized by the addition of a polymeric adsorbate such as gelatine or polyvinyl alcohol, which protects the particles from intimate contact on collision. At low concentrations of neutral adsorbates however there is slow formation of initially small aggregates, which remain dispersed, and thus a slow change in the color of the sol without precipitation.

Electron microscopic examination of gold or silver colloids slowly aggregated with low concentrations of pyridine show that in some cases they consist of strings of particles rather than of globular clusters.[5] This may be seen from Figure 1 which shows micrographs of a gold-citrate sol with particles of 17 nm diam. and total gold concentration $3 \times 10^{-4}M$, before and respectively 4 and 26 hr after making $8 \times 10^{-5}M$ in pyridine. Also shown in Figure 1 is a micrograph of a silver-borohydride sol aggregated by addition of pyridine. The good sphericity and uniformity of size of the gold particles prepared by the citrate method may be clearly seen from these micrographs. The formation of "pearl strings" rather than globular clusters has a large effect on the optical properties of the sols (see below). This mode of aggregation may be accounted for in terms of the opposing effects of short range attractive forces, which hold the particles together on contact, and the relatively large longer range Coulomb repulsive forces between the particles under these conditions of low pyridine concentrations due to the displacement of relatively few adsorbed ions from the particles by adsorbed pyridine.

EXTINCTION SPECTRA

Provided the particles are small compared to the wavelength of light and are approximately spherical and unaggregated, silver and gold aqueous colloids are respectively yellow and wine red, with a single extinction band at ca. 385 and 520 nm, respectively. This extinction is due to the resonant excitation of plasma oscillations in the confined electron gas of the particles. For a very small sphere (diameter < ca. 20 nm) illuminated by visible light, the electromagnetic field across the particle is approximately constant

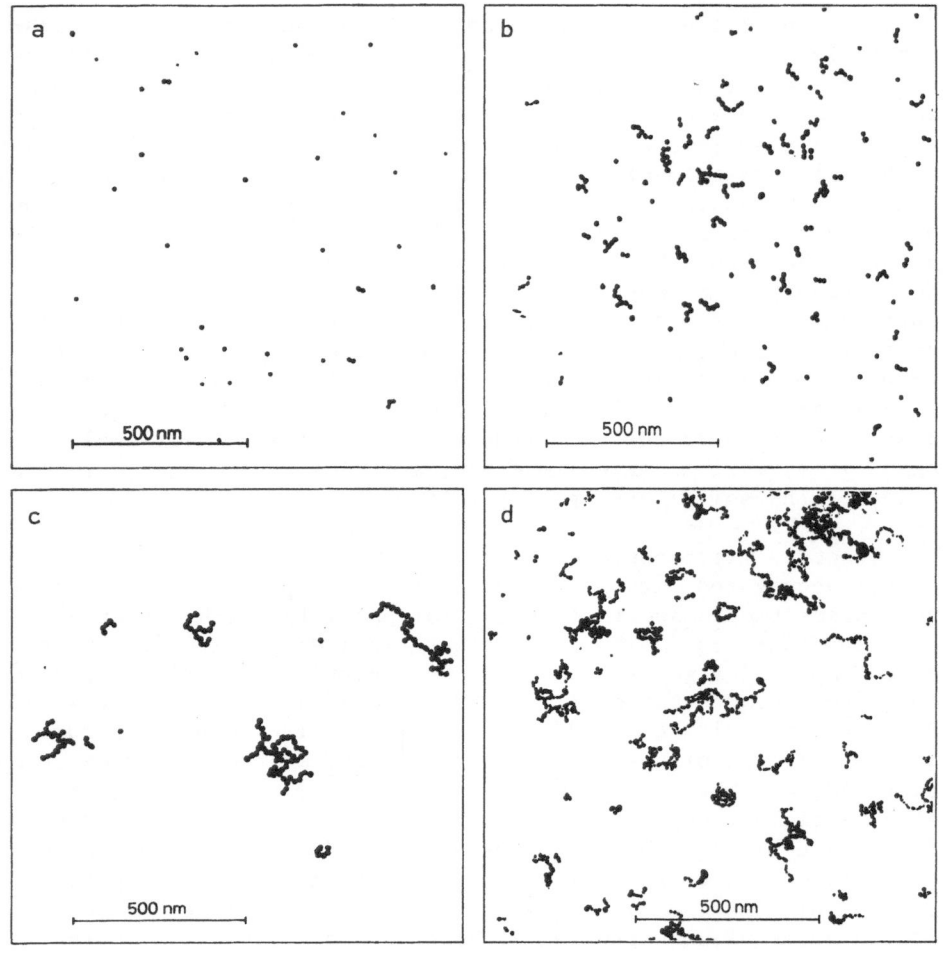

Fig. 1.   Transmission electron micrographs of gold and silver
          sols. a-c show gold-citrate sols respectively before and
          4 and 26 hr. after adding pyridine, and d shows a silver-
          borohydride sol aggregated by pyridine. From ref. 5.

and only the dipolar plasma mode is significantly excited. The
absorption contribution to the extinction cross-section per particle for spheres which are sufficiently small to be within this
Rayleigh scattering regime is given in terms of the radius a by

$$C_{abs} = \frac{8\pi^2 a^3}{\lambda} \left| Im \left(\frac{\varepsilon - 1}{\varepsilon + 2} \right) \right|$$

where $\epsilon = \epsilon_1 + i\epsilon_2$ is the dielectric constant of the metal at the optical frequency relative to that of the surrounding medium and λ is the wavelength in that medium. For a colloidal dispersion consisting of N small spheres per unit volume, the absorption A, measured in units of absorbance, for a path length ℓ is given by

$$A = NC_{abs}\ell \ .$$

The absorption thus passes through a maximum at the wavelength at which $\epsilon_1 = -2$. Since ϵ_2 is a slowly varying function for silver and gold whereas ϵ_1 varies rapidly with wavelength in the visible range, it may be shown that the absorption band shape is approximately Lorentzian, with C_{abs} at the band center equal to $24\pi^2 a^3 / \lambda \epsilon_2$ and with the full band width at half height given by $\Delta\lambda = 2\epsilon_2(\partial\lambda/\partial\epsilon_1)$. To this approximation ϵ_1 therefore determines the wavelength of the resonance independent of particle size, while for an efficient resonance ϵ_2 must be small. Dielectric data for silver and gold have been tabulated,[8,9] and for these metals in water $\epsilon_1 = -2$ occurs at 384 and 520 nm respectively. These are very close to the observed wavelengths of the extinction maxima for spheres of these metals within the Rayleigh regime, though in practice there is a small variation in the wavelength of the maximum, and in the half-width, with particle size.[10]

In addition to the absorption the other contribution to the extinction is that due to elastic scattering, and for small spheres within the Rayleigh regime the elastic scattering cross section per particle is given by

$$C_{sca} = \frac{128\pi^5 a^6}{3\lambda^4} \left| \frac{\epsilon - 1}{\epsilon + 2} \right|^2 \ .$$

It may be seen that the elastic scattering, like the absorption, peaks at the wavelength at which $\epsilon_1 = -2$, and increases rapidly with increase in particle radius, but within the Rayleigh regime the contribution of scattering to the extinction is very small.

It will become apparent below that in the context of surface enhanced Raman scattering by colloidal particles there is particular interest in situations which shift the extinction resonance to longer wavelengths than those for isolated spheres within the Rayleigh regime. In order to achieve this result, it is necessary that the spheres are of a size which takes them beyond this regime, or that the particles are aspherical or are very close together. The first of these situations is illustrated in Figure 2, which compares the calculated extinction and scattering cross sections for gold spheres of 20 and 100 nm radius. For the larger sphere the greater part of the extinction is no longer the result of ab-

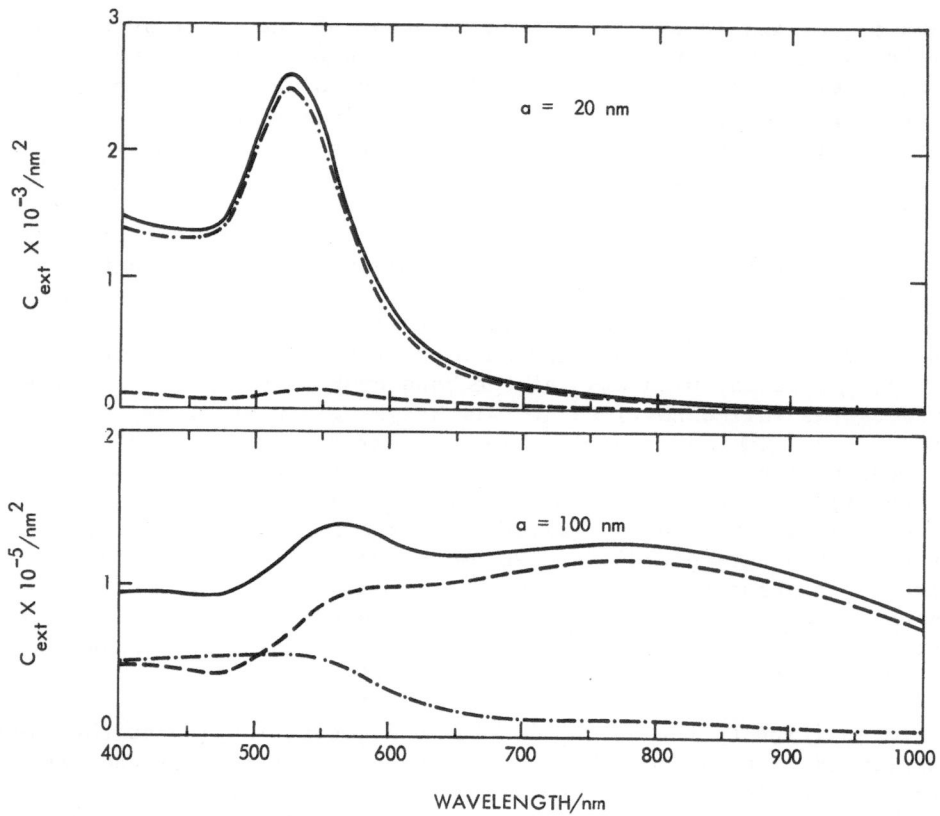

Fig. 2.   Calculated extinction (———), elastic scattering (----),
          and absorption (—·—·—·) of isolated gold spheres of 20
          and 100 nm radius.

sorption but is rather due to scattering, and the dipole resonance
contribution has shifted out to near 800 nm, while there is also a
peak near 550 nm due mainly to quadrupole excitation.   These curves
in Figure 2 were computed from the dielectric data for gold of
Hagemann et al.[8] using the full equations of Mie theory,[11] and are
in good agreement with similar curves calculated by Messinger et al.[12]
from the dielectric data of Johnson and Christy.[9]

   More pronounced in its effect on the extinction spectra than
increasing the size of spherical particles is that of changing
their shape.   Figure 3 illustrates this by showing the absorption
spectra, calculated in the dipole approximation, of small prolate
gold spheroids of various axial ratios.[5]   To this approximation
the absorption and elastic scattering cross sections are given by[11]

$$C_{abs} = (8\pi^2/3\lambda) \text{ Im } (\alpha_\ell + 2\alpha_t)$$

$$C_{sca} = (128\pi^5/9\lambda^4)(|\alpha_\ell|^2 + 2|\alpha_t|^2) \ .$$

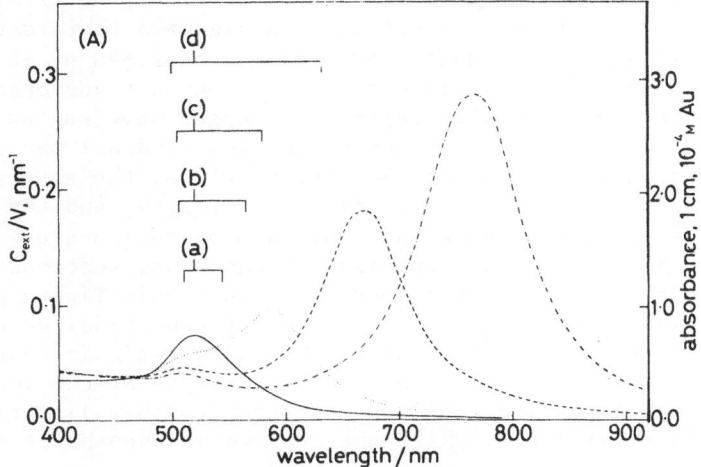

Fig. 3.   Absorption cross section per unit particle volume calcu-
          lated for gold spheroids in water. Minor radius of sphe-
          roids 10 nm, major radius 10 (——), 20 (·········), 30 (----),
          40 nm (-·-·-·). Also shown are the wavelengths of the ab-
          sorption maxima for linear strings of two (a), three (b),
          four (c), and an infinite number (d) of gold spheres of
          10 nm radius in contact in water, assuming dipole coupling
          between the spheres. From ref. 5.

Here the longitudinal and transverse polarizabilities α_ℓ and α_t of
the spheroid are given by

$$\alpha_{\ell,t} = \frac{V(\varepsilon - 1)}{4\pi + (\varepsilon - 1)P_{\ell,t}}$$

where V is the volume of the spheroid whose axial ratio is r, and

$$P_\ell = \frac{4\pi}{r^2 - 1}\left[\frac{r}{\sqrt{(r^2 - 1)}}\ln[r + \sqrt{(r^2 - 1)}] - 1\right]$$

$$P_t = (4\pi - P_\ell)/2 .$$

It is seen that there are now two maxima at the wavelengths at
which

$$\varepsilon_1 = 1 - 4\pi/P_{\ell,t}$$

these being the wavelengths at which the two components of the polarizability come into resonance. In Figure 3 therefore, the shorter wavelength component which remains near 520 nm is due to the mode in which the dipole oscillates along a transverse axis, while the band which shifts rapidly to longer wavelengths with increase in the axial ratio is due to the longitudinal oscillation. As for spheres within the dipole approximation, the absorption and elastic scattering maxima coincide in wavelength, and the resonance wavelengths for spheroids within this approximation are also independent of spheroid size. In Figure 3 the cross sections are divided by the volume of the speroid so that this Figure gives the variation in absorbance of a dispersion of spheroidal particles of fixed total metal concentration as the axial ratio is changed. This Figure therefore shows that the absorbance of the longitudinal mode for such a changing colloid increases roughly linearly with increase in axial ratio, while the transverse absorbance decreases.

Finally there is also a shift and splitting of the extinction resonances of small particles if the particles are within a few diameters or less of each other, on account of coupling of the induced dipoles or higher multipoles. The effect of dipole coupling on absorption has been discussed by Clippe et al.[13] for small spheres in contact in aggregates of various shapes, and it may be shown[5] from their results that the coupled resonances occur at the wavelengths at which

$$\epsilon_1 = (\lambda - 2)/(\lambda + 1)$$

where λ is here a geometrical constant tabulated by Clippe et al.[13] for various aggregate shapes. Figure 3 shows the wavelengths of the coupled dipole resonances which give the most intense absorption, for gold spheres aggregated into linear strings of various lengths. As for spheroidal particles there is a splitting of the dipole resonance of the individual spheres into two components. These are a longitudinal component which moves to longer wavelengths with increase in the string length, in which the coupled dipoles are in phase and parallel to the string axis, and a transverse in-phase coupled component which remains near the isolated sphere resonance wavelength. However this splitting for linear strings of dipole-coupled particles is considerably less than it is for spheroids of similar axial ratios, as may be seen from Figure 3, and the data of Clippe et al.[13] show that the splittings for more compact aggregates of dipole-coupled particles is even less than it is for linear strings containing the same number of particles.

ENHANCED RAMAN SCATTERING

Surface enhanced Raman scattering by metal particles was first reported by Creighton, Blatchford, and Albrecht, who made measurements on pyridine adsorbed on aqueous silver and gold colloids.[3] The silver colloids were prepared by borohydride reduction of silver nitrate solution. Before addition of pyridine the sols were yellow with a sharp extinction band centered at 380 nm, showing the particles to be roughly spherical and of size within the Rayleigh limit. Upon addition of pyridine the extinction spectrum changed as shown in Figure 4, and the color of the sol progressed from yellow through red to blue-grey. It is now clear that this spectral change was due to aggregation in which there was formation of particle strings of increasing length, the longitudinal resonance shifting out to beyond 550 nm in Figure 4, while the transverse resonance is seen at 340 nm as a shoulder on the residual single particle extinction band. Thus Figure 1d reproduces a transmission electron micrograph of an aggregated silver sol (the sol of Fig. 3b of ref. 3) taken from this work.

It was found that the aggregated silver sols exhibited strong SER scattering of incident light in the green-yellow region, and that the SER spectra (Figure 5) were characteristic of adsorbed pyridine. The prominent bands due to pyridine ring vibrations observed at 1038 and 1010 cm^{-1} are within experimental error the same as those observed in the SER spectra of pyridine adsorbed at roughened silver electrodes, and are different from the frequencies for pyridine in solution. There was also a strong band at 228 cm^{-1} in the SER spectrum of the colloids, due perhaps to stretching of the Ag-pyridine bond. As observed for silver electrode surfaces, the 1038 and 1010 cm^{-1} SERS bands of the colloids also had remarkably high depolarization ratios, in view of the very low ratios ($\rho <$ 0.05) for the corresponding 1035 and 1003 cm^{-1} bands of pyridine in aqueous solution. The intensities of the SERS bands of the colloids were up to 5 times more intense than the corresponding bands

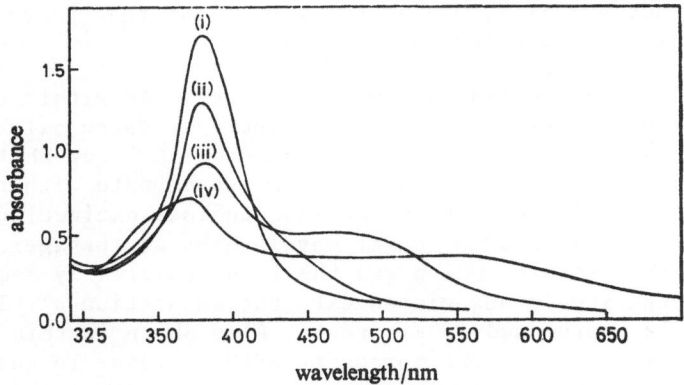

Fig. 4.   Extinction spectrum of silver-borohydride sol (i) freshly prepared, and (ii-iv) at increasing intervals after adding pyridine. From ref. 3.

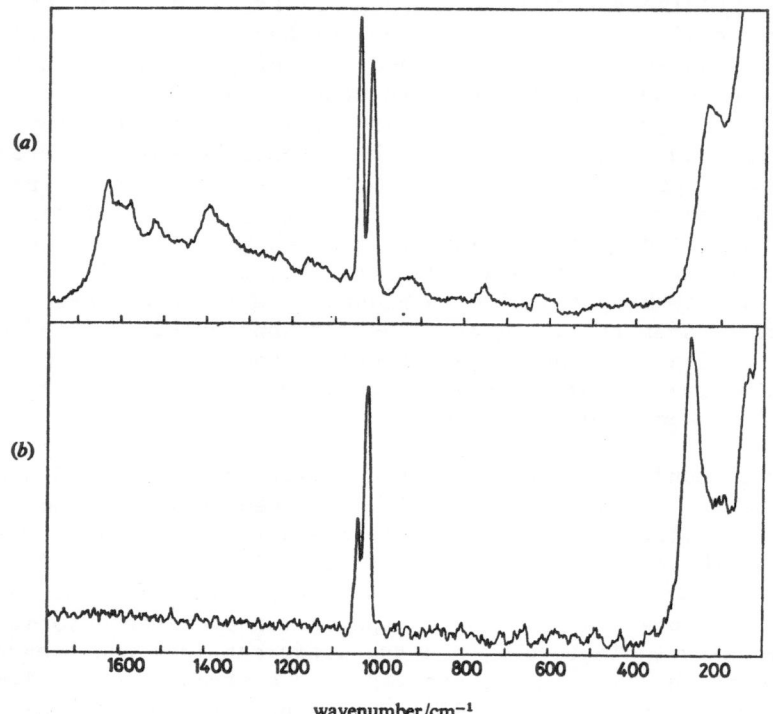

Fig. 5. Raman spectra of (a) silver-borohydride sol, (b)
         gold-borohydride sol, both with added pyridine. From
         ref. 3.

of 0.1M aqueous pyridine, and it was thus clear that there was a
substantial Raman intensity enhancement.

The excitation wavelength used in Figure 5 was within the range
of the longitudinal extinction band of the aggregated silver sol,
and the most striking result of this work[3] was to show that the
SER excitation profile peaked at the same wavelength within the
accuracy of the measurements as the longitudinal extinction band,
and moved with this band to longer wavelengths as the aggregation
proceeded. This result, which was the first to clearly demonstrate
that SER scattering is associated with the excitation of plasma
resonances, is reproduced in Figure 6. Also shown in this Figure
is the large increase in SER intensity with increase in the resonance wavelength, and thus with increase in the aggregate string
length, for the silver sols.

Preliminary results in SER scattering by pyridine adsorbed on
a gold sol, prepared by borohydride reduction of $K[AuCl_4]$ solution,

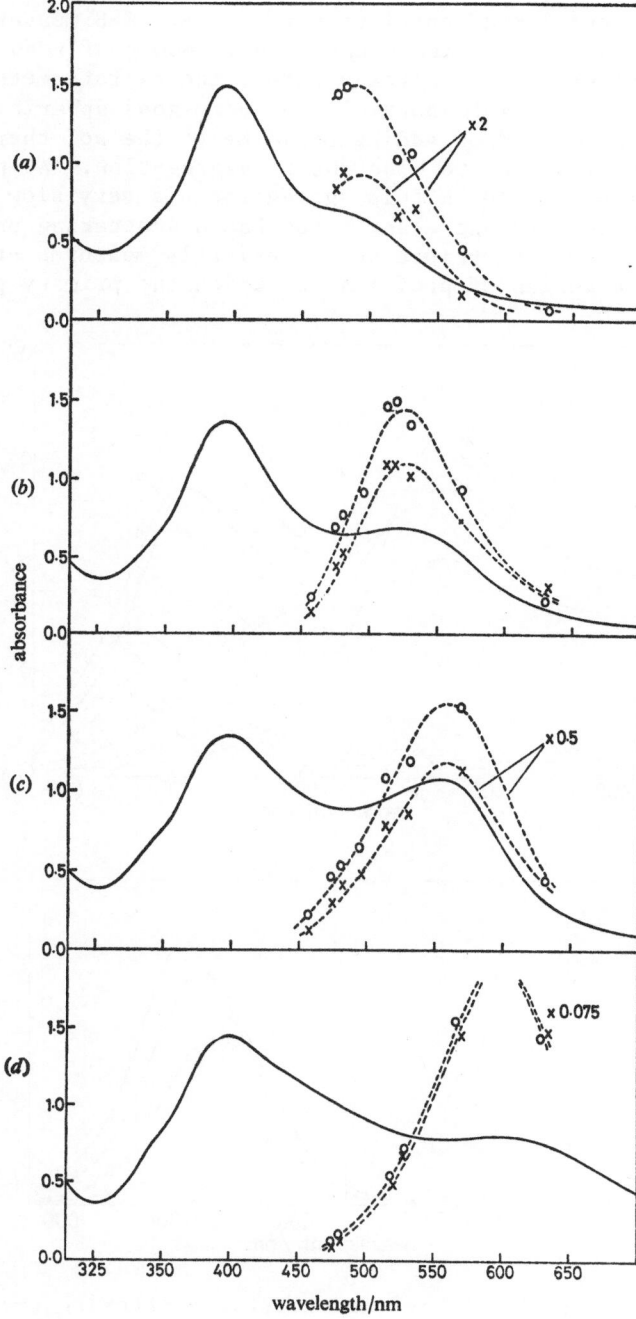

Fig. 6.   Extinction spectra (———·) and Raman excitation profiles
          (— — —) for silver-borohydride sols with added pyridine.   O
          and X denote intensities of 1038 and 1010 cm^{-1} bands of
          adsorbed pyridine.   From ref. 3.

were also reported,[3] and constituted the first SERS measurements on
gold surfaces, but fuller measurements have now confirmed and extended these results.[5] As already noted, the citrate method provides a dispersion of gold particles of very good sphericity and
uniformity of size, and on adding pyridine to the sol there is a
change of color from red to blue due to aggregation. At pyridine
concentrations of ca. $10^{-5}M$ this aggregation is very slow, and
the extinction spectra and elastic and Raman scattering profiles
for a single sol may therefore be conveniently measured at various
times to give a series of profiles for which the primary particle

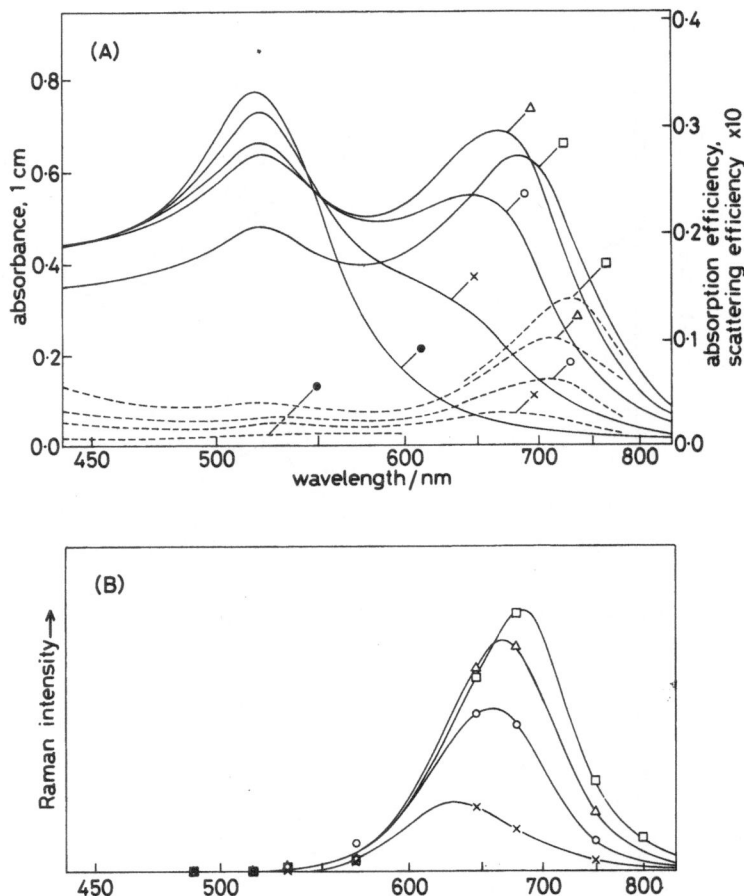

Fig. 7.   (A) Extinction (——) and elastic scattering (----) of
           gold-citrate sol, and (B) the excitation profile for the
           1014 cm^{-1} Raman band of pyridine adsorbed on the colloid
           particles. Measurements before (•), and 3 hr. (x), 8 hr.
           (0), 24 hr. (□), and 58 hr. (Δ) after adding pyridine.
           From ref. 5.

size and concentrations are constant and only the average string
length varies. Figure 7 shows such a series of extinction and
scattering profiles for a single gold sol, and electron micrographs
are also given for the same sol (Figure 1 a-c) to show the development of the aggregation. Initially the extinction spectrum closely
resembles the spectrum calculated for an isolated small gold sphere
(Figure 3), with no evidence of a shoulder on the long wavelength
side indicative of aggregation. As the aggregation proceeds the
longitudinal extinction band of the linear aggregates appears and
grows in intensity at the expense of the single particle peak,
moving slowly to longer wavelengths as for the silver sols. It
may be seen (Figure 3) that these changes are well reproduced by a
superposition of the calculated extinction spectrum of a single
sphere upon that of a gold spheroid of increasing axial ratio. The
excitation profiles for the 1014 cm^{-1} SERS band of adsorbed pyridine is also shown in Figure 7. Comparison of the extinction and
elastic scattering profiles in Figure 7 shows that the measured
elastic scattering cross sections were much less than the extinction cross sections, and the extinction spectra are thus essentially the absorption spectra of the sols. It is therefore clear
from Figure 7 that it is the adsorption contribution to the extinction rather than the elastic scattering which is more closely
followed by the SERS excitation profiles.

Figure 7 also shows that no SER scattering was detected for
excitation under the transverse resonance/single particle peak for
the gold colloids, and this result has also been observed for a
silver-borohydride sol aggregated with pyridine, as demonstrated
in Figure 8. As also observed for the silver sols (Figure 6), the

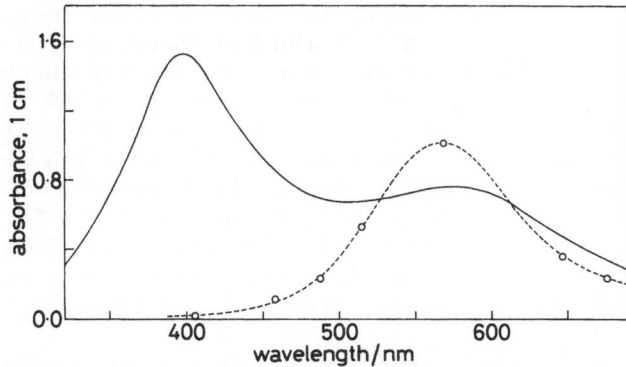

Fig. 8.   Extinction (——) and Raman excitation profile (----, 1008
          cm^{-1} band) for silver-borohydride sol with added pyridine,
          showing low Raman intensity for excitation under 400 nm
          extinction resonance. From ref. 5.

SERS intensity for the gold sols increases steeply with increase in the longitudinal resonance wavelength, and thus in the aggregate string length. This increase is steeper than the simultaneous increase in the height of the longitudinal extinction band, and Figure 9 shows that, at least for moderate degrees of aggregation, there is a near-quadratic relationship between the SERS intensity and the absorbance at a given excitation wavelength within the longitudinal extinction band.

The SERS enhancement factor for the most aggregated gold sol prepared in this work[5] was determined by measuring the amount of adsorbed pyridine by a method which involved centrifugation and spectrophotometry, and comparing the SERS intensity with that of normal Raman scattering from a pyridine solution of known concentration. The longitudinal extinction resonance for the sol was beyond 800 nm, and the enhancement factor measured for 752.5 nm excitation was 4.7×10^5. This is comparable with or bigger than the total enhancement factors found for molecules adsorbed at roughened silver electrodes[14] or at silver diffraction gratings.[15]

SERS excitation profiles for adsorbates on metal colloids have also been measured by Kerker et al.[7] and by von Raben and co-workers.[6] Kerker et al. prepared silver sols by the Carey Lea method in which ferrous ions reduce Ag^+ in the presence of citrate ions. The sols were not investigated by electron microscopy, but their extinction peaked sharply at 400 nm, from which it was assumed the particles were mostly essentially spherical. Their mean radius was estimated to be 21 nm by comparing the wavelength of this extinction maximum with that calculated for various particle radii by Mie theory.

These sols were found to show strong SER scattering of incident light in the green-red region, and the adsorbed species was identified as initally citrate ions, which the spectra (Figure 10) showed to undergo slow chemical change over the course of several days. Although the extinction peaked in the blue however, the SERS intensity increased with increase of the excitation wavelength from 350.7 to 647.1 nm, with the enhancement factor for 647.1 nm excitation estimated to be 6×10^5. This increase in SER scattering to the red is similar to the result for the silver-borohydride sols after considerable aggregation, and it is therefore of obvious relevance to note that the extinction spectra of the Carey Lea sols showed considerable extinction extending out on the long wavelength side of the maximum at 400 nm, indicative of the presence in the colloids of some aggregated or aspherical particles.

In von Raben et al.'s study[6] gold colloids were prepared by borohydride reduction of $K[Au(CN)_2]$ or $K[AuCl_4]$ solutions. The particles were examined by electron microscopy, which showed the primary particles to be approximately spherical and mostly within

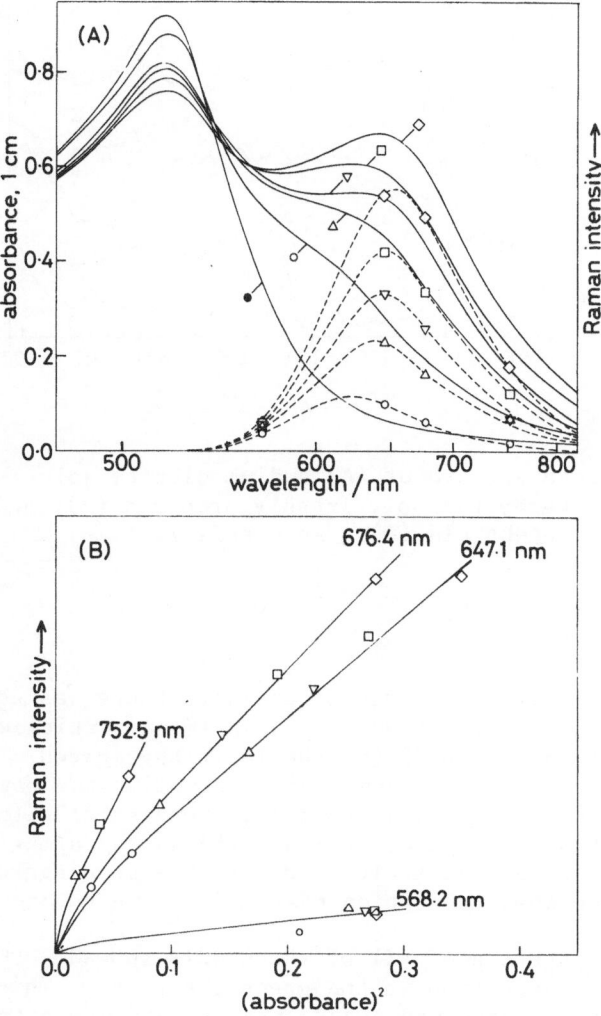

Fig. 9. (A) Extinction (———), and excitation profile(————) for
the 1014 cm^{-1} Raman band of pyridine adsorbed on a gold-
citrate sol, and (B) the Raman intensity vs. the square of
the absorbance at various excitation wavelengths, measured
as aggregation proceeds. (●) indicates before, and (0)
11 hr., (Δ) 25 hr, (▽) 35 hr., (□) 49 hr. and (◇) 73 hr.
after adding pyridine. From ref. 5.

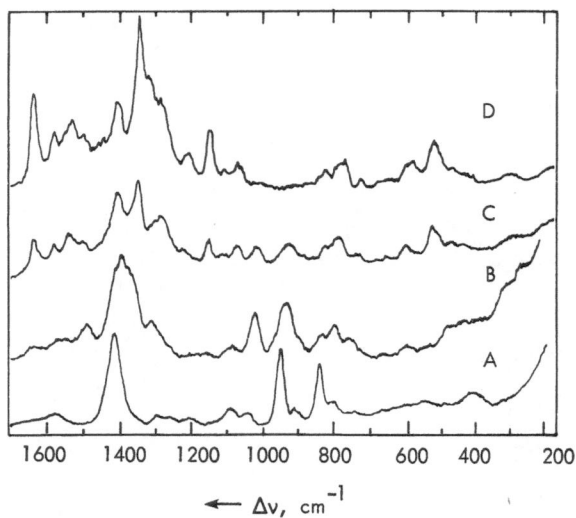

Fig. 10.   Raman spectra of (A) sodium citrate solution, and (B-D)
           of Carey Lea sol, freshly prepared (B), and one (C) and
           two weeks old (D).   From ref. 7.

the Rayleigh limit. For each of the sols there was some aggregation, amounting to almost 60% of the primary particles for the most
aggregated sol but only 7% for the least aggregated. In each sol
the aggregates were of random shape and unlike the particle strings
of the gold- or silver-pyridine sols. Consistent with these results the extinction spectra (Figure 11) showed a maximum near 520
nm due to the single particles, and for the more aggregated sols
there was also some absorption near 700 nm due to aggregates.

The Raman spectra of all of these sols showed a SER band at
2138 cm^{-1} for excitation in the green-red range. This band is
clearly due to an adsorbed cyano-species, and was attributed to
$[Au(CN)_2]^-$, whose frequency in solution is 2164 cm^{-1}. Superimposed
on this band was a broad background due to re-emission from the
metal, which could be seen to peak at 530 nm for excitation at
514.5 nm or shorter wavelengths. Raman bands of water were also
apparent in all of the spectra, but it was demonstrated by spectral
subtraction that these were due to the bulk water of the colloids,
and that there was no detectable enhancement of the spectrum of
water by the sol particles. Figure 11 shows the excitation profiles of the 2138 cm^{-1} SERS band measured for each of the sols.
In each case there was maximum SERS intensity for excitation in
the red region, despite the lack of extinction in this region for

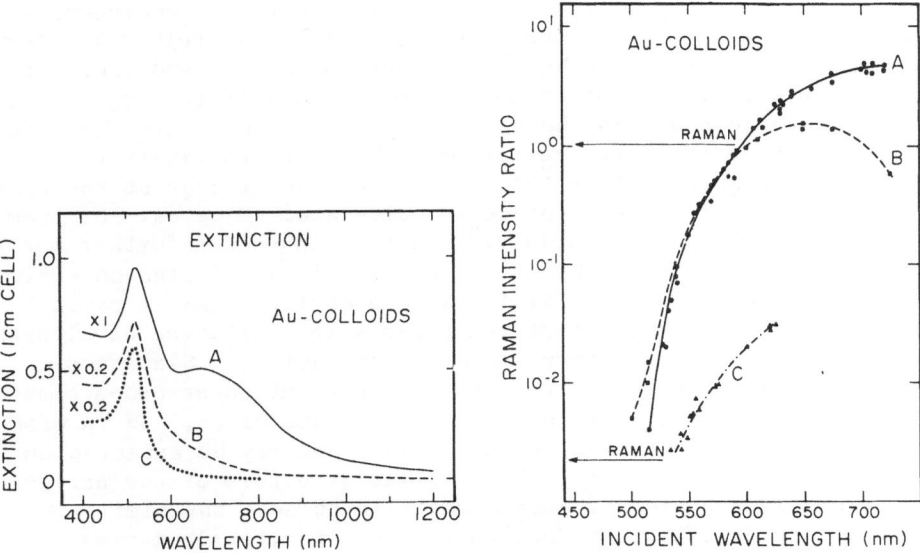

Fig. 11. Extinction and Raman excitation profiles (2138 cm^{-1} band)
of gold-borohydride sols containing [CN]$^-$. Colloids A-C
contain respectively 57, 31, and 7% aggregated particles.
From ref. 6.

the least aggregated sol, and particularly for this sol therefore,
the extinction and SERS excitation profiles were rather different.

A theoretical description of the enhancement of Raman scattering by molecules adsorbed at the surface of isolated metal
spheres and spheroids has recently been developed by Kerker and
coworkers.[16,17] In this description the SERS effect owes its high
intensity to an enhancement of the electromagnetic fields at the
metal surface, due to the resonant response of the particle to the
incident light and a further resonant response to the outgoing
Raman scattered light. The theoretical analysis includes spheres
of size beyond the Rayleigh regime. In the simplest case of
spheres within the dipole limit the Raman enhancement factor G is
given by[16]

$$ G = \left| \left(1 + \frac{2(\varepsilon_i - 1)}{\varepsilon_i + 2} \right) \left(1 + \frac{2(\varepsilon_r - 1)}{\varepsilon_r + 2} \right) \right|^2 $$

where ε_i and ε_r are the values of the metal complex dielectric function relative to the surrounding medium at the incident
and Raman scattered frequencies. The SERS enhancement is seen from
this to peak at an incident wavelength very close to (slightly
shorter than) that of the extinction maximum for small spheres. In

the case of small spheroids it was shown[17] that this enhancement
maximum splits into two peaks which again approximately coincide
in wavelengths with the spheroid extinction maxima, and thus the
longitudinal SERS excitation maximum moves rapidly to longer wavelengths with increase in the spheroid axial ratio, while the wavelength of the transverse excitation maximum is relatively unchanged. The magnitude of the SERS enhancement factor at the peak
(520 nm) for a small gold sphere is calculated to be ca. 10^3 from
the data of Johnson and Christy.[9] This is increased further for a
prolate spheroid for excitation at the longitudinal plasmon excitation wavelength, reaching ca. 10^6 at 670 nm for an axial ratio 3,
while the SERS enhancement at the transverse excitation wavelength
quickly diminishes with increase of axial ratio.[17] Since two
resonances are involved in the SERS enhancement these enhancement
factors are critically dependent on the values of ε_2, and in practice somewhat smaller enhancements than these may be expected on
account of imperfections in the metallic structure of the particles.
The point of significance here however is to note the similarity in
the wavelengths of the extinction and SERS excitation maxima expected on theoretical grounds for spheres and spheroids, and the
rapid increase in the SERS enhancement at the longitudinal resonance wavelength for prolate spheroids with increase in axial
ratio.

It may be seen that, as was also true of the extinction, these
theoretical results for the SERS enhancement factors for spheres
and spheroids account quite well for several of the observations
already described for the particle strings. Thus the approximate
coincidence of the wavelengths of the longitudinal extinction and
SERS excitation maxima, the rapid increase in SERS intensity with
increase in string length, and the failure to observe SERS scattering from excitation under the transverse resonance/small particle extinction peak of the aggregated colloids, are rationalized
by the Kerker theory. On the other hand the experimental data of
Kerker[7] and Chang[6] and coworkers, where there are large differences between the extinction and SERS excitation profiles, appear
to be contrary to the theory. However these differences may also
be rationalized in a way which is consistent with Kerker's theoretical results if account is taken of the quite wide range of particle sizes present in the sols used in the latter work. Both for
elastic and SER scattering the cross sections increase much more
rapidly with increase in particle size than the extinction cross
section. For a sol consisting mainly of small particles but with
a small fraction of significantly larger aggregates, as in Chang's
least aggregated sol, the extinction spectrum may thus be dominated
by the extinction of the more numerous small particles, whereas the
SER (and elastic) scattering may be mainly due to the larger aggregates. It is clear that in sols with this diversity of particles
it would be advantageous to measure the elastic scattering spectrum

as well as the extinction spectrum for comparison with the SERS excitation profile.

Since none of the SERS excitation profiles so far discussed maximize near the small sphere extinction maximum the question arises whether aggregation is essential for the SER scattering to reach a detectable level. In this connection two recent reports of SERS excitation profiles which over a limited wavelength range resemble the small sphere extinction spectra are thus of considerable interest. Wetzel and Gerischer[18] prepared aqueous silver and gold colloids by reduction of silver nitrate or $H[AuCl_4]$ solution with ethylenediaminetetraacetate,[4] and in the case of the gold sols agar was added to inhibit aggregation. These sols were apparently stable to aggregation by pyridine, as judged by their extinction spectra. Thus the freshly prepared sols showed a single extinction peak near the small sphere resonance wavelength, and there was no change on adding pyridine to the sols. Raman excitation of the silver sol near this wavelength gave SER spectra consisting of the characteristic 1010-1036 cm^{-1} bands of pyridine, showing pyridine to be adsorbed, and these bands increased considerably in intensity on addition of chloride ions, as has previously been noted for pyridine adsorbed on silver electrodes. The SERS intensity was found to increase as the exciting radiation was changed from red to blue, and thus the SERS excitation profile resembled the extinction spectrum over the range of wavelengths of the measurements, as shown in Figure 12. For the gold sol however the SERS excitation profile peaked in the yellow-red and thus at a significantly longer wavelength than the extinction maximum of the gold sol.

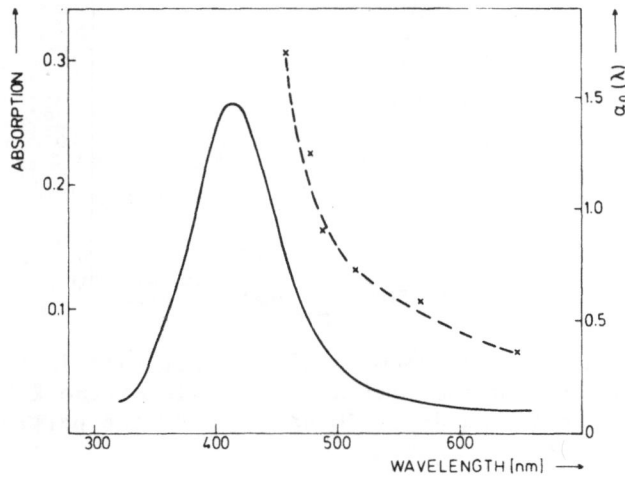

Fig. 12. Extinction (———) and Raman excitation profile (----) for silver-EDTA sol with added pyridine. From ref. 18.

In the other measurements due to Moskovits et al.[19] the colloid
consisted of silver particles of about 10 nm diam. dispersed in a
solid matrix of CO at 4°K. The silver particles were prepared by
evaporation of silver into a stream of argon gas at a few Torr
pressure, where nucleation and growth occurred in the gas phase.
The particles were then swept through an aperture into a separate
chamber and the argon removed by pumping before the particles were
co-condensed with CO. As discussed by Abe and coworkers[10] the
particles produced by this technique are well rounded and fairly
spherical, and it is possible by appropriate choice of evaporation
conditions to control the particle size, though the range of sizes
is fairly wide. The extinction maximum for the CO matrix of silver
particles prepared by this technique was at 376 nm. The Raman spectrum of the matrix was found to consist of bands of good intensity

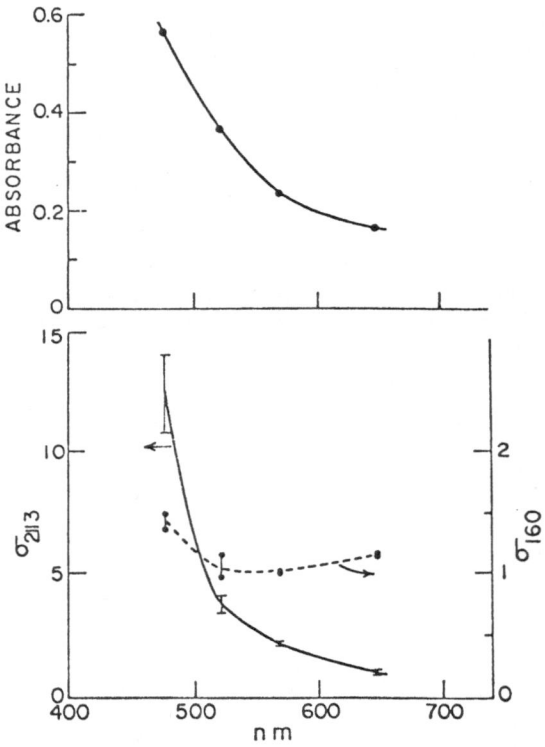

Fig. 13. Extinction spectrum of silver particles in a solid CO
          matrix, and the excitation profile of the 2113 and 160
          cm^{-1} Raman bands of CO adsorbed on the particles. From
          ref. 19.

at 2113, 160, and 64 cm^{-1}, and these are clearly SERS bands due to CO molecules chemisorbed at the surface of the particles. The excitation profiles of the 2113 and 160 cm^{-1} bands and the extinction spectrum are reproduced in Figure 13. It is seen that the two bands have different excitation profiles, and this interesting observation was discussed by Moskovits et al.,[19] but the point at issue here is that the profile for the 2113 cm^{-1} band rises towards the blue, while for the 160 cm^{-1} band there is also not the large profile rise to the red observed for the SERS bands of other colloids discussed earlier.

In neither of these latter studies, however, were the colloid particles investigated by electron microscopy. In the work of Wetzel and Gerischer[18] the extinction maximum for the silver sol was ca. 30 nm to longer wavelengths than expected for small silver spheres within the Rayleigh limit, while in the work of Moskovits et al.[19] the matrix concentration may have been too high to ensure complete particle isolation. Although the excitation profiles suggest that these may be reports of SER scattering from single small isolated spheres, there is therefore room for further work to fully establish this point.

CONCLUSION

The emphasis in this chapter has been on the excitation profiles and particle shapes for SER scattering from metal colloids, and in this respect it reflects the interest in the mechanism of SERS as a whole which has dominated the early phase of the development of the phenomenon at all types of surface. Measurements on colloids have contributed to this development, both in demonstrating the significance of plasma resonances in SERS, and in showing the importance of resonator shape. These matters are relevant to SER scattering at all types of metal surface, and thus the role of localized particle-like resonances in SER scattering from highly roughened electrodes was early recognized by Moskovits,[20] and the importance of the shape of particles[21] or of protrusions on rough electrodes[22] was raised by Gersten and Nitzan. Related to the work discussed here on metal colloids is that on island films or particle arrays discussed elsewhere. All types of particulate surface have the experimental advantage over randomly rough continuous surfaces that the wavelength of the plasma resonances may be readily measured by normal techniques of absorption spectrophotometry. There is also some control of particle size or shape, and hence of the resonance wavelengths, by appropriate preparative procedures.

Although determination of excitation profiles has been a major part of published SERS measurements on colloids, there are other details of the work which should be noted. The flow of charge between adsorbate and metal in the course of the vibrations of CO adsorbed on silver particles has been discussed in a proposed mechanism of SER scattering which accounts for the different excitation profiles of different vibrational modes of this adsorbate;[19] the relative intensities of the bands of adsorbed pyridine have been used to estimate the ζ-potential of a silver sol;[18] a procedure for the preparation of silver sols has been reported which results in a large number of SERS bands due to adsorbed impurities or biproducts of the preparation.[23] These are matters of chemical interest, and serve as a reminder of the likely potential of Raman spectroscopy as a probe of surface chemistry. Although there has been much work done on the mechanism of SERS, its development as a chemical probe both of particle surfaces and of metal surfaces in general is still almost untouched.

ACKNOWLEDGEMENT

The author is grateful to Profs. R. K. Chang and M. Kerker and Dr. J. C. Tsang for sending copies of papers cited here prior to publication. Hospitality from Exxon Research and Engineering during the writing of this chapter is gratefully acknowledged.

REFERENCES

1. J. Turkevich, P. C. Stevenson, and J. Hillier, A Study of the Nucleation and Growth Processes in the Synthesis of Colloidal Gold, Disc. Faraday Soc. 11:58 (1951).
2. G. Frens, Controlled Nucleation for the Regulation of the Particle Size in Monodisperse Gold Suspensions, Nature Physical Science 241:20 (1973).
3. J. A. Creighton, C. G. Blatchford, and M. G. Albrecht, Plasma Resonance Enhancement of Raman Scattering by Pyridine Adsorbed on Silver and Gold Solid Particles of Size Comparable to the Excitation Wavelength, J. Chem. Soc. Faraday II 75:790 (1979).
4. A. Fabrikanos, S. Athanassiou, and K. H. Lieser, Darstellung stabiler Hydrosole von Gold und Silber durch Reduktion mit Athylendiamintetraessigsaure, Z. Naturforsch, 18B:612 (1963).
5. J. A. Creighton, C. G. Blatchford and J. R. Campbell, The Effect of Aggregation on the Plasma Resonance Enhanced Raman Scattering by Adsorbates on Gold Colloids, J. Chem. Phys., submitted.
6. K. U. von Raben, R. K. Chang, and B. L. Laube, Surface Enhanced Raman Scattering of Au(CN)$_2^-$ Ions Adsorbed on Gold Colloids, Chem. Phys. Lett. 97:465 (1981).
7. M. Kerker, O. Siiman, L. A. Bumm, and D.-S. Wang, Surface Enhanced Raman Scattering (SERS) of Citrate Ion Adsorbed on Colloidal Silver, Appl. Opt. 19:3253, 4137 (1980).
8. H. J. Hagemann, W. Gudat, and C. Kunz, Optical Constants from the Far Infrared to the X-ray Region. Magnesium, Aluminum,

Copper, Silver, Gold, Bismuth, Carbon and Aluminum Oxide, J. Opt. Soc. Amer. 65:742 (1975).

9. P. B. Johnson and R. W. Christy, Optical Constants of the Noble Metals, Phys. Rev. B 6:4370 (1972).

10. H. Abe, W. Schulze and B. Tesche, Optical Properties of Silver Microcrystals Prepared by Means of the Gas Aggregation Technique, Chem. Phys. 47:95 (1980).

11. M. Kerker, "The Scattering of Light and Other Electromagnetic Radiation", Academic Press, New York (1969).

12. B. J. Messinger, K. U. von Raben, R. K. Chang, and P. W. Barber, Local Fields at the Surface of Noble Metal Microspheres, Phys. Rev. B, (in press).

13. P. Clippe, R. Evrard, and A. A. Lucas, Aggregation Effect on the Infrared Adsorption Spectrum of Small Ionic Crystals, Phys. Rev. B 14:1715 (1976).

14. R. P. Van Duyne, in "Chemical and Biological Applications of Lasers", Vol. 4, ed. by C. B. Moore, Academic Press, New York (1978) p. 101; C. C. Busby and J. A. Creighton, Factors Influencing the Enhancement of Raman Spectral Intensity from a Roughened Silver Surface: the Adsorption and Surface Raman Scattering of 2-amino 5-nitro Pyridine, J. Electroanal. Chem., submitted.

15. J. C. Tsang, J. R. Kirtley and S. S. Jha, Surface Enhanced Raman Scattering from Characterized Metal Surfaces, Proceedings of Second International Conference on Vibrations at Surfaces, Namur, Belgium, Sept. 1980, ed. by A. Lucas, R. Caudano, and J. M. Gilles, Plenum, 1981, to be published.

16. M. Kerker, D-S. Wang, and H. Chew, Surface Enhanced Raman Scattering (SERS) by Molecules Adsorbed at Spherical Particles, Appl. Opt. 19:4159 (1980).

17. D-S. Wang and M. Kerker, Enhanced Raman Scattering (SERS) by Molecules Adsorbed at the Surface of Colloidal Spheroids, Phys. Rev. B, submitted.

18. H. Wetzel and H. Gerischer, Surface Enhanced Raman Scattering from Pyridine and Halide Ions Adsorbed on Silver and Gold Sol Particles, Chem. Phys. Lett. 76:460 (1980).

19. H. Abe, K. Manzel, W. Schulze, M. Moskovits, and D. P. DiLella, Surface-enhanced Raman Spectroscopy of CO Adsorbed on Colloidal Silver Particles, J. Chem. Phys. 74:792 (1981).

20. M. Moskovits, Surface Roughness and the Enhanced Intensity of Raman Scattering by Molecules Adsorbed on Metals, J. Chem. Phys. 69:4159 (1978).

21. J. I. Gersten, The Effect of Surface Roughness on Surface Enhanced Raman Scattering, J. Chem. Phys. 72:5779 (1980).

22. J. I. Gersten and A. Nitzan, Electromagnetic Theory of Enhanced Raman Scattering by Molecules Adsorbed on Rough Surfaces, J. Chem. Phys. 73:3023 (1980).

23. M. E. Lippitsch, Observation of Surface Enhanced Raman Spectra by Adsorption to Silver Colloids, Chem. Phys. Lett. 74:125 (1980).

INELASTIC MIE SCATTERING FROM ROUGH METAL SURFACES

D. A. Weitz, T. J. Gramila and A. Z. Genack

Exxon Research and Engineering Company
P. O. Box 45
Linden, New Jersey 07036

INTRODUCTION

One of the most widely debated questions concerning surface-enhanced Raman scattering (SERS) is the nature of the surface morphology required to obtain the large enhancements. While it is generally agreed that surface roughness is important, the specific nature of this roughness, and its exact role in the enhancement has been a subject of considerable debate. In this chapter, we discuss experimental results that address this question directly. We study an extremely low frequency Raman mode which we attribute to Raman scattering (RS) from the rough metal surface itself.[1,2] From the low frequency Raman scattering we can determine the nature of the roughness which contributes to SERS. From the anomalous behavior of this Raman scattering, we can determine the role of this roughness in the large enhancement of the RS cross section of adsorbed molecules.

We have discovered that a characteristic of virtually all surfaces that exhibit SERS is a very intense low frequency maximum in the RS, peaked at a frequency shift, of $\Omega \simeq 10$ cm^{-1}. It is never observed from smooth metal surfaces from which we do not observe SERS, but only from rough ones from which SERS is observed. Although the RS intensity of the mode increases as the surface coverage of a specifically adsorbed organic molecule is increased, the presence of such a molecule is not required for its observation. Furthermore, the frequency shift of the mode is generally independent of the specific adsorbate. Thus the mode is characteristic of RS from the rough metal surface itself. The most unusual feature of this Raman mode is the fact that the frequency shift from the laser line changes as the excitation frequency, ω_{ex}, is varied.

Furthermore, Ω also changes when n_s, the index of refraction of the medium surrounding the surface, is changed. To our knowledge, this is the only example of RS exhibiting this anomalous behavior.

We attribute the low frequency peak to RS from acoustic vibrations of the metal surface, localized by the surface roughness. The change in Ω with excitation frequency suggests that a resonant excitation is involved in the RS process, while the change in Ω when n_s is varied suggests that this excitation is a collective electronic resonance or surface plasmon localized on some roughness feature of the surface. Thus we believe that each laser excitation frequency resonantly excites surface plasmons localized on surface roughness features of a given size and shape. The size of the region of plasmon excitation determines the vibration frequency of this roughness feature and is observed as a low frequency peak in the Raman spectrum. Thus the frequency of the mode gives the characteristic dimension of the roughness directly, which is typically ~ 100 Å. These observations are the first direct evidence of the involvement of some form of resonant excitation in RS from these surfaces, and we believe this same excitation is also responsible, at least in part, for SERS from adsorbed molecules.

These results suggest an explanation for the broad excitation spectrum of SERS, which extends from the blue to the deep red for RS from Ag electrodes. They indicate that this width is not characteristic of RS from a single molecule, but rather results from overlapping plasmon resonances of different roughness features of the surface. This hypothesis, and the critical role of localized surface plasmons in SERS, can be tested by examining SERS from a surface with more uniform roughness than is found on electrode surfaces. We expect that the SERS excitation spectrum should be narrowed for these surfaces. We, therefore, present data on RS from molecules adsorbed on silver-island films. These surfaces are known to exhibit SERS[3] and also exhibit an anomalous absorption that has been studied in detail[4] and is due to the electronic resonance of the islands. This localized-surface-plasmon-induced adsorption has a distinct frequency dependence and we show that SERS from the adsorbates has an excitation-frequency dependence which can be fully determined from the absorption. These results confirm the "electromagnetic" theory for SERS that has been proposed by a number of authors[5-8] and demonstrate the importance of the localized surface plasmon excitation in SERS, which is suggested by the low frequency RS data.

Finally, we also discuss a simple, although somewhat idealized, model for the low-frequency RS from the metal. We consider isolated metal spheroids whose lowest-order Mie-scattering resonances correspond to the localized surface plasmon excitation. The acoustic vibration is modeled as a mechanical vibration of the spheroid that changes its shape and hence its resonance frequency.

This will produce Raman sidebands on the Mie scattering, leading us to denote these modes as inelastic Mie scattering. A comparison of the prediction of the model to the data provides some information about the morphology of the electrode surface.

The remainder of this chapter is organized as follows. In the next section we describe the experimental apparatus used to perform these studies. The following section contains a detailed discussion of our experimental results and their interpretation. This is followed by a section in which the theoretical model is presented. Next we briefly discuss and refute an alternate hypothesis for the low frequency RS. Finally, we close with a brief summary of our conclusions.

EXPERIMENTAL APPARATUS

All of our RS measurements were performed with a Spex 14018 double spectrometer equipped with 1800 l/mm holographic grating and a "spatial filter" between the two spectrometers. Because the rough surfaces had very intense diffuse scattering and because we were looking so close to the laser line, a third spectrometer, scanned in tandem with the double, was crucial for these experiments. To obtain the required rejection of the elastic scattering, the entrance and exit slits were typically set at 20 μm for 514.5-nm excitation. This allowed us to observe Raman scattered light to within ~2 cm^{-1} of the elastic peak. Slits with a similar spectral bandpass were used at other excitation frequencies. We note that the very narrow slits required to obtain the necessary rejection meant that a relatively intense RS signal was required. Standard photon counting electronics were used.

Most of our experiments were performed using a backscattering optical arrangement. A single lens focused ~50-100 mW from an Ar$^+$ or Kr$^+$ laser onto the surface and collected and collimated the scattered light. A second lens focused the collimated scattered beam into the spectrometer slits. The best collection efficiency was obtained using a point focus of ~20 μm diameter; however, when the power density was a problem, a cylindrical lens was used prior to the collection-focusing lens. This caused the laser beam to diverge in one direction resulting in a line focus at the sample.

The electrochemical experiments were performed in a standard cell with a ~1 mm^2 polycrystalline silver working electrode, a platinum counter electrode and a saturated calomel electrode (SCE) as a reference. The electrolyte typically consisted of 0.1 M salt and 0.05 M pyridine in distilled, deionized water. The silver surface was mechanically polished to a mirror-like finish and then roughened electrochemically using a "mild" oxidation-reduction cycle[9] consisting of stepping from -0.6 V vs. SCE to 0.2 V to col-

lect ∿50 mC/cm^2 followed by a step back to -0.6 V. The solution
was deoxygenated before and during the experiment with nitrogen gas.

To determine the nature of the low frequency mode, we studied
the temperature dependence of the scattering down to temperatures
for which kT < hcΩ, where k, h, and c are Boltzman's and Planck's
constants and the speed of light, respectively. This required
cooling to liquid helium temperatures. A silica cuvette was used
for this purpose and the Ag electrode was first electrochemically
roughened in the standard way and then forced flush against the
flat window and held in place with a spring. Potentiostatic control of the electrode voltage was removed and the remaining electrodes and most of the electrolyte withdrawn. The whole cell was
then mounted in a Dewar. Since the light path through the electrolyte was limited to the thickness trapped between the electrode and
the cell window, scattering of the light by the frozen electrolyte was not a problem. For the very low temperature experiments,
a line focus was used to minimize heating of the surface by the
laser.

The other forms of silver surfaces used in the experiments were
made in the standard fashions. The silver sols used were produced
using the method of Creighton et al.[10] The silver island films
were made by a slow vacuum evaporation of silver from a resistively
heated source onto silica substrates. We typically evaporated
total mass thicknesses of ∿50 Å at ∿0.5 Å/sec. High resolution
electron micrographs of these films showed them to consist of islands with approximately circular cross sections, ∿200 Å diameter
and separated by ∿200 Å.

RESULTS AND DISCUSSION

A typical scan of the low frequency RS showing both the anti-
Stokes (negative frequency shift) and Stokes (positive frequency
shift) spectra is shown in Fig. 1. The electrolyte used contained
0.1 M KBr and 0.5 M pyridine, and the excitation wavelength was
λ_{ex} = 514.5 nm. The complete RS spectrum in this case showed the
usual molecular modes of a SERS spectrum of pyridine.[9] However,
the peak shown here at Ω ≃ 8 cm^1 was the most intense feature
of the spectrum, and was ∿25 times as intense as the pyridine mode
at 1008 cm^{-1}. The existence and frequency of the peak is not dependent on the nature of the adsorbed RS molecule. Under similar
experimental conditions, we have observed the peak at the same Ω
for nearly every adsorbed molecule we have tried. These have included pyridine, 4-cyanopyridine, deuterated pyridine and cyanide.
Thus we conclude that the mode is not characteristic of the adsorbed organic RS molecule.

In fact the presence of the specific organic adsorbate is not

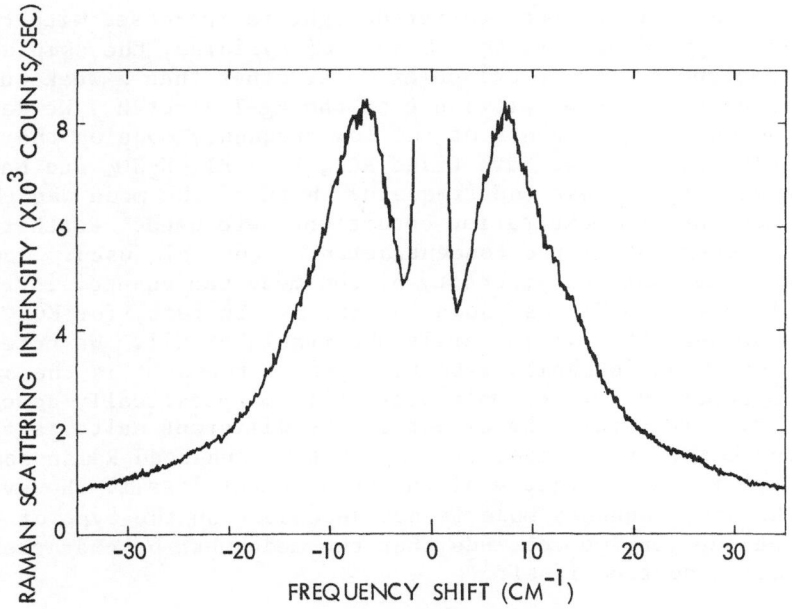

Fig. 1: Low frequency Raman spectrum.

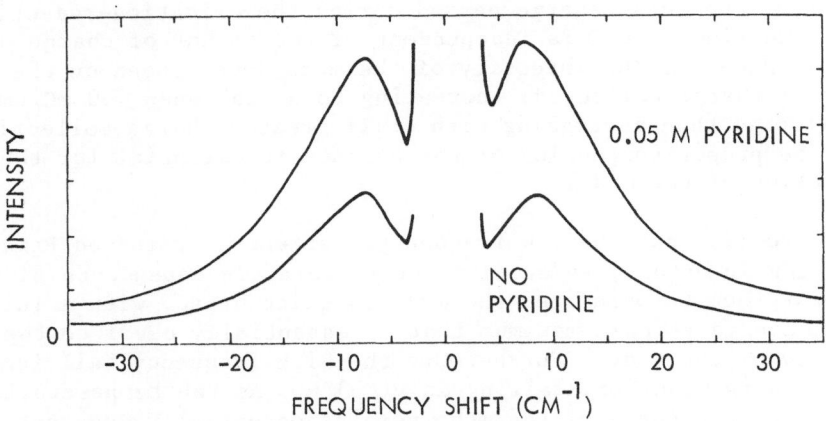

Fig. 2: Low frequency RS with and without 0.05M pyridi...
         in electrolyte.

essential to observe the mode. Figure 2 shows two scans taken under identical conditions except that one electrolyte contained 0.1 M KI and 0.05 M pyridine, while the second contained only 0.1 M KI. The low frequency peak is observed in both cases at the same Ω. However the intensity of the scattered light is increased with the presence of pyridine. In the absence of pyridine, the complete Raman spectrum showed no molecular modes other than a weak one at $\Delta\upsilon \simeq 115$ cm^{-1}, which we attribute to the Ag-I stretch. We have also studied the dependence of the low frequency mode on the salt in the electrolyte. We have tried KCl, KBr, KI, K_2SO_4 and NaCl and in each case found that the frequency shift of the mode was the same, when the same excitation conditions were used. Furthermore, Ω was independent of the concentration of the salt used. However, we find that the intensity of the mode can change, increasing from KCl to KBr to KI, as shown in Fig. 3. In fact, for KCl, the mode is so weak that it is barely observable at all. We note that this ordering of intensity with the type of the salt is the same as has been reported for the molecular RS from specifically adsorbed molecules.[9] Therefore the effect of the different salts is simply to change either the number of contributing enhanced Raman scatterers or the actual magnitude of the enhancement itself. However, since the low frequency mode is not dependent on the type of adsorbed halide ion, we conclude that the mode must be characteristic of the metal surface itself.[7]

Surface roughness is crucial to the observation of the low frequency mode. We have never seen it in RS from a smooth surface. In an electrochemical cell, an oxidation-reduction cycle is essential for the observation of the mode. In Fig. 4, we show the effect of increasing the total charge passed during the oxidation-reduction cycle. We find that Ω is independent of the amount of charge collected. However, the intensity of the mode does depend on the amount of charge collected, increasing to a peak when ~ 50 mC/cm^2 is collected then decreasing with still greater charge collection. Thus, the proper roughening of the surface is essential for the observation of the mode.

To verify that the low frequency scattering is indeed Raman scattering in nature, we studied the temperature dependence of the mode. At room temperature, the mode is quite broad, with a full width measured at half maximum that is essentially equal to the frequency of the mode. Furthermore the high frequency tail is Lorentzian in behavior, falling as $\sim(\Delta\upsilon)^{-2}$. As the temperature is lowered, the shape of the mode remains essentially constant suggesting both the shape and width are determined by inhomogeneous broadening effects rather than a homogeneous broadening, which might be expected to decrease with temperature. However, the peak heights do scale with temperature as expected for a mode which is vibrational in nature, with the Stokes side varying as $n(\Delta\upsilon)+1$ and the anti-Stokes side as $n(\Delta\upsilon)$, where $n(\Delta\upsilon)$ is the Bose-Einstein

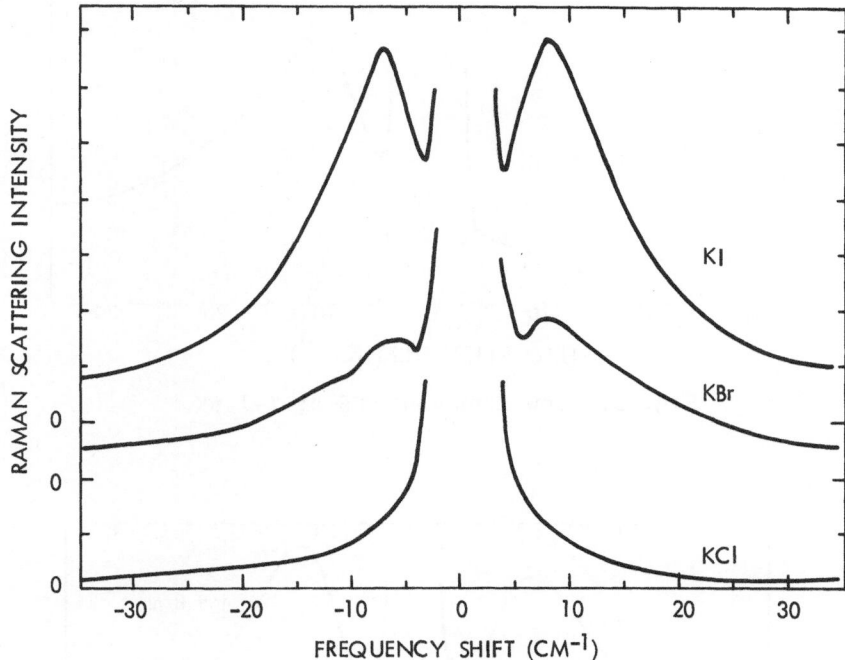

Fig. 3: Effect of halide ion in electrolyte on low frequency RS.

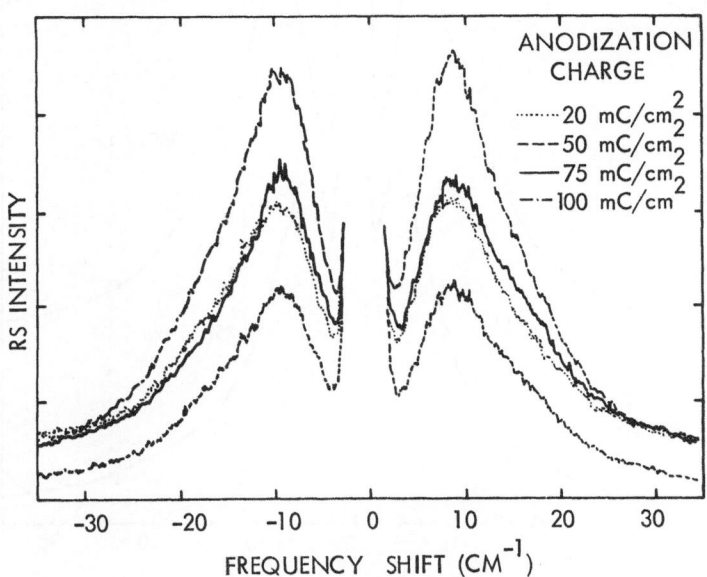

Fig. 4: Effect of total charge collected in oxidation-reduction
          cycle on intensity of low frequency RS.

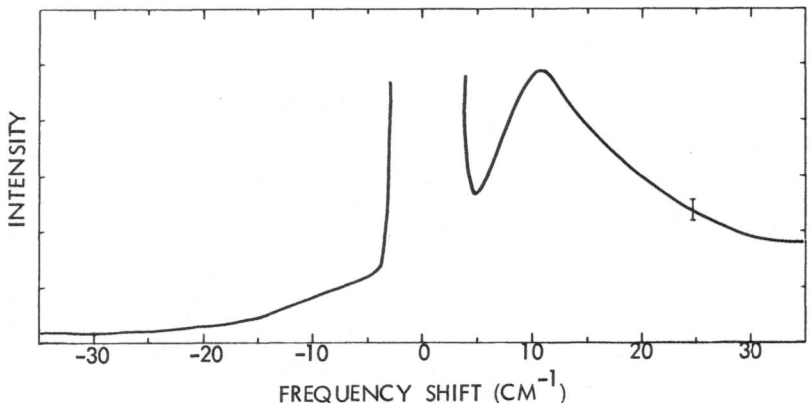

Fig. 5: Low frequency RS at T=1.6K.

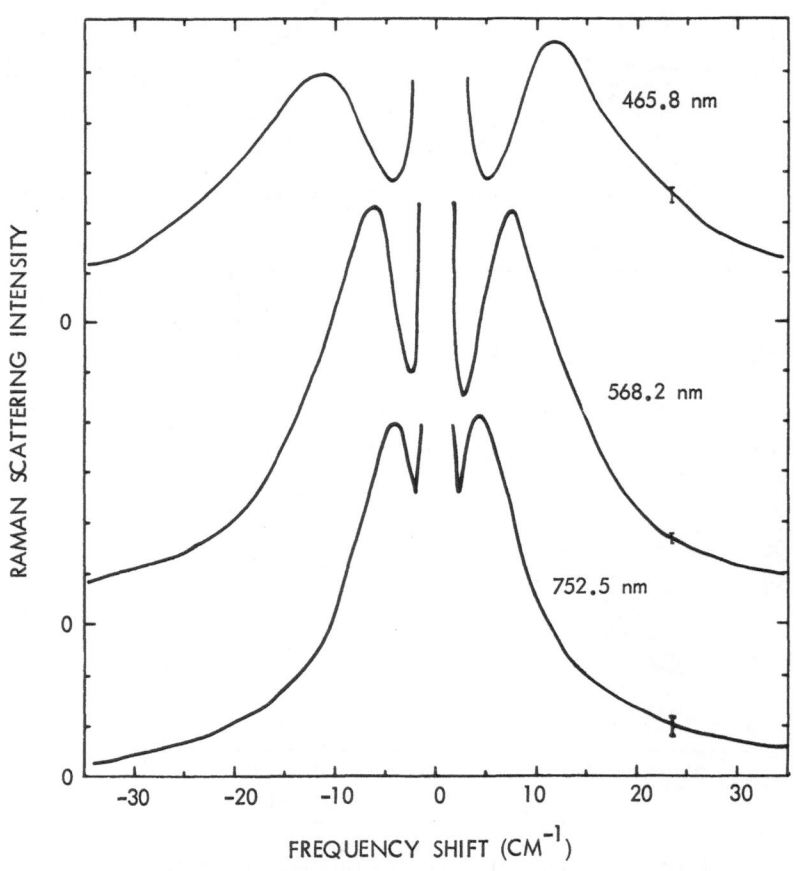

Fig. 6: Low frequency RS with different excitation wavelengths,
         showing shift in Ω.

factor. The lowest temperature achieved in these experiments was
$T \simeq 1.6$ K, in which case $kT < hc\Omega$ and the anti-Stokes scattering virtually disappeared, as shown in Fig. 5. The anti-Stokes side of
this figure also confirms our success in rejecting the elastically
scattered light.

The results presented above suggest that the low frequency mode
is characteristic of RS from the rough electrode surfaces that exhibit SERS. It does not depend on the adsorbed molecule or anion,
but does require some degree of roughness. We have also seen the
mode in RS from other surfaces that exhibit SERS, including silver
sols,[10] silver-island films,[3] and electrochemically roughened
copper electrodes.[11]

The most remarkable aspect of this low frequency RS is its behavior when the laser excitation frequency is varied. We find that
Ω increases as the excitation frequency is increased. This is illustrated in Fig. 6 which shows scans of the low frequency RS of a
surface, prepared in an electrolyte containing 0.1 M KBr and 0.05 M
pyridine, and taken using three different excitation wavelengths.
As shown in Fig. 7, Ω increases monotonically from 3.7 cm^{-1} at an
excitation frequency of ω_{ex} = 12,502 cm^{-1} (λ_{ex} = 799.9 nm) to
11.5 cm^{-1} at ω_{ex} = 21,839 cm^{-1} (λ_{ex} = 457.9 nm). Although we looked
at higher ω_{ex}, the intensity of the mode became too weak for us
to measure the frequency shift of its peak. This is similar to SERS
from the molecular modes which also becomes extremely weak at λ_{ex}
shorter than 457.9 nm on silver electrodes. To check whether this
anomalous behavior with excitation frequency was dependent on the
scattering k vector, we repeated the measurements at a number of
excitation frequencies using a scattering geometry with the k vectors of the incident and scattered radiation separated by 90°.
We found no change in Ω between the 90° scattering geometry and a
backscattering geometry. We also changed the angle between the
surface and the scattering k vector and again found no change in Ω.
Thus we conclude that the anomalous dependence of Ω on ω_{ex} is not
a wave-vector-dependent effect.

A second unusual aspect of the low frequency mode is its behavior when the index of refraction of the medium surrounding the
surface is changed. We first prepared a silver surface in an electrochemical cell in our standard fashion and took a scan of the low
frequency RS. Then the electrolyte was forced out with nitrogen
gas and the electrode dried and maintained under a nitrogen atmosphere thus changing n_s from 1.33 to 1.0. A second scan of the
low frequency RS was taken and typical results for the RS are shown
in Fig. 8, for a silver surface prepared in an electrolyte containing 0.1 M KI and excited with λ_{ex} = 514.5 nm. The frequency of the
mode shifts to a lower value, from 8.0 cm^{-1} with the electrolyte
present, to 6.5 cm^{-1} without the electrolyte. Similar results were

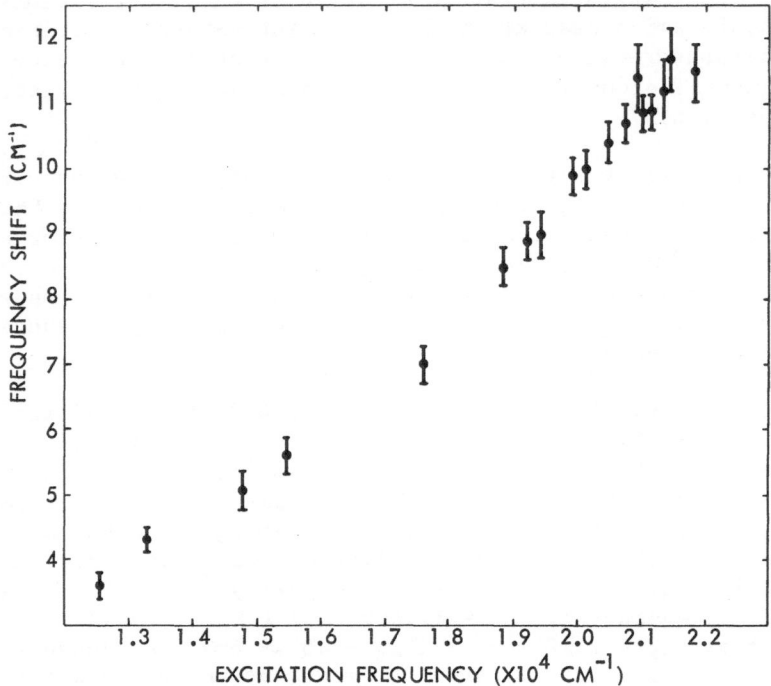

Fig. 7: Dependence of frequency shift on excitation frequency.

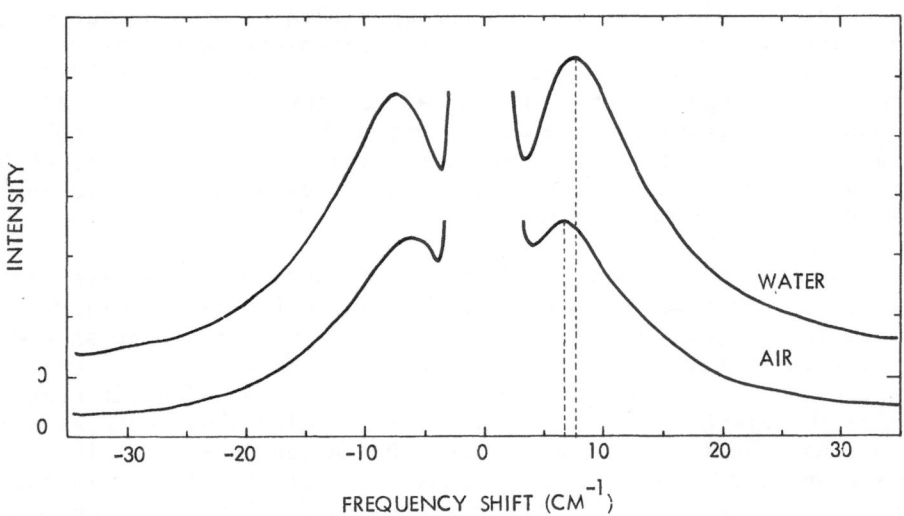

Fig. 8: Low frequency RS with different indices of refraction of
the surrounding medium, showing shift in Ω with n_s.

obtained for a surface prepared in an electrolyte containing 0.1 M
K_2SO_4 and 0.01 M KCN.

The shift in Ω when ω_{ex} or n_s are varied is highly unusual, and
is the first observation of such behavior in RS when there is no
k-vector dependence. The question we now address is the origin of
the mode and its anomalous behavior. Since the low frequency mode
is characteristic of the metal itself, we attribute it to RS from
an acoustic vibration of the metal surface. The high degree of
roughness on the surface will cause the vibration to be localized
on some roughness feature on the surface. We expect that the fundamental frequency of this vibration would be approximately $\Omega \simeq v/cd$,
where v and c are the speeds of sound and light respectively and d
is the characteristic dimension of the roughness feature. Inelastic
neutron scattering from phonons on small particles confirms that
there is a large peak in the density of phonon states when the
phonon energy is such that its wavelength is approximately the same
size as the particle dimension.[12] For the rough silver surfaces,
d $\simeq$ 100 Å for v = 3 x 10^3 m/sec and $\Omega \simeq 10$ cm^{-1}. The localized
nature of the vibration will also relax the requirements of momentum
conservation, allowing it to couple to the optical radiation.
Structure in the Raman spectrum at higher frequencies has recently
been identified as scattering from shorter wavelength bulk acoustic
phonons on silver, coupled to the radiation field by the surface
roughness.[13]

The shift in Ω with ω_{ex} implies that the characteristic size
or shape of the roughness feature causing the RS is different for
each excitation frequency. This implies that only some particular
roughness features on the surface are resonantly excited at each
ω_{ex}, and is direct evidence for the involvement of some form of
resonant intermediate state in the RS process. We can critically
examine the nature of this resonance using the observed behavior of
the low frequency mode. The change in Ω with n_s implies that the
resonant state is not spatially localized at a surface-adsorbed
molecule, but rather is localized over a larger region and has the
characteristics of a surface plasmon. For optimum coupling the
spatial extent of the electronic excitation should be approximately
that of the vibrational excitation, which is on the order of
$\sim$100 Å. Thus, we believe that the resonant intermediate state is a
localized surface plasmon. These excitations are the collective
oscillations of the electrons confined in a small metal particle,
or on a surface roughness feature, responding resonantly to an
optical field. The contribution of these resonances to SERS from
adsorbed molecules has been discussed by Gersten and Nitzan,[5] as
well as other authors.[6-8] The frequency of the localized surface
plasmon will depend on the size and shape of the roughness feature
as well as on the effects of the field of the surrounding surface.
The very rough surfaces used to obtain SERS will contain a random
inhomogeneous array of surface roughness features, each of which
may have a localized surface plasmon resonance at some frequency.[14]

Laser excitation of RS at a given ω_{ex} will resonantly select some subset of these roughness features. The resonant acoustic vibration of each surface roughness feature will also depend on the size and shape of the feature. Thus, as ω_{ex} is changed, different roughness features are resonantly excited, giving different Ω for the low frequency RS. The particular dependence of Ω on ω_{ex} will be determined both by the relationship between the frequencies of the acoustic and electronic excitations as well as by the distribution of roughness features on the surface.

There is a second mechanism that can produce additional RS intensity at exactly the same frequency as the acoustic vibration. The electromagnetic field at the adsorbed molecules is modulated by the acoustic vibrations on the surface since the field enhancement at the surface depends on the shape of the roughness feature. Thus the Rayleigh scattering from the adsorbed molecules will be modulated at the acoustic frequency characteristic of the roughness feature, resulting in an inelastic component in the scattered light. We label the contribution of the metal inelastic Mie scattering, and that of the molecules, inelastic Rayleigh scattering. We attribute the observed increase in intensity of the low frequency mode upon addition of a specifically adsorbed organic molecule, as shown for example in Fig. 2, to inelastic Rayleigh scattering. A more detailed example is shown in Fig. 9, which shows the RS spectrum of an electrode, prepared in an electrolyte containing 0.1 M KBr and increasing amounts of pyridine. The amount of pyridine adsorbed on the surface increases with electrolyte concentration, as measured by the intensity of the 1008 cm^{-1} molecular mode. In this particular case, the contribution of the inelastic molecular scattering appears to dominate. This is also the case for electrolytes containing 0.1 M KCl, where the low frequency mode is almost unobservable (see Fig. 3) without an organic adsorbate.

The results on the low frequency RS give us an understanding of the role of roughness in SERS in general. Roughness is required for both the observation of the low frequency mode and the largest enhancements in the RS from adsorbates because plasmons, localized on surface roughness features, serve as the intermediate state in light scattering from surfaces. The change in Ω with ω_{ex} is direct evidence that a resonance is involved in SERS, and that only some parts of the surface contribute at each excitation energy. Furthermore, roughness is important because the localized plasmon frequencies on roughness features lie below the plasmon frequency for a smooth surface. This enhances the electromagnetic interaction both because the plasmon resonance is narrowed since the imaginary part of the dielectric function, ε'', is reduced in many metals, and because the field strength on the surface is enhanced because the absolute value of the real part, ε', is increased. Finally, from the Raman frequency itself, we can

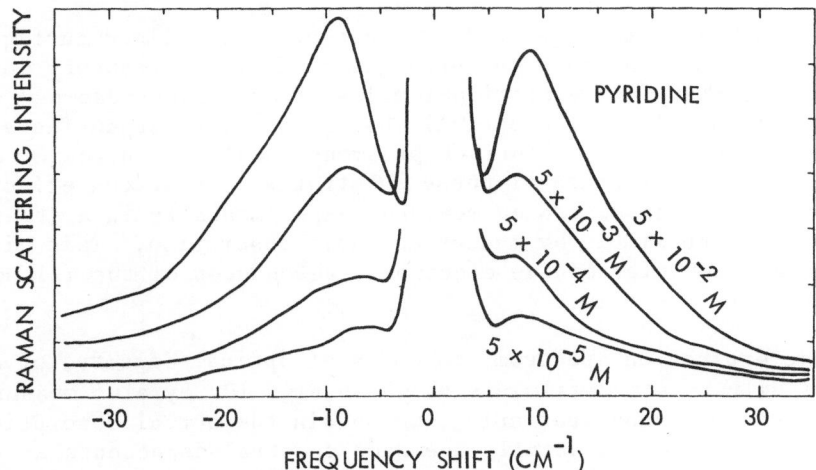

Fig. 9: Effect of the concentration of a specific organic
         adsorbate on the intensity of the low frequency RS.

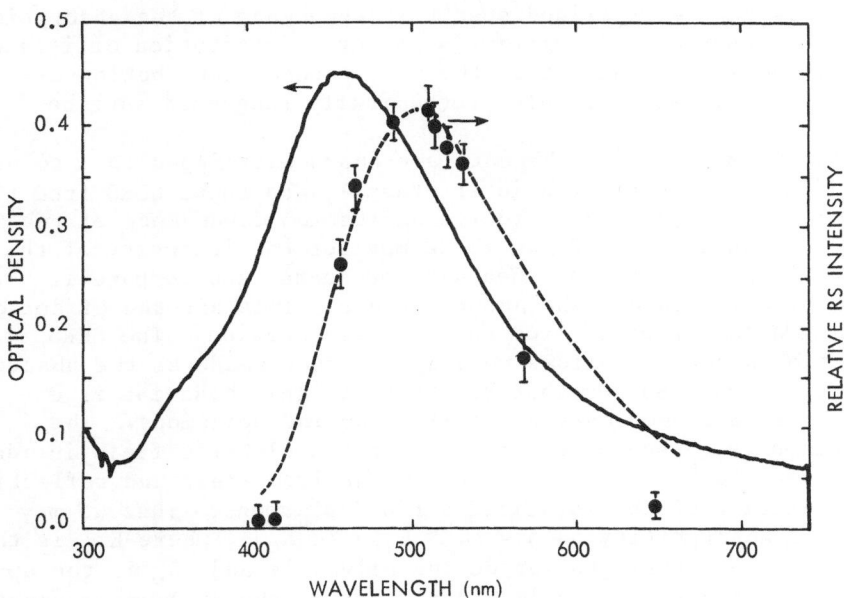

Fig. 10: SERS excitation spectrum for nitrobenzoate on a silver-
          island film compared to its extinction spectrum (solid
          line) and the predicted spectrum (dotted line).

determine the characteristic dimension of the roughness that is
important, and find this to be on the order of 100 Å.

To determine the precise contribution of localized surface
plasmons to SERS, we have measured the excitation frequency dependence of SERS from vibrational modes of molecules adsorbed on
silver-island films. The optical absorption of these surfaces is
due to the excitation of surface plasmons localized on the metal
islands.[4] The advantage of these substrates is that the effects of
the optical resonance can be measured experimentally in a straightforward fashion simply by measuring their absorption. This gives
the average behavior of the electronic resonances of the islands
that comprise the film.

The extinction spectrum, in units of optical density, of a
typical silver-island film, is shown in Fig. 10. By also measuring
the reflectivity and scattering, we obtain the actual absorption of
the film. It has essentially the same spectral dependence as the
extinction which is shown. The absorption peak is relatively
narrow since the islands are rather uniform in size and shape. The
peak position and width are determined by the distribution of
shapes of islands and the island-island and island-substrate
electromagnetic interaction.[4] For our purposes, the important
feature is the well-defined spectral dependence of the absorption,
which is caused by the relatively uniform distribution of island
shapes, and is in contrast to the very random distribution of
roughness features on an electrochemically roughened surface.

The film for which the data are shown was dipped in a solution
of 10^{-3} M p-nitrobenzoic acid in ethanol, and shows SERS from the
adsorbed molecules. The relative excitation dependence of the SERS
is also shown in Fig. 10, where we monitor the intensity of the
1650 cm^{-1} band of the adsorbed nitrobenzoate, and compare it to
the RS of a liquid cyclohexane reference. This has the effect of
removing a factor of ω^4 from the RS cross section. The SERS
excitation spectrum follows roughly the same shape as the absorption but is shifted somewhat to the red. This behavior is expected for RS caused by the localized surface plasmons. The
absorption is caused by an increase in the electric field inside
the silver due to its response to the applied field and reflecting
the excitation of the localized surface plasmons. Thus we may
write the absorptivity as $A \propto \omega_L \varepsilon'' |E_{in}|^2/c |E_o|^2$, where E_{in} is the
increased field strength inside the silver island, E_o is the ap-
plied field and ε'' is the imaginary part of the dielectric constant
of silver, $\varepsilon(\omega) = \varepsilon' + i\varepsilon''$. We can use the measured absorptivity to
obtain some average value of the increased field strength inside
the silver islands. The RS, however, depends on the field strength
at the adsorbed molecules outside of the silver islands. The continuity of the displacement field implies that $E_{out} = \varepsilon E_{in}$ for the
component of the field normal to the surface. Because $|\varepsilon'|$ is quite

large, this component will dominate. The interaction with the localized surface plasmons not only enhances the local fields, but also increases the emission dipole.[5] Thus the intensity of the RS light is related to the product of the enhanced incident and scattered intensities which can be expressed in terms of the absorptivities as

$$I_{RS} \propto |\epsilon_L|^2 \lambda_L A_L^2 |\epsilon_s|^2 \lambda_s A_s^2 \Big/ \epsilon_L'' \epsilon_s'' . \tag{1}$$

Here the subscript L refers to the quantity measured at the laser excitation frequency while s refers to the Stokes frequency. All of the quantities in this expression can be determined experimentally with ϵ measured for bulk silver.[15] The RS intensity calculated in this manner, is compared to the data in Fig. 10. The discrepancy in the red is due to our inability to accurately measure the very small absorptivities of the film in that spectral region. Nevertheless, the agreement with the data is excellent.

These data prove that, for molecules adsorbed on these surfaces, the frequency dependence of SERS can be completely determined by the contribution of the localized surface plasmons to the enhancement of the RS from adsorbates of these surfaces. If there is any other enhancement mechanism which is contributing, it must have a very flat frequency dependence, which is unlikely for any resonance effect, such as resonance RS scattering of e-h pairs or local surface states.[16,17] We conclude therefore that localized surface plasmon resonances will cause SERS from molecular modes of adsorbates as well as from the acoustic vibration of the surface itself.

Finally, we note that the origin of the red shift of the peak in I_{RS} with respect to the absorption peak comes about because of the inclusion of term $|\epsilon_L|^2 |\epsilon_s|^2$, which is a very rapidly increasing function as ω decreases. This suggests that if the localized surface plasmon resonance can be shifted to the red, the increase in the field strength outside the silver is greater, thus improving the SERS from the adsorbates. Furthermore, this presumably explains our failure to observe the low frequency mode from dilute spherical silver sols,[2] which have their absorption resonance at rather high frequencies where $|\epsilon|$ is rather low, resulting in a decreased intensity of the RS.

THEORY

In this section we discuss a simple model that can account for the behavior of the low frequency mode in more detail. We choose a geometry for which a solution can be obtained for both the electronic and the vibrational excitations. Because of the large degree of surface roughness, we choose a local model which is based

on the theory for SERS formulated by Gersten and Nitzan and presented in a separate chapter of this book.[5,18] We consider surface roughness in the shape of spheroidal or hemispheroidal protrusions. While this is certainly an oversimplified model for a randomly rough surface, its success in interpreting our data leads us
to believe that it does reflect much of the essential physics.

We begin by considering an isolated prolate spheroid whose
major and minor axes are a and b, and whose dielectric constant is
$\varepsilon(\omega)$. We take a, b $<< \lambda_{ex}$ and so neglect retardation effects.
The incident electric field vector, E_0, is taken to be along the
major axis. To include the effects of molecules on the low frequency RS, we consider a molecule of polarizability, α, a distance
H from the surface and along the major axis of the spheroid.
Gersten and Nitzen[5] have obtained the total electric dipole
moment of this system,

$$D = \frac{f^3 E_0 \xi_0 [\varepsilon(\omega) - \varepsilon_s]}{3\Delta} + \frac{\alpha E_0}{1-\Gamma} \left(1 + \frac{\xi_0 Q_1'(\xi_1)[\varepsilon_s - \varepsilon(\omega)]}{\Delta} \right)^2 . (2)$$

Here $f = (a^2 - b^2)^{1/2}$, $\xi_0 = a/f$, $\xi_1 = \xi_0 + H/f$, $Q_n(\xi_1)$ is a Legendre
function of the second kind and ε_s is the dielectric constant of
the medium surrounding the spheroid. The term represented by Γ
reflects the effect of the image of the molecular dipole in metal
acting back on the molecular dipole. It has been shown[5] to be
small for most cases of interest and is neglected. The denomina-
tor $\Delta = \varepsilon(\omega) Q_1(\xi_0) - \xi_0 \varepsilon_s Q_1'(\xi_0)$ becomes small at some value of the
aspect ratio and frequency, leading to a large increase in the di-
pole and corresponding increase in the scattered intensity. In Eq.
2, the first term represents the dipole induced on the spheroid
itself and thus will result in Mie scattering. The second term
represents the dipole induced on the molecule and will result in
Rayleigh scattering. The actual scattering cross sections can be
obtained by inserting the dipole given by Eq. 2 into Larmor's
formula and normalizing to unit light intensity.[5]

The condition for this resonance, $Re\Delta = 0$, corresponds to the
condition for the localized surface plasmon, and is identical to
the condition for a resonance in the Mie scattering from the
spheroid. The frequency for which this condition is satisfied is
shown as a function of aspect ratio, a/b, for silver, copper, and
gold in Fig. 11, where we have used the values of $\varepsilon(\omega)$ reported in
the literature for measurements on bulk samples.[15] As can be
seen, the resonance frequency is lowered if the aspect ratio is
increased or if the index of refraction of the surrounding medium
($n_s = \varepsilon_s^{1/2}$) is increased. When this resonance condition is satisfied, there is a large increase in both terms of Eq. 2, which will
result in a resonance in both the Mie scattering from the spheroid
and the Rayleigh scattering from the molecule. The amplitude of
this resonance will be limited by ε''.

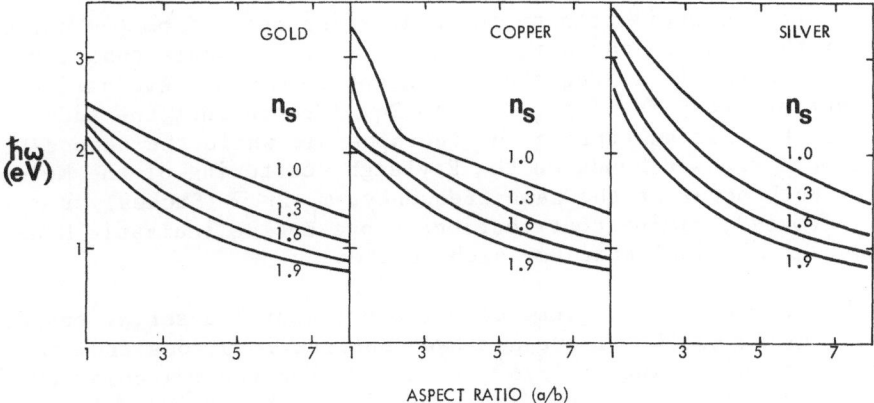

Fig. 11: Frequency of localized dipolar surface plasmon as a
function of aspect ratio for different indices of
refraction of the surrounding medium for an isolated
spheroid.

To consider an acoustic vibration of the spheroid, we allow
the dimensions of the spheroid to oscillate periodically in time.
The largest effect on the scattering will occur when the aspect
ratio changes. This corresponds to a fundamental mode of oscillation, the spheroidal vibration, in which the semi-major axis oscillates as $a = a_0 + A \cos 2\pi\Omega ct$ and the semi-minor axis lags by $180°$.
The frequency of the spheroidal vibration is given by $\Omega = v/ca$ from
simple dimensional considerations. In fact for a sphere, an analytic solution for the spheroidal vibration gives[19]

$$\Omega = kv/2ac \qquad\qquad\qquad (3)$$

where $k = 0.84$. Thus we might expect Eq. 3 to be valid for a
spheroid as well, with k, a constant that is on the order of
unity. Although there are many other modes of vibration of the
spheroid, all the others will result in a small change in the
aspect ratio and will couple to the radiation more weakly, and
therefore are neglected here. Thus, we differentiate Eq. 2 with
respect to a, and use the resulting dipole to calculate the inelastic scattering cross section[2]

$$\frac{d\sigma}{d\omega} \approx \frac{\pi A^2}{6} \left(\frac{\omega}{c}\right)^4 \left|\frac{\partial\Delta}{\partial a}\right|^2$$

$$\times \left|\frac{f^3\xi_0(\epsilon-\epsilon_s)}{3\Delta^2} + \frac{2\alpha(\epsilon-\epsilon_s)^2\xi_0^2[Q_1'(\xi_1)]^2}{\Delta^3}\right|^2 . \quad (4)$$

The dominant contribution to the scattering comes from the derivative of the denominator Δ in Eq. 2 and we retain only that term. The term $|\partial\Delta/\partial\alpha|$ is a frequency dependent term whose evaluation is straightforward.[2] The first term in Eq. 3 represents the sidebands on the Mie scattering due the spheroid while the second term represents the sidebands on the Rayleigh scattering of the molecule. Both occur at the same frequency, $\omega_L - 2\pi c\Omega$ (Stokes) or $\omega_L + 2\pi c\Omega$ (anti-Stokes), giving contributions representing inelastic Mie scattering and inelastic Rayleigh scattering.

We can account for some of our experimental observations directly, using Eq. 4. There are two contributions, one from the spheroid itself going as f^3/Δ^4, a second from the molecular adsorbate, going as α^2/Δ^6. The determination of the dominant contribution will depend on the details of the surface. However, the molecular term accounts for our observation of the increase in the RS intensity with the increasing coverage of surface adsorbed molecules (Figs. 2 and 9). If n_s is increased by replacing the electrolyte with gaseous N_2, and all other parameters are held constant, spheroids of a larger aspect ratio will be on resonance with the laser excitation frequency. Excitation of these spheroids in water would require a lower ω_{ex}. Thus we would expect Ω to decrease upon decrease of n_s, as is observed experimentally (Fig. 8).

To obtain the dependence of Ω on ω_{ex}, we must find the relationship between the acoustic and electronic excitations. However, within our model, Ω depends on a dimension, a, while the condition for the electronic resonance, $\mathrm{Re}\,\Delta = 0$, depends on a ratio of dimensions, a/b. Thus, to relate the two in this model, we need an additional assumption about the distribution of spheroid shapes and sizes on the surface. We obtain the good agreement shown in Fig. 12 if we assume that the spheroids that contribute are prolate and have a distribution of semi-minor axes, b, that is strongly peaked at some value, b_p. The distribution of semi-major axes is assumed to be independent of, and broader than, the distribution of b. In this case, we have

$$\Omega_p = \frac{v(\xi_p^2-1)^{1/2}}{2cb_p\xi_p} \tag{5}$$

where $\xi_p = a/(a^2-b_p^2)^{1/2}$. From a fit to the data, $b_p \simeq 50$ Å, consistent with the characteristic dimension we expect from our naive assumption about Ω in the previous section. This particular distribution of shapes is consistent with dendritic growth of roughness features on the surface. They might be expected to grow to some average width, but of varying length. We note, however, that this prediction of the dependence of Ω on ω_{ex} is sensitive to the

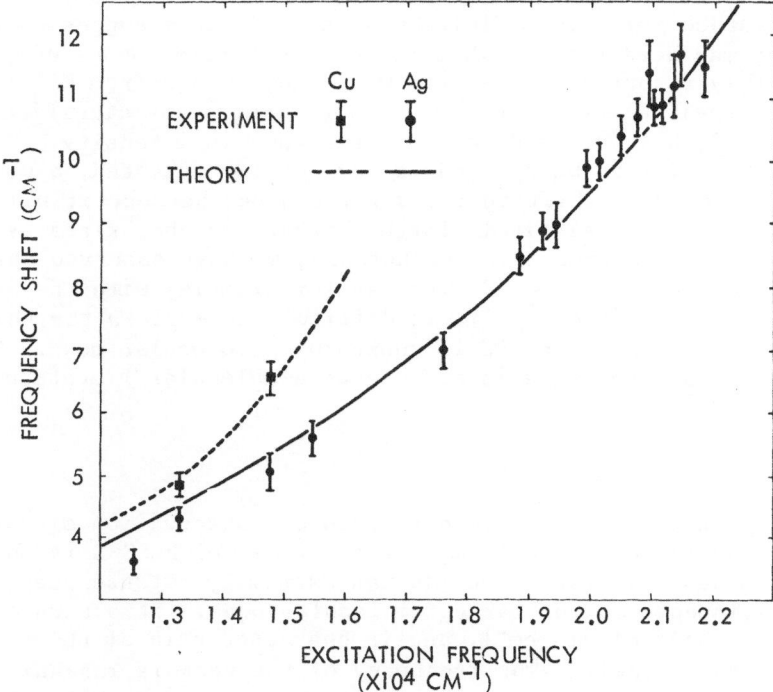

Fig. 12: Comparison of model with data for the variation in Ω with
          excitation frequency for copper and silver.

particular model chosen for the surface and therefore any conclusions drawn about the specific nature of the surface roughness must
be regarded as tentative. Nevertheless, the qualitative agreement
between the model and the experiment is excellent.

MOLECULAR LIBRATIONS

     Based on some of our preliminary data for the low frequency RS,
we originally suggested that the mode may be due to a libration of
the adsorbed molecule as a whole.[20] This was supported by calculations[21] which suggested that a molecule bound by the attraction
between its dipole and its image in the metal might have a libration
at these low frequencies. However, this hypothesis can not account
for the occurence of the same frequency for all adsorbed molecules,

or for our observation of the mode in the absence of any specifically adsorbed molecule. Fleischman et al.[22] have suggested that
the mode may be due to a libration of the halide ion-Ag complex itself. They report a change in Ω when they change from KCl to KBr
to KI in their electrolyte. We do not observe any significant
change in Ω; however we do observe a change in intensity. We note
that if the elastic background is not totally rejected, a weaker
mode can appear to shift to a lower frequency because it will be
superimposed on a relatively larger background that rises as the
frequency shift decreases. Furthermore, we have observed the mode
in the absence of any halide ion, as for example, with the Ag sol
or island films. Finally, it is difficult to explain the shift in
Ω with ω_{ex} and n_s if the RS is caused by a molecular mode. Thus
we believe that the mode is not due to a molecular libration.

SUMMARY

In this chapter we have discussed our observation of rather
unusual low frequency RS from surfaces from which SERS is observed
from adsorbed molecules. We find an extremely intense peak, $\sim$25
times as intense as the largest molecular peak. It occurs at a
frequency shift of $\sim$10 cm^{-1}, and is characteristic of the rough
metal surface itself. The frequency of the peak is insensitive to
the specific adsorbed molecule, and the peak occurs even in the
absence of any particular organic adsorbate, although at lower
intensity. However, the frequency shift varies when the excitation frequency is changed and when the index of refraction of the
surrounding medium is changed. Besides being an example of a very
unusual form of RS, these results provide important insight into
both the nature of the surface required to obtain a large enhancement of the RS cross section, and the origin of SERS from molecular adsorbates.

The mode is attributed to an acoustic vibration of the surface
itself, localized by the roughness. Its frequency shift thus suggests that the characteristic dimension of the effective roughness
is typically $\sim$100 Å. The change in Ω with ω_{ex} and with n_s is direct
evidence that a localized surface plasmon is resonantly excited in
the SERS process. This suggests that at each ω_{ex} roughness features of a particular size and shape are selectively excited and
have a characteristic acoustic frequency. This leads to the dependence of Ω on ω_{ex}. These localized surface plasmon excitations also
lead to enhanced RS from adsorbates as the two effects are always
observed together. This is confirmed by a study of the excitation
frequency dependence of SERS on silver-island films. Finally, we
present a simple model for the low frequency RS and show that there
are two contributions to the intensity, which we label inelastic
Mie scattering and inelastic Rayleigh scattering.

ACKNOWLEDGEMENTS

    We thank Joel Gersten for his many, pivotal contributions to-
ward the understanding of these results.

REFERENCES

1. D. A. Weitz, T. J. Gramila, A. Z. Genack and J. I. Gersten,
    Anomalous Low Frequency Raman Scattering from Rough Metal
    Surfaces and the Origin of Surface-Enhanced Raman Scattering,
    Phys. Rev. Lett. 45:355 (1980).
2. J. I. Gersten, D. A. Weitz, T. J. Gramila and A. Z. Genack,
    Inelastic Mie Scattering--Theory and Experiment, Phys. Rev. B
    22:4562 (1980).
3. C. Y. Chen, E. Burstein and S. Lundquist, Giant Raman Scat-
    tering by Pyridine and CN⁻ Adsorbed on Silver, Solid State
    Commun. 32:63 (1979).
4. S. Yoshida, T. Yamaguchi and A. Kinbara, Optical Properties
    of Aggregated Silver Films, J. Opt. Soc. Am. 61:62 (1971).
5. J. I. Gersten and A. Nitzan, Electromagnetic Theory of
    Enhanced Raman Scattering by Molecules Adsorbed on Rough
    Surfaces, J. Chem. Phys., 73:3023 (1980).
6. M. Kerker, D. S. Wang and H. Chew, Surface-Enhanced Raman
    Scattering by Molecules Adsorbed at Spherical Particles:
    Errata, Appl. Opt. 19:4159 (1980).
7. C. Y. Chen and E. Burstein, Giant Raman Scattering by
    Molecules at Metal Island Films, Phys. Rev. Lett. 45:1287
    (1980).
8. S. L. McCall, P. M. Platzman and P. Wolff, Surface-Enhanced
    Raman Scattering, Phys. Lett. A77:381 (1980).
9. D. L. Jeanmaire and R. P. Van Duyne, Surface Raman
    Spectroelectrochemistry. Part I. Heterocyclic, Aromatic and
    Aliphatic Amines Adsorbed on the Anodized Silver Electrode,
    J. Electroanal. Chem. 84:1 (1977).
10. J. A. Creighton, C. G. Blatchford and M. G. Albrecht, Plasma
    Resonance Enhancement of Raman Scattering by Pyridine
    Adsorbed on Silver or Gold Sol Particles of Size Comparable
    to the Excitation Wavelength, J. Chem. Soc. Faraday Trans. II
    75:790 (1979).
11. U. Wenning, B. Pettinger and H. Wetzel, Angular-Resolved
    Raman Spectroscopy of Pyridine on Copper and Gold Electrodes,
    Chem. Phys. Lett. 70:49 (1980).
12. D. Richter and L. Passell, Neutron Scattering as a Probe of
    Small Particle Dynamics in Hydroxylated Amorphous Silica,
    Phys. Rev. Lett. 44:1593 (1980)
13. I. Pockrand and A. Otto, Surface Enhanced and Disorder
    Induced Raman Scattering from Silver Films, Solid State
    Commun. 37:109 (1980).

14. R. Ruppin, Electric Field Enhancement Near a Surface Bump, preprint (1981).

15. P. B. Johnson and R. W. Christy, Optical Constants of the Noble Metals, Phys. Rev. B 6:4370 (1972).

16. E. Burstein, C. Y. Chen and S. Lundquist, Giant Raman Scattering by Molecules Adsorbed on Metals: An Overview, in "Light Scattering in Solids," J. L. Birman, H. Z. Cummins and K. K. Rebane, ed., Plenum, New York (1979).

17. A. Otto, J. Timper, J. Billmann and I. Pockrand, Enhanced Inelastic Light Scattering from Metal Electrodes Caused by Adatoms, Phys. Rev. Lett. 45:46 (1980).

18. J. I. Gersten and A. Nitzan, Electromagnetic Theory--A Spheroidal Model, this volume.

19. A. E. H. Love, "Mathematical Theory of Elasticity," 3rd ed., Cambridge University Press, Cambridge (1920).

20. A. Z. Genack, D. A. Weitz and T. J. Gramila, Very Low Frequency Surface-Enhanced Raman Scattering, Surf. Sci. 101: 381 (1980).

21. H. Morawitz and T. R. Koehler, A Model for Raman-active Librational Modes on a Metal Surface: Pyridine and CN^- on Silver, Chem. Phys. Lett. 71:64 (1980).

22. M. Fleischmann, P. J. Hendra, I. R. Hill and M. E. Pemble, Enhanced Raman Spectra from Species Formed by the Coadsorption of Halide Ions and Water Molecules on Silver Electrodes, J. Electronanal. Chem. 117:243 (1981).

RAMAN SCATTERING AND LUMINESCENCE BY MOLECULES ADSORBED

AT METAL ISLAND FILMS

G. Ritchie and C. Y. Chen*

Physics Department and Laboratory for Research on
the Structure of Matter
University of Pennsylvania
Philadelphia, PA 19104

INTRODUCTION

It is now well established that surface roughness on a sub-microscopic scale plays an important role in the enhancement of the Raman scattering (RS) by molecules adsorbed on metals.[1-4] Moreover, recent work indicates that the optical absorption and luminescence by molecules adsorbed at a rough metal surface can also be enhanced.[5-8]

An important contribution towards the elucidation of the role played by surface roughness was made by Creighton et al,[9] who suggested that the enhanced RS by molecules adsorbed on a Ag electrode is a direct result of the excitation of the collective electron oscillations of structures (e.g. nodules, cavities) on the surface. (A similar interpretation was proposed independently by Moskovits.[10]) In a study of the RS by pyridine adsorbed on submicroscopic Ag and Au aqueous sol particles, Creighton et al[9] observed an enhanced RS by the adsorbed molecules, which was most intense at wavelengths corresponding to the Mie extinction maximum for the metal particles. They suggested that the enhanced RS by the molecules on the sol particles had an origin similar to that for molecules adsorbed on an electrochemically processed Ag electrode, namely the excitation of the collective electron resonances of the sol particles, or of structures on the electrode surface.

We summarize here the results of experiments (mostly at Penn)

---

*Current address: Universal Energy Systems, Dayton OH 45432

designed to study the RS and luminescence by molecules adsorbed at thin metal (Ag, Au) island films. In contrast to the three-dimensional spatial distribution of sol particles in solution, the metal island films consist of a two-dimensional distribution of submicroscopic metal particles on a surface.[11,12] The island films are simple to prepare, and the characterization of their optical properties is relatively straightforward. As a consequence, these films are useful as model systems for the study of the RS and luminescence by molecules adsorbed at rough metal surfaces in general.

The optical properties of noble metal island films have been studied extensively. Evaporated island films of Ag, Au, and Cu exhibit a broad absorption band in the visible and near infrared, which corresponds to the excitation of the transverse collective electron resonance of the metal islands (i.e. electronic oscillations in a direction parallel to the plane of the film.) The spectral position, width and strength of the resonance depend on the film mass thickness, substrate temperature during evaporation, deposition rate, etc. Ag island films also exhibit a narrow absorption band in the near ultraviolet, due to the excitation of the perpendicular collective electron resonance. For Au or Cu island films, the perpendicular resonance is strongly damped by interband electron-hole pair excitations, and is not apparent in the optical absorption spectra for these films.

It is understood[13] that the enhanced RS by molecules adsorbed at a metal island film is largely due to the combined effects of (a) the sizable increase in the effective electric field at the molecules, which occurs when the transverse collective electron resonance is excited by the incident electromagnetic field, and (b) the large radiating electric moments that are induced in the metal islands by the coulomb fields of the Raman-excited molecules. The metal islands thus serve as "resonant amplifiers" for both the incident and the scattered radiation.

The luminescence by molecules located near a metal surface has been rather extensively studied, for systems where the luminescent molecules are separated from the metal surface by an intermediate, dielectric (e.g. fatty acid, solid argon) layer.[14,15] These investigations showed that for a molecule-metal separation $d \lesssim 100\text{Å}$, the intensity and lifetime of the luminescence decrease monotonically with decreasing d. These effects are attributed to the de-excitation ("quenching") of the molecules via nonradiative energy transfer from the molecules to the metal.

This luminescence-quenching effect is particularly pronounced when the molecules are adsorbed directly at the metal surface, which limits the feasibility of using luminescence as a probe of the properties of the adsorbed molecules. The luminescence by a monolayer of dye molecules (e.g. fluorescein, rhodamine 6G) adsorbed at a smooth evaporated Ag film surface is more than three orders of magnitude weaker than that for the corresponding molecules adsorbed on a glass substrate.[16]

When the molecules are adsorbed at a Ag island film, however, the luminescence can be an order of magnitude more intense than that for the molecules on glass, and contains "hot" as well as "relaxed" luminescence components.[16] Moreover, the luminescence is accompanied by an enhanced Raman scattering by the adsorbed molecules which is typically comparable in intensity to the luminescence. As in the case of the RS, the enhancement of the luminescence is attributed to the "amplification" of the incident and emitted electromagnetic fields by the metal islands. The field enhancements effectively offset the luminescence-quenching effect of the metal substrate, making the luminescence easily detectable under many conditions.

EXPERIMENTS AND RESULTS

Raman scattering by nonresonantly excited
molecules adsorbed at Ag and Au island films

Thin (approximately 50 Å mass thickness) Ag and Au island films were prepared by evaporating the metal onto a glass or sapphire substrate in a vacuum of 2 X 10^{-7} torr, at a rate of 0.2 to 0.4 Å per second. The mass thickness of the film was monitored by a quartz crystal oscillator mounted close to the sample substrate. The films were then removed from the vacuum, immersed momentarily in an aqueous solution of the molecules to be studied, rinsed with distilled water and air-dried. The Raman spectra were measured in a backward scattering geometry, using about 50 mW argon or krypton laser excitation.

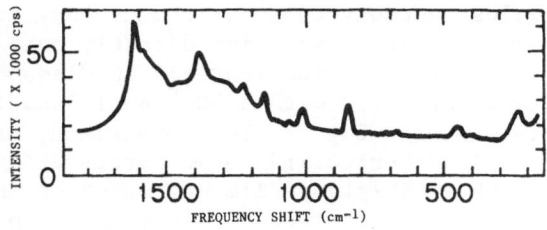

Fig. 1.   Raman spectrum for isonicotinie acid adsorbed onto a
          50 Å Ag island film substrate (50 mW, 5145 Å excitation).

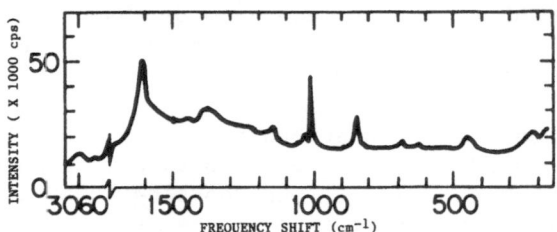

Fig. 2.   Raman spectrum for benzoic acid adsorbed onto a 50 Å Ag
          island film substrate (50 mW, 5145 Å excitation).

We found, on the basis of the observed molecular RS, that although isonicotinic acid and benzoic acid molecules are readily
adsorbed from solution onto the Ag island films, pyridine is not adsorbed to any appreciable extent. We believe that this is due to an
oxide layer that is formed when the highly chemically reactive Ag
islands are exposed to air. The oxide layer allows chemisorption
of the isonicotinic acid and benzoic acid molecules via the carboxylate group (COO$^-$), but prevents the bonding of pyridine to Ag via
the nitrogen lone pair. On the other hand, all three molecular
species are adsorbed to about the same extent from aqueous solution
onto the Au island films, indicating that the surface of the Au island is not appreciably oxidized when exposed to air.

Figs. 1 and 2 show, respectively, the Raman spectrum for isonicotinic acid and for benzoic acid adsorbed onto a Ag island film. The
strong peaks at approximately 1380 cm^{-1} in both spectra are due to
the symmetric stretching mode of the carboxylate group that is bonded
to the oxide layer on the Ag islands. The absence of a peak in the
spectra in the vicinity of 1700 cm^{-1}, which corresponds to the
stretching vibration mode of C=O, indicates that the molecules are
chemisorbed, and that the coverage is at most a monomolecular layer.

Certain molecules (notably those with a carboxylic acid group)
can be adsorbed from an aqueous solution directly onto a glass substrate. The Raman spectrum for isonicotinic acid adsorbed on glass,
which was subsequently overlaid with a 50 Å Ag island film and then
measured in air, is shown in Fig.3. For comparison, we show in Fig.
4 the Raman spectrum for isonicotinic acid adsorbed on glass, which
was then overlaid with a Ag island film and measured in high vacuum.
The shoulder at 202 cm^{-1} is attributed to the Ag-N vibration. The
additional peaks at 230, 811, 930, and 1274 cm^{-1} in Fig. 3 are due to

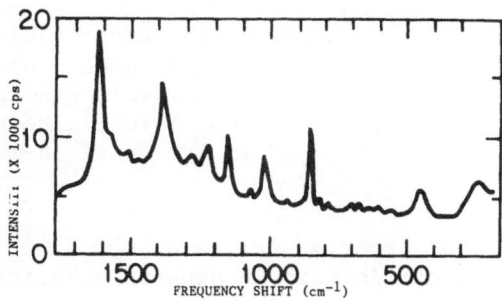

Fig. 3.   Raman spectrum for isonicotinic acid adsorbed at a 50 Å
          Ag island film overlayer (60 mW, 4880 Å excitation).

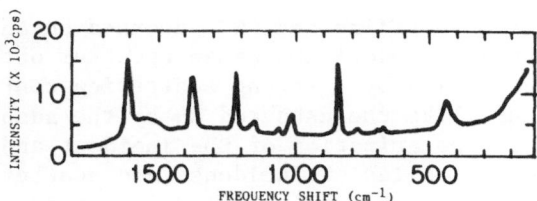

Fig. 4.   Raman spectrum for isonicotinic acid adsorbed at a Ag
          island film overlayer in high vacuum (60 mW, 4880 Å).

contaminants adsorbed from the atmosphere. Some of these contaminants are apparently removed from the Ag surface on immersing the
Ag island film into the aqueous solution of isonicotinic acid, since
some of these peaks do not appear in Fig. 1. The immersion of the
Ag island film into the aqueous solution also results in an increase
in the background in the spectrum between 1200 and 1600 cm^{-1}, which
has been attributed to the adsorption of carbon layers.[17,18]

We were not able to observe the RS by benzoic acid when it was
initially adsorbed onto the glass substrate, and then overlaid with
a Ag island film. Based on this result, it was proposed[19] that
chemisorption plays an important role in the Raman enhancement,

since benzoic acid exhibits an enhanced RS when it is chemisorbed
on the Ag islands via COO^-, but the RS is undetectable when the
benzene ring is in close proximity, but presumably not chemically
bonded, to the Ag. Isonicotinic acid, however, can apparently be
bonded to the Ag islands via COO^- or the N lone electron pair, and
therefore also exhibits enhanced RS in the overlaid-molecule
configuration.

On the basis of our experimental data, the Raman cross section
per molecule for isonicotinic acid in aqueous solution is approximately one-fifth of that for pyridine in aqueous solution. By comparing
the RS intensity for isonicotinic acid adsorbed on a 50 Å Ag island
film with the intensity measured for pyridine adsorbed on an optimally
anodized Ag electrode, we are able to estimate the enhancement factor
for the Ag island films to be 10^5, since the enhancement factor for
pyridine adsorbed at a Ag electrode has been experimentally determined
to be 10^5 to 10^6.

As in the case of pyridine and CN^- adsorbed on an electrochemically processed Ag electrode,[1] the Raman spectrum obtained for a Ag
island film is accompanied by a strong scattering continuum. The
scattering continuum, like the enhanced RS by the adsorbed molecules,
is insensitive to the polarization of the incident and scattered
radiations, and to the angles of incidence and scattering. A strong
continuum is also observed for a Ag island film prepared and studied
under ultra-high vacuum, without the adsorbed molecules.

The scattering continuum has been attributed to inelastic light
scattering by electron-hole pair excitations in the surface region of
the metal. We believe[1] that the nature of the enhancement of the
scattering continuum is similar to that of the enhancement of the
RS by the adsorbed molecules. The shoulder at ~ 1600 cm^{-1} in Fig. 5
is observed for different excitation wavelengths, e.g. 5145 Å, 4880 Å,
and 4765 Å, and hence results from a Raman effect. Recently, this
broad structure has been attributed to an amorphous carbon contaminant
on the Ag surface.[18] Our preliminary results indicate that its
wavelength dependence is quite different from that of the enhanced RS
by adsorbed molecules. This suggests that the contaminant has an
optical absorption band in the visible region, i.e. the broad structure may correspond to resonant Raman scattering.

The RS spectra (6471 Å excitation) for isonicotinic acid and
benzoic acid adsorbed on Au island films are about two orders of
magnitude weaker than the corresponding spectra for the molecules
adsorbed on the Ag island films. The accompanying scattering
continuum from the Au island film is also weaker than that from the
Ag island film. The two strong peaks at 1022 and 1612 cm^{-1} in the
Raman spectrum for isonicotinic acid adsorbed on the Au island film
(Fig. 6) are due to the ring stretching modes which involve carbon-

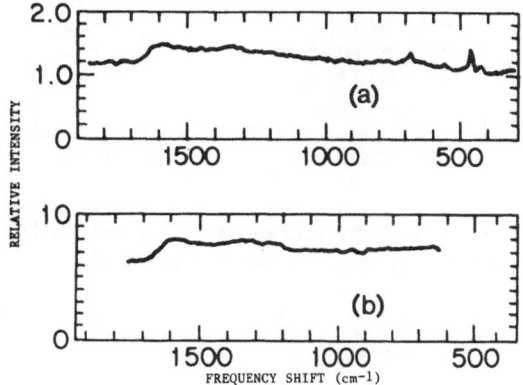

Fig. 5. (a) Spectrum of inelastic scattering by a 42 Å Ag
             island film on sapphire in ultra-high vacuum. The
             low frequency peaks in the spectrum are due to the
             sapphire substrate.
         (b) Spectrum of inelastic scattering by an electro-
             chemically anodized Ag electrode in 0.1 M KCl.

carbon bonds. In contrast, the strong peak at 845 cm^{-1}, (which
corresponds to the C-H vibration) in the Raman spectrum of isonicotinic acid adsorbed on the Ag islands becomes a very weak peak in
the Raman spectrum for isonicotinic acid adsorbed on the Au island
film. The differences in the relative strengths of the Raman peaks
may be due to the differences in chemisorption and in the orientation of the molecules at the different metals.

In order to gain further insight into the nature of the enhancement mechanism, we studied the wavelength dependence of the RS
by isonicotinic acid adsorbed on Ag and Au island films. The 241
cm^{-1} Raman band (corresponding to the TO phonon) of a BaF_2 window
placed near the sample surface was used as the standard in these
measurements. Normalizing the data with respect to the intensity
of this standard eliminates the ω^4 factor in the RS intensity, and
compensates for the wavelength dependence of the spectrometer system sensitivity.

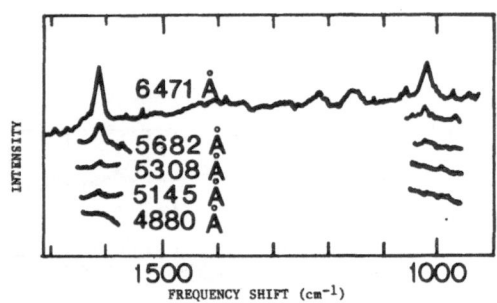

Fig. 6. Wavelength dependence of RS by isonicotinic acid
         adsorbed on a 50 Å Au island film.

Fig. 7 shows the Raman excitation profiles for the 443, 847, and
1010 cm^{-1} peaks for isonicotinic acid adsorbed on the Ag island film.
The normalized Raman peak intensity increases in magnitude by about
a factor of 10 on increasing the excitation wavelength from 4765 Å
to 6471 Å. The Raman peak intensity also depends on the scattered
wavelength, as can be seen from the variation in the ratio of the
intensity of the 443 cm^{-1} peak to that of the 1016 cm^{-1} band.

Fig. 8 shows the transmission spectrum for a Ag island film on
glass, with adsorbed isonicotinic acid. The corresponding spectrum
for a Au island film is given in Fig. 9. The broad absorption bands
in both spectra are due the excitation of the transverse collective
electron resonance of the films. The absorption maximum occurs near
7000 Å for the Ag island film, and near 7500 Å for the Au island film.

The RS intensity for isonicotinic acid adsorbed at the Ag or Au
island film increases as the excitation wavelength approaches the
absorption maximum of the film. We note that for 5145 Å excitation
the low frequency vibrational modes of isonicotinic acid adsorbed at
the Au islands are not observable, whereas the high frequency modes
(e.g. 1612 cm^{-1} in Fig. 6) can be observed. The 5145 Å wavelength
corresponds to excitation beyond the onset (~5200 Å) of interband
transitions in Au. This indicates that an appreciable enhancement can
be obtained for excitation outside the absorption resonance of the
island film, if the scattered wavelength lies within the resonance.
Moreover, the fact that the RS by the low frequency molecular modes
is not observable for an excitation wavelength below 5200 Å suggests

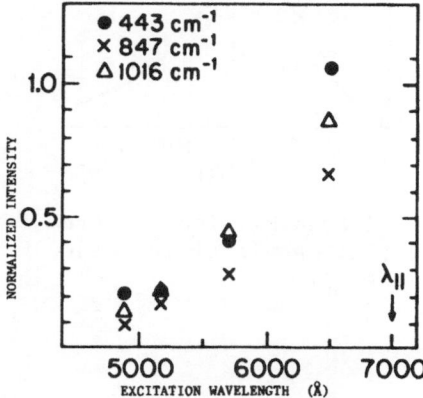

Fig. 7. Wavelength dependence of Raman peaks for
isonicotinic acid adsorbed on Ag island film
with transverse collective electron resonance
at 7000 Å. The Raman peak intensity was
normalized with respect to the TO phonon
peak of BaF_2 at 241 cm^{-1}.

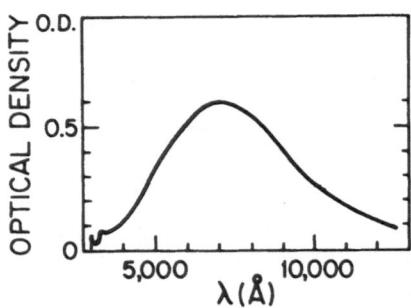

Fig. 8.   Normalized transmission spectrum for a 50 Å Ag
          island film.   Angle of incidence θ = 55°.

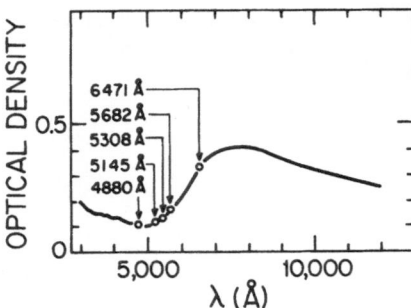

Fig. 9.   Normalized transmission spectrum for a 50 Å Au
          island film.

that interactions between the adsorbed molecules and interband electron hole pair excitations in the metal substrate do not play an
important role in the enhancement of the RS. The intensity of the
inelastic scattering continuum is increased, however, when the excitation wavelength is below 5200 Å. This behavior is attributed to
photoluminescence by the Au, which involves the radiative recombination of electrons below the Fermi level in the conduction band with
holes generated in the d band by optical interband transitions.[20]

Luminescence and Raman scattering by
dye molecules adsorbed at Ag island films

The data presented in this section were obtained using organic
dye molecules (fluorescein isothiocyanate [FITC] and rhodamine 6G
[R6G]) which absorb and luminesce in the visible. The secondary light
emission (luminescence and RS) spectra obtained for FITC molecules
adsorbed from an ethanol solution onto a 50 Å Ag island film, for
three different excitation wavelengths, are shown in Fig. 10. The
inset in the figure gives the absorption (maximum near 4930 Å) and
luminescence (largely relaxed fluorescence, peaked near 5200 Å)
spectra for FITC in ethanol. The spectra (solid curves) corresponding
to resonant excitation (4880 Å, 5145 Å) of the molecules exhibit RS
peaks superimposed on a broad luminescence band. The dashed curve
below each spectrum represents an estimate of the luminescence component of the total spectrum. The apparent luminescence maximum for
the adsorbed FITC molecules occurs near 5310 Å. Excitation at a wavelength (6471 Å) above the FITC absorption band yields RS which is
nearly two orders of magnitude weaker, and there is no apparent
molecular luminescence.

Fig. 11 shows the normal-incidence optical transmission spectra
for several Ag island films, which differ only in mass thickness. The
transverse collective electron resonance shifts towards longer wavelengths and is broadened as the film thickness is increased. On a
qualitative basis, one expects the maximum luminescence and RS by the
adsorbed molecules to occur when the resonance band of the film
overlaps the absorption and emission wavelengths in an "optimum" manner, i.e. there should be an optimum mass thickness.

In order to study the dependence of the luminescence and RS spectra on the Ag film thickness, it is advantageous to first adsorb the
molecules onto the glass substrate and subsequently overlay them with
Ag. In this manner, the number of adsorbed molecules remains reasonably constant while the Ag film thickness is varied. Monolayers of R6G
adsorbed on glass were prepared by immersing glass slides in a 10^{-4}
molar solution of the dye in ethanol for 30 minutes. The slides were
then rinsed under a stream of pure ethanol for several minutes, and
blown dry with compressed Freon. The optical absorption spectrum for
the R6G layer on glass, obtained by measuring the transmission through
ten such slides in tandem, exhibits a broad band at 5300 Å. The cor-

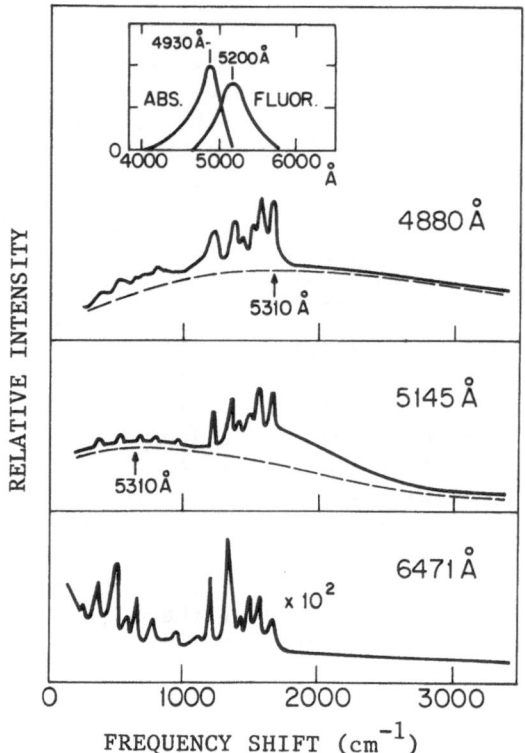

Fig. 10.   Luminescence and Raman spectra for FITC adsorbed at a
           50 Å Ag island film, for three different excitation
           wavelengths. Dashed curves represent an estimate of
           the luminescence component of the spectrum. Inset:
           absorption and luminescence spectrum for FITC in
           solution.

responding luminescence spectrum peaks near 5700 Å. Using data on the
integrated absorption by the R6G monolayer, we estimate the coverage
to be 0.4. The R6G monolayers were then overlaid with Ag island films
ranging from 5 Å to 160 Å in thickness. We note that the transmission
spectra for the R6G monolayers overlaid with the Ag island films do
not perceptively differ from the corresponding spectra for the Ag island films without the R6G monolayers (Fig. 11). On the other hand,
changes in the absorption spectra are clearly evident when the dye
molecules are adsorbed from solution onto the Ag islands. This suggests that considerably more than a monolayer of molecules is adsorbed
onto the Ag island film when it comes into contact with the dye
solution.

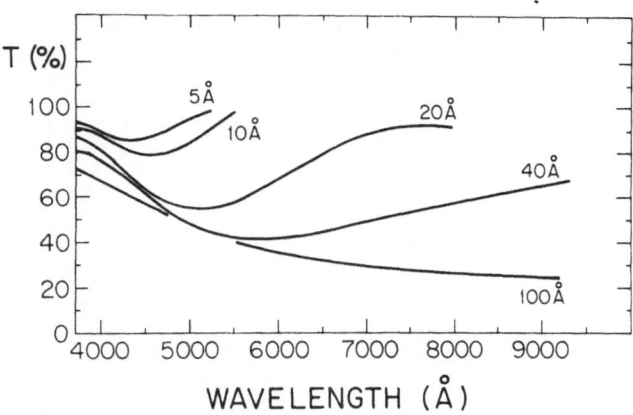

Fig. 11.   Normal-incidence transmission spectra for Ag island films
of different mass thickness.

      Fig. 12 shows the luminescence and RS spectra obtained (4765 Å
continuous laser excitation, 5 mW) for a R6G monolayer on glass, before and after deposition of a 40 Å Ag overlayer. The corresponding
spectrum (inelastic light scattering continuum) for a 40 Å Ag island
film on glass, without the R6G molecules, is also shown for comparison. The luminescence spectrum has a maximum at 5550 Å, and is 6
times more intense than that for the R6G monolayer on glass alone.
The spectrum also exhibits peaks at 1170, 1340, 1570, and 1650 cm^{-1}
due to RS by the vibration modes of the adsorbed R6G molecules.[21] We
interpret the appreciable light intensity at wavelengths close to the
laser line to be "hot luminescence"[22] by the overlaid molecules. The
fact that the hot luminescence and the relaxed luminescence become
comparable in intensity for molecules adsorbed at the Ag islands implies that the lifetime of the thermally relaxed state of the excited
R6G molecules is as much as several orders of magnitude smaller than
that for the molecules adsorbed on glass.[16] Weitz et al[23] have recently reported the direct observation of a large reduction in the
lifetime of the excited state of Eu^{3+} ions in a thin chelate layer
applied onto a silver island film substrate. The lifetime reduction
is due in part to energy transfer from the excited molecules to the
metal, and in part to the increased rate of radiative emission by the
molecules when they are adsorbed at the metal islands.[16]

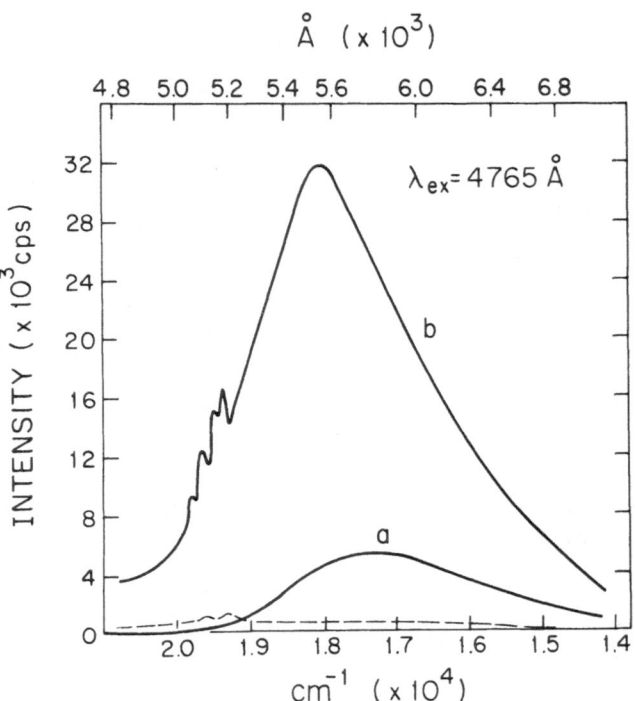

Fig. 12. Curve (a) - the luminescence spectrum for a R6G monolayer
          on glass; curve (b) - luminescence and RS spectrum for
          R6G monolayer overlaid with a 40 Å Ag island film; dashed
          curve - inelastically scattered light spectrum for a 40 Å
          Ag island film without the R6G monolayer.

     The intensity of the luminescence peak at 5550 Å, and the inten-
sity of the 1650 cm^{-1} RS peak above the luminescence background are
graphed in Fig. 13 as functions of Ag film overlayer mass thickness.
Both the luminescence and the RS are most intense when the overlayer
thickness is approximately 40 Å. Glass et al.[7] have reported similar
dependences on film mass thickness of the luminescence by dye molecules (rhodamine B, nile blue) adsorbed from solution onto Ag, Au and
Cu island films. From Fig. 11 we see that the 40 Å Ag island film
exhibits a pronounced collective electron resonance band which overlaps the excitation and emission wavelengths. The incident and emit-

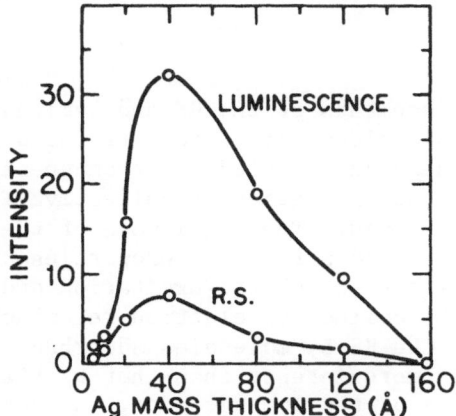

Fig. 13.   Intensities of luminescence peak (5550 Å) and 1650 cm^{-1}
           RS peak for R6G monolayer overlaid with Ag, as functions
           of Ag overlayer mass thickness (4765 Å excitation).

ted electromagnetic fields are therefore strongly coupled to the
resonance, and the optimum field enhancements are obtained with this
Ag film thickness.

     For a thickness greater than approximately 150 Å, the Ag film
becomes continuous, and the luminescence and RS intensities become
undetectable, i.e. they are more than 3 - 4 orders of magnitude weaker
than those obtained for the 40 Å Ag overlayer. The luminescence by
the overlaid R6G monolayer is effectively enhanced, relative to that
for R6G adsorbed on glass alone, only for Ag film thicknesses within
the range 20 Å < d < 120 Å. This indicates that the field enhancements more than offset the tendency of the metal to quench the luminescence for Ag island films in this thickness range. The deposition
of an overlayer with a thickness outside this range causes an apparent
decrease in the overall R6G luminescence intensity. Finally, we note
that for the 5 Å Ag island film overlayer, the observed luminescence

and RS intensities are reduced by only an order of magnitude relative
to those obtained for the optimum, 40 Å overlayer. This indicates
that surface roughness on an extremely small scale can give rise to
relatively large local field enhancements, since the Ag particles
which compose the 5 Å film are typically tens of Å in diameter.[24]

SUMMARY

The observed enhancements of the RS and luminescence by molecules adsorbed at metal island films are attributed to the "amplification" of the incident and emitted electromagnetic fields, which
results from the resonant excitation of collective electron oscillations in the metal islands. The magnitudes of the enhancements
depend on the dielectric and structural properties of the islands,
as well as on the extent to which the excitation and emission
frequencies overlap the collective electron resonance band. Under
"optimum" conditions, the RS by molecules adsorbed at a Ag island
film can be $\sim 10^5$ times more intense than that by the "isolated"
molecules. In the case of the luminescence, the corresponding
intensity ratio (e.g. for molecules adsorbed at Ag islands versus
for molecules adsorbed on glass) is much smaller. The nonradiative
de-excitation of the luminescent states of the molecules by the
metal offsets the effects of the field amplifications, thereby
limiting the apparent enhancement of the luminescence.

In discussing the Raman and luminescence data, we have focussed
on the roles played by the collective electron resonances associated
with metal surface roughness and, in the case of the luminescence,
the nonradiative de-excitation of the molecules by the metal. Other
factors can also be important. These include modifications in the
electronic structure and transitions of the molecules due to adsorption at the metal surface, and the coupling of the adsorbed molecules
to electron-hole pair excitations in the metal substrate. These
latter effects should become more important in the luminescence and
RS by molecules adsorbed at a smooth metal surface, where the
additional effects of localized collective electron resonances are
not present.

ACKNOWLEDGEMENTS

We would like to express our gratitude for the guidance provided by E. Burstein during the execution of the work described
here. We would also like to acknowledge helpful discussions with
H. Talaat, D. Whittle and R. Hochstrasser. This work was supported
in part by the NSF MRL program under grant no. 7923647, by ARO,
and by ONR.

REFERENCES

1. C. Y. Chen, E. Burstein, and S. Lundquist, Giant Raman scattering by pyridine and CN⁻ adsorbed on silver, Solid State Commun. 32:63 (1979).

2. J. E. Rowe, C. V. Shank, D. A. Zwemer, and C. A. Murray, Ultra-high-vacuum studies of enhanced Raman scattering from pyridine on Ag surfaces, Phys. Rev. Lett. 44:1770 (1980).

3. J. I. Gersten and A. Nitzan, Electromagnetic theory of enhanced Raman scattering by molecules adsorbed on rough surfaces, J. Chem. Phys. 73:3023 (1980).

4. E. Burstein and C. Y. Chen, Raman scattering by molecules adsorbed at metal surfaces. The role of surface roughness, in: "Proceedings of the VIIth International Conference on Raman Spectroscopy," Ottawa, Canada, North Holland (1980).

5. G. Ritchie, E. Burstein, and R. B. Stephens, Secondary light emission by molecules at metal surfaces with axially symmetric bumps, Bull. Amer. Phys. Soc. 26:359 (1981).

6. A. Hartstein, J. R. Kirtley, and J. C. Tsang, Enhancement of the infrared absorption from molecular monolayers with thin metal overlayers, Phys. Rev. Lett. 45:201 (1980).

7. A. M. Glass, P. F. Liao, J. G. Bergman, and D. A. Olson, Interaction of metal particles with adsorbed dye molecules: absorption and luminescence, Opt. Lett. 5:368 (1980).

8. C. Y. Chen, I. Davoli, G. Ritchie and E. Burstein, Giant Raman scattering and luminescence by molecules adsorbed on Ag and Au metal island films, in: "Proceeding of the International Conference on Non-Traditional Approaches to the Study of the Solid-Electrolyte Interface," T. E. Furtak, K. L. Kliewer, and D. W. Lynch, eds., 24-27 Sept. 1979, North-Holland (1980); Surf. Sci. 101:363 (1980).

9. J. A. Creighton, C. G. Blatchford and M. G. Albrecht, Plasma resonance enhancement of Raman scattering by pyridine adsorbed on silver or gold particles of size comparable to the extinction wavelength, J. Chem. Soc. Faraday II 75:790 (1979).

10. M. Moskovits, Surface roughness and the enhanced intensity of Raman scattering by molecules adsorbed on metals, J. Chem. Phys. 69:4159 (1978).

11. R. P. Rouard and A. Messen, Optical properties of thin films, Prog. in Opt. XV:79 (1977), and the references therein.

12. T. S. Yamaguchi, S. Yoshida and A. Kinbara, Optical effect of the substrate on the anomalous absorption of aggregated silver films, Thin Sol. Films 21: 173 (1974).

13. C. Y. Chen and E. Burstein, Giant Raman scattering by molecules at metal island films, Phys. Rev. Lett. 45:1287 (1980).

14. R. R. Chance, A. Prock, and R. Sibley, Molecular fluorescence and energy transfer near interfaces, in: "Advances in Chemical Physics 37," I. Prigogine and S. A. Rice, eds., Wiley (1978).

15. A. Campion, A. R. Gallo, C. B. Harris, H. J. Robota and P. M. Whitmore, Electronic energy transfer to metal surfaces: a test

of classical image dipole theory at short distances, Chem. Phys. Lett. 73:447 (1980).

16. G. Ritchie and E. Burstein, Luminescence by molecules adsorbed at a Ag surface, Phys. Rev. B (in press).

17. R. P. Cooney, M. R. Mahoney and M. W. Howard, Intense Raman spectra of surface carbon and hydrocarbons on silver electrodes, Chem. Phys. Lett. 76:448 (1980).

18. J. C. Tsang, J. E. Demuth, P. N. Sanda and J. R. Kirtley, Enhanced Raman scattering from carbon layers on silver, Chem. Phys. Lett. 76:54 (1980).

19. E. Burstein, C. Y. Chen, and S. Lundquist, Giant Raman scattering by molecules adsorbed on metals: an overview, in: "Light Scattering in Solids," J. L. Birman, Z. Cummins, and K. K. Rebane, eds., p. 479, Plenum (1979).

20. A. Mooradian, Photoluminescence of metals, Phys. Rev. Lett. 22:185 (1969).

21. R. Konig, A. Lau, and H. J. Weigmann, Vibrational spectra of the excited electronic state of rhodamine molecules obtained by CARS spectroscopy, Chem. Phys. Lett. 69:87 (1980).

22. K. K. Rebane and P. Saari, J. Luminesc. 16:223 (1978).

23. D. A. Weitz, S. Garoff, and C. D. Hanson, The effect of rough silver on fluorescent lifetimes, Bull. Amer. Phys. Soc. 26:339 (1981).

24. In very small particles, quantum size effects must be considered. See E. Burstein, S. Lundqvist, and D. L. Mills, plus J. I. Gersten and A. Nitzan in this volume.

SILVER STRUCTURES PRODUCED BY MICROLITHOGRAPHY

P. F. Liao

Quantum Electronics Research Department
Bell Telephone Laboratories
Holmdel, N.J.   07733

INTRODUCTION

The observation of surface enhanced Raman scattering (SERS) of molecules adsorbed on metal surfaces generally requires that the metal surface be roughened. Certainly all cases in which giant enhancement of 10^4-10^6 have been observed have been on roughened silver surfaces. In this chapter we shall discuss the use of lithographic techniques to produce controlled surface "roughness". These surfaces consist of arrays of isolated submicron silver particles which are uniform in shape and size. Modern techniques of microlithography are used to produce these surfaces. With the help of these surfaces we are able to elucidate the mechanism responsible for SERS and can design the surface for maximum enhancement. We have, to date, found[1] enhancements of 10^7 on such surfaces.

Roughened surfaces have been produced by a number of techniques. The most common techniques are electrochemical roughening[2,3] and the deposition of island films.[4] Roughening via a photochemical process and evaporation of silver on initially rough CaF_2 surfaces has also provided silver surfaces showing very large enhancement of Raman signals. Electronmicroscope photographs[5] of both electrochemically and photochemically roughened surfaces show the surfaces to consist of randomly sized and shaped microscopic silver spheroidal shaped bumps or protrusions with typical dimension of $\sim$ 500-1000Å. Micrographs of silver island films show that the films which show the largest Raman signals are composed of isolated islands of approximately 200-400Å dimensions.[4]

The requirement of surface roughness has given rise to several theories for the Raman enhancement. In 1979, J. G. Bergman et al., demonstrated[6] that the role of roughness was <u>not</u> to increase the silver sample's surface area. By using radioactive molecules containing carbon 14 they were able to monitor the coverage of an electrochemically roughened silver electrode by simply monitoring the radioactivity of the surface. They found that as the surface was increasingly roughened by the electrochemical anodization process, the Raman intensity of the $2144cm^{-1}$ stretch mode of CN adsorbed on the surface increased by almost five orders of magnitude. On the other hand, the molecular coverage, as determined by the radioactive count, measured with a Geiger counter, only increased by a factor of two. All measurements were taken after the electrode was first removed from the electrochemical cell and allowed to dry. Their results demonstrated unambiguously that some mechanism was responsible for a real enhancement of the Raman effect.

The requirement of surface roughness, combined with the observation[7] of enhancement for more than the first layer of the molecules has led to electromagnetic models[8-12] for SERS in which plasmon resonances of the microscopic bumps or the islands of an island film act to increase the local field at the molecules and to amplify the re-radiated field of the Raman active molecule.

The electromagnetic theory was first proposed by Moskovits[8] and has been elaborated on by several groups.[9-12] The enhancing surface is generally modeled as covered with a random array of metal spheroids. The main features of these theories are easily seen by considering a single dielectric ellipsoid with an external laser field, E_L directed along the principal axis of the ellipsoid, and a nearby molecule also located on the principal axis. If the ellipsoid major (a) and minor (b) axis dimensions are such that $(a, b << \lambda)$ the problem can be solved in an electrostatic approximation.[9-12] To further simplify the problem we replace the ellipsoid by a point dipole of magnitude[13] $\vec{\mu}_E = \alpha_E \vec{E}$ with

$$\alpha_E = -\frac{ab^2 \varepsilon_o}{3} \frac{1-\varepsilon/\varepsilon_o}{1-(1-\varepsilon/\varepsilon_o)A} \quad . \tag{1}$$

Here ε and ε_o are the dielectric constants of the ellipsoid material and of the surrounding medium respectively; A is a depolarization factor given by

$$A = \frac{ab^2}{2} \int \frac{ds}{(s+a^2)^{3/2}(s+b^2)} \quad .$$

This integral has been tabulated.[14] For a sphere, A = 1/3; for a 3:1 aspect ratio ellipsoid, A = 0.1. Equation (1) shows a resonance at $\varepsilon \simeq \varepsilon_o(1-1/A)$. For a sphere this resonance occurs if $\varepsilon \simeq -2\varepsilon_o$.

For $\varepsilon_0 = 1$, a silver sphere would have a resonance at 3.5eV i.e., in the ultraviolet part of the spectrum near 350nm. A 3:1 ellipsoid is resonant at $\varepsilon = -8.25\varepsilon_0$ which would imply the resonance is shifted to $\sim 5000\text{\AA}$ for such silver ellipsoids when $\varepsilon_0 = 1$. The resonances are quite intense since the imaginary part of ε (i.e., ε_2) is near zero in the visible region of the spectrum. These examples show that as the ellipsoid is made more needlelike, the particle plasmon resonance corresponding to electron motion along the major axis is shifted toward longer wavelengths. Increasing ε_0, for example by dipping the surface into inert liquids, will also shift the resonance to longer wavelengths. Both of these predictions are borne out by our lithographically produced samples.

At resonance the dipolar field of the ellipsoid becomes very large and induces a large Raman molecular polarization. The resulting molecular dipole moment, μ_m oscillating at the Stokes frequency ω_S is given by $\mu_m(\omega_S) = 2\alpha_R\alpha_E(\omega_L)E_L/\varepsilon_0 r^3$. Here ω_L is the laser frequency and r is the distance from the center of the ellipsoid to the molecule. The field of the molecular dipole in turn polarizes the ellipsoid to produce an ellipsoid dipole at the Stokes frequency, $\mu_E(\omega_S) = 2\alpha_E(\omega_S)\mu_m(\omega_S)/\varepsilon_0 r^3$, which is larger than the usual Raman molecular dipole by the factor

$$f = \frac{4}{9} \frac{[1-\varepsilon(\omega_s)/\varepsilon_o]}{[1-(1-\varepsilon(\omega_s)/\varepsilon_o)A]} \frac{[1-\varepsilon(\omega_L)/\varepsilon_o]}{[1-(1-\varepsilon(\omega_L)/\varepsilon_o)A]} \left[\frac{ab^2}{r^3}\right]^2 . \qquad (2)$$

The net enhancement of the Raman intensity is given by $|f|^2$. This expression is essentially the same as that given in reference 12. The major difference between the result given in equation (2) and the solution[12] to the correct boundary value problem for the case of an ellipsoid is the enhancement caused by the concentration of the field around the tips of the ellipsoid by what Gersten and Nitzan have referred to as the "lightning rod" effect. Recently Liao and Wokaun[15] have shown that this effect can be expressed as a simple factor

$$\gamma^2 = \left\{\frac{3}{2}(\frac{a}{b})^2(1-A)\right\}^2$$

with which equation (2) should be multiplied. This "lightning rod" effect is an important one. For a 3:1 aspect ratio ellipsoid it is responsible for $\gamma^4 = 2\times10^4$ of the total enhancement for molecules located at the ellipsoid tips.

Although equation (2) predicts resonant behavior as a function of excitation wavelength, experiments have generally shown only a slowly increasing enhancement as the excitation wavelength is varied from 450nm to 650nm. Since the resonant frequency is shape dependent through the factor A, this non-resonant behavior can be attributed

to the wide range of shapes and sizes of the metal protrusions which
are found on a rough surface. Since ε increases in magnitude with
increasing wavelength, the factors of ε found in the numerator of
Eq. (2) cause any sample with any significant distribution of shapes
to exhibit increasing enhancement for longer wavelengths. The lightning rod factor also tends to weight the long wavelength resonant
ellipsoids, which are also the most needlelike, more heavily than
the more spherical particles which have resonances at shorter wavelengths.

To obtain a good comparison with the particle plasmon theory we
have developed microlithographic techniques to produce regular arrays
of isolated, uniformly sized and shaped silver particles of 100nm
dimension. These particles can then be designed to maximize the
Raman enhancement.

MICROLITHOGRAPHY

Modern techniques of lithography are capable of producing extremely small structures. Structures as small as 50Å have been made
by electron beam writing on ultrathin substrates.[16] X-ray lithography has been shown to be capable of creating tungsten features
of 175Å dimension.[17] Laser interference patterns can be used to
produce periodic patterns of ∿ 2000Å period. With such tools it
should be possible to directly produce silver structures of dimensions and shapes which would be capable of testing the electromagnetic model for SERS. Indeed new studies with specially designed
surfaces will be possible.

The main difficulty with using lithography for production of
Raman enhancing silver surfaces is the possibility of contamination
of the silver surface with the chemicals used in the lithographic
processing steps. We also have the requirement that a rather large
patterned area is required in order to have sufficient Raman signal.
Typically one wishes to have a minimum area of 1mm x 5mm.

To circumvent these difficulties we have chosen not to directly
pattern the silver but instead to produce a large area patterned
substrate upon which we can evaporate fresh silver under vacuum conditions. Although the experiments which we have reported thus far[1]
have not utilized the ultra pure capabilities of UHV systems, such
studies with our patterned substrates are underway. The large patterned areas (1 cm^2) of our substrates are obtained by using holographic exposure of photoresist followed by several etching steps.

The substrate pattern consists of an array of tall SiO_2 posts
on a silicon wafer. By evaporating silver at grazing incidence onto
this substrate we obtain isolated silver particles on the tops of
the posts.

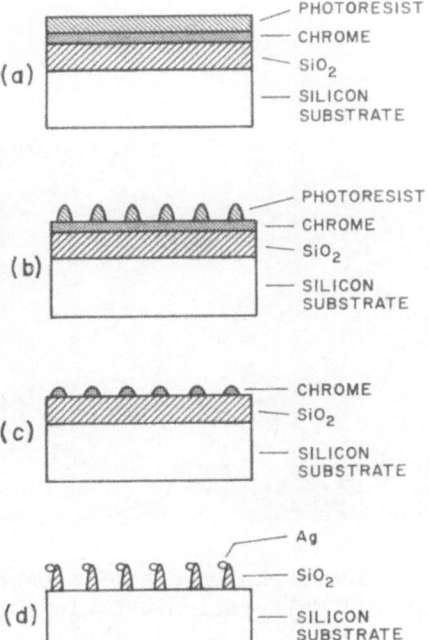

(a)
- PHOTORESIST
- CHROME
- SiO₂
- SILICON SUBSTRATE

(b)
- PHOTORESIST
- CHROME
- SiO₂
- SILICON SUBSTRATE

(c)
- CHROME
- SiO₂
- SILICON SUBSTRATE

(d)
- Ag
- SiO₂
- SILICON SUBSTRATE

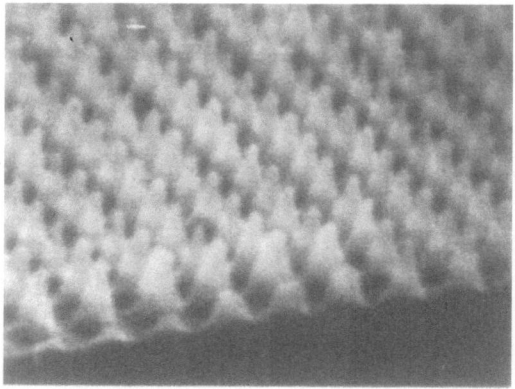

Fig. 2. Electron microscope photograph of substrate before evapora-
         tion of silver. Calibration bar is 1000Å long.

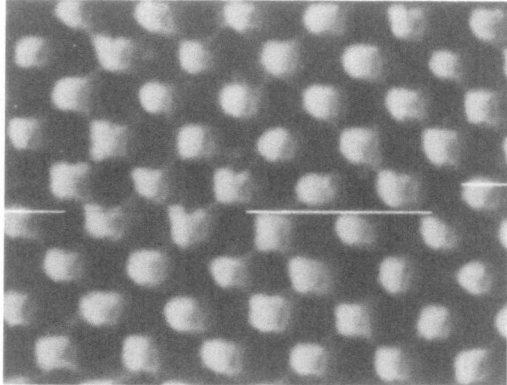

Fig. 3. Top view of substrate after silver evaporation. Calibra-
         tion bar is 1 micron in length.

must be evaporated onto the substrate along a nonchanneling direction
so that the posts shadow each other. In Figure 4 the silver particles are seen to be uniform in size and as a first approximation

Fig. 4. Side view of substrate after silver evaporation. Calibration bar is 1 micron in length.

can be considered to be ellipsoids with a 3 to 1 aspect ratio. The silver particles in Figure 5 have an aspect ratio of approximately 2 to 1. Samples such as these are ideal for testing the electromagnetic model of SERS. We note the substrates are quite durable and can be chemically cleaned and reused many times.

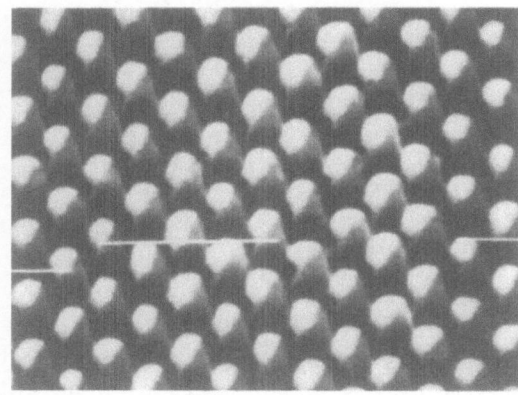

Fig. 5. Side view of substrate after silver evaporation. Calibration bar is 1 micron in length.

RESULTS

After silver is deposited onto the substrate, CN molecules are adsorbed onto the silver by exposing the sample to HCN vapor. Raman light, Stokes shifted by the 2144 cm^{-1} CN stretch mode, is then excited with 20-100mW of laser light from either an argon ion or Rh6G dye laser and measured through a double monochromator, photon counting system set for 8 cm^{-1} resolution. The samples are mounted at 60^0 to the horizontally polarized incident laser beam.

In Fig. 6 the variation of the measured normalized peak Raman intensity versus excitation photon energy is shown for the sample in a nitrogen atmosphere. The normalization eliminates the ω^4 density of states factor in the expression[2] for the Raman cross section, the laser intensity and the detector sensitivity. Data are shown for two different aspect ratio particles. The open circles correspond to the ∿ 3:1 aspect particles shown in Fig. 4, while the solid points were taken with ∿ 2:1 aspect ratio particles of Fig. 5. In each case one clear resonance is observed. Evaluation of the complete expression for the enhancement, including the lightning rod effect, would predict two narrow peaks separated by ω_{vib} = 0.27eV. One peak corresponds to the resonance of the incident light and one to the resonance of the emitted Raman light. However, if one broadens the theoretical resonances by increasing ϵ_2 to 2.0, or if one assumes our particles have a slight distribution (±12%) of aspect ratios,

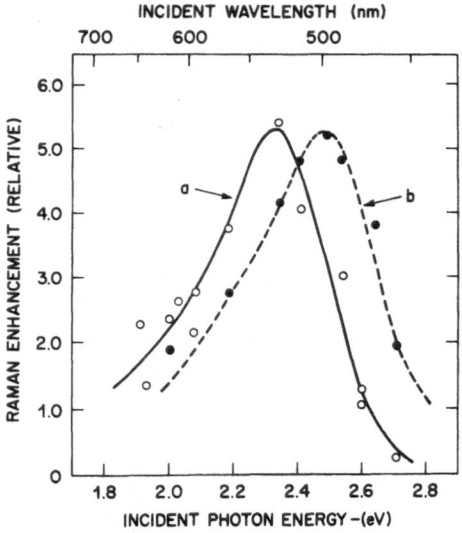

Fig. 6. Dependence of the Raman signal on the aspect ratio of the silver ellipsoids. The normalized Raman intensity of the CN (2144 cm^{-1}) vibration in nitrogen is shown as a function of incident photon energy. (a) 3:1 aspect ratio ellipsoids (Fig. 4). (b) 2:1 aspect ratio ellipsoids (Fig. 5).

the two peaks coalesce into one, and the width, location and shift of the peak can be fit to the data yielding an aspect ratio of 3.9:1 for the open circle data and 3.2:1 for the solid points. These values compare favorably with the observed ratios, especially considering that we have neglected the effects of interactions between particles, the presence of the SiO_2 posts, and deviations from ellipsoidal shape. Furthermore, the $\sim$ 100nm size of the particles is only moderately small compared to λ and the dipolar approximation is not strictly valid. Mie resonances of small particles are known to broaden as particle size increases.[9]

Calibration of the absolute detection sensitivity of our apparatus against the known output of a standard tungsten lamp indicates a $\sim 10^7$ enhancement of the CN Raman cross section at the resonant peak (assuming a Raman cross section of 3×10^{-30} cm^2 and one monolayer coverage[4] of 10^{15} cm^{-2}). Using a 12% distribution in aspect ratios and the dielectric function of bulk silver,[19] we calculate an enhancement of 1.3×10^8 in reasonable agreement with the measured value.

A further test of the particle plasmon model can be made by varying the dielectric constant of the surrounding medium. As we

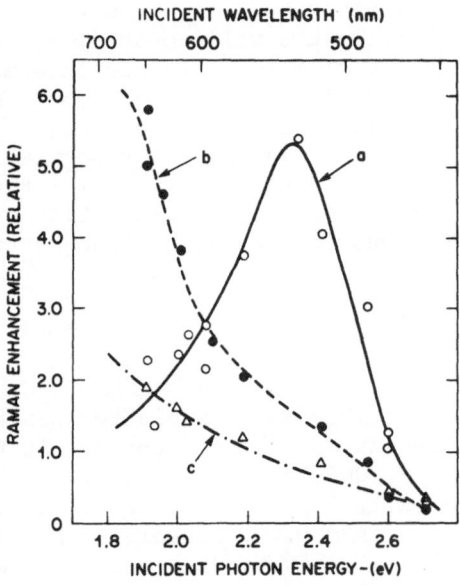

Fig. 7. Dependence of the CN Raman enhancement on the dielectric constant (ε_0) of the surrounding medium, for 3:1 ellipsoids. (a) Nitrogen ($\varepsilon_0 = 1$), (b) H_2O ($\varepsilon_0 = 1.77$), (c) cyclohexane ($\varepsilon_0 = 2.04$).

discussed earlier, if ε_O is increased, the resonance should shift to the red. In Figure 7 we see that this behavior is indeed verified as the sample of Figure 4 was immersed in either water (ε_O=1.77) or cyclohexane (ε_O=2.04). Unfortunately our dye laser could not be tuned sufficiently to completely resolve the resonances. The predicted resonance for 3.9:1 aspect ellipsoids with a 12% distribution would be at 1.9eV in water and 1.5eV in cyclohexane. The relative amplitudes of the three sets of data could be repeatedly checked by removing the samples from the liquids into nitrogen atmosphere.

CONCLUSIONS

Lithography can be used to create substrates to produce a surface consisting of uniform arrays of isolated silver particles. These surfaces produce Raman enhancements of order 10^7 and have been used to test the electromagnetic model of SERS. The dependence on particle shape and surrounding dielectric constant is found to agree with the electromagnetic model and strongly supports the particle plasmon theories of enhanced Raman scattering.

The general utility of these surfaces is immediately evident. Clearly one can design them to enhance the particular wavelengths of interest. One can "tune" the particle resonances to increase the Raman efficiency. Gersten and Nitzan have predicted enhancements as large as 10^{11} for properly sized and shaped particles. Because the spacings between the particles are well controlled, these samples are also ideal for the study of inter-particle interactions. Such studies are underway. Other electromagnetic processes will also be enhanced by the intense fields near the particles,[20] and for these processes lithographic samples have unique properties. For example, in recent studies of second harmonic generation,[21] the grating like character of particle arrays resulted in generation of second harmonic radiation in directions into which no fundamental radiation is diffracted.

ACKNOWLEDGEMENTS

The work which is described in this chapter has been the result of the efforts and collaboration of many people. At Bell Laboratories J. G. Bergman, D. S. Chemla, A. Glass, C. V. Shank, T. H. Wood and A. Wokaun made especially important contributions. Microfabrication advice was received from L. D. Jackel, E. Hu and R. E. Howard of Bell Laboratories. The microstructures were made during a visit by the author to the microstructures laboratory at M.I.T. where J. Melngailis, A. Hawryluk and N. P. Economou directly contributed to the final results.

REFERENCES

1. P. F. Liao, J. G. Bergman, D. S. Chemla, A. Wokaun, J. Melngailis, A. M. Hawryluk and N. P. Economou, Surface enhanced Raman scattering from microlithographic silver particle surfaces, Chem. Phys. Lett., to be published 1981.

2. M. Fleischman, P. J. Hendra and A. J. McQuillan, Raman spectra of pyridine adsorbed at a Ag electrode, Chem. Phys. Lett. 26:163 (1974); see also R. P. Van Duyne in "Chemical and Biochemical Applications of Lasers," Vol. 4, ed. C. B. Moore, Academic Press, NY, 1978).

3. C. Y. Chen, E. Burstein and S. Lundquist, Giant Raman scattering by pyridene and CN⁻ adsorbed on Ag, Sol. State Commun. 32:63 (1979).

4. J. G. Bergman, D. S. Chemla, P. F. Liao, A. M. Glass, A. Pinczuk, R. M. Hart and D. H. Olson, Relationship between surface enhanced Raman scattering and the dielectric properties of aggregated silver films, Opt. Lett. 6:33 (1981).

5. Surface-enhanced Raman effect, Physics Today April 1980 p. 19.

6. J. G. Bergman, J. P. Heritage, A. Pinczuk, J. M. Worlock and J. H. McFee, Cyanide coverage on silver in conjunction with surface enhanced Raman scattering, Chem. Phys. Lett. 68:412 (1979).

7. J. E. Rowe, C. V. Shank, D. A. Zwemer and C. A. Murray, Ultra-high-vacuum studies of enhanced Raman scattering from pyridine on Ag surfaces, Phys. Rev. Lett. 44:1770 (1980).

8. M. Moskovits, Surface roughness and the enhanced intensity of Raman scattering by molecules adsorbed on metals, J. Chem. Phys. 69:4159 (1978).

9. S. L. McCall, P. M. Platzman and P. A. Wolff, Surface enhanced Raman scattering, Phys. Lett. A 77:381 (1980).

10. M. Kerker, D. S. Wang, H. Chen, Surface enhanced Raman scattering (SERS) by molecules adsorbed at spherical particles, Appl. Optics 19:4159 (1980).

11. C. Y. Chen and E. Burstein, Giant Raman scattering by molecules at metal-island films, Phys. Rev. Lett. 45:1287 (1980).

12. J. I. Gersten and A. Nitzan, Electromagnetic theory of enhanced Raman scattering by molecules adsorbed at rough surfaces, J. Chem. Phys. 73: 3023 (1980).

13. See for example, C. J. F. Bottcher, "Theory of Electric Polarization," Vol. 1 (Elsevier Sci. Publ. Co., NY 1973) p. 79.

14. E. C. Stoner, Demagnetizing factors for ellipsoids, Phil. Mag. 36:803 (1945).

15. P. F. Liao and A. Wokaun, to be published.

16. H. P. Zingsheim, STEM as a tool in the construction of two-dimensional molecular assemblies, in "Scanning Electron Microscopy," Vol. 1, O. Johari, ed. ITT Research Institute (1977).

17. D. C. Flanders, Replication of 175Å lines and spaces in polymethylmethacrylate using x-ray lithography, Appl. Phys. Lett.

17. D. C. Flanders, Replication of 175 lines and spaces in poly-
     methylmethacrylate using x-ray lithography, Appl. Phys. Lett.
     36:93 (1980).
18. H. W. Lehmann and R. Widner, Fabrication of deep square wave
     structures with micron dimensions by reactive sputter etching,
     Appl. Phys. Lett. 32:163 (1978).
19. P. B. Johnson and R. W. Christy, Optical constants of the noble
     metals, Phys. Rev. B 6:4370 (1972).
20. A. M. Glass, P. F. Liao, J. G. Bergman and D. Olson, Interaction
     of metal particles with adsorbed dye molecules:absorption and
     luminescence, Opt. Lett. 5:368 (1980).
21. A. Wokaun, J. G. Bergman, J. P. Heritage, A. M. Glass, P. F.
     Liao and D. H. Olson, Surface second harmonic generation
     from metal island films and microlithographic structures,
     Phys. Rev. B (1981).

NONLINEAR OPTICAL EFFECTS

J. P. Heritage and A. M. Glass

Bell Telephone Laboratories

Holmdel, N. J.   07733

INTRODUCTION

Experimental studies of surface enhanced nonlinear optical effects are reviewed within the framework of electromagnetic theories of enhancement. Nonlinear optical effects are catalogued by the well known power series expansion of the generalized polarization in terms of the applied optical fields and local field factors. Surface enhanced optical effects that are described by the first, second and third order polarization are in general agreement with simple models of the local field enhancement which arises from plasmon resonances of small metallic particles. Nonlinear optical phenomena studied on surfaces with controlled morphology promise to yield additional information concerning local field effects.

The discovery (1) of the anomalous enhancement of Raman scattering from molecular monolayers on silver surfaces has led to a renewed interest in the optical properties of metal surfaces, films and microscopic structures, both with and without molecular overlayers. The recent observation of enhancement of a variety of linear and nonlinear optical processes as well as Raman scattering at a metallic surface provides tests for the theory of surface enhanced light scattering, as well as yielding new surface sensitive optical spectroscopies. New insights into old effects that were not previously well understood may well emerge as the large body of work on optical properties of metallic films and dielectric particles is reinvestigated in the light of theories of enhancement. One striking example of this is the S1 photocathode which is based on enhanced photoemission (2).

In this chapter we review experimental enhanced nonlinear optics at a metallic interface with a particular emphasis on the electromagnetic theory of enhancement. The electromagnetic theory describes the enhancement in terms of the local excitation fields experienced by microscopic polarizable objects (atoms, molecules, microstructures) which can be much larger than the applied fields. Furthermore, outgoing fields may be more efficiently radiated when the object is near a metallic microscopic particle. These effects occur when the fields are resonant with the local surface plasma frequency of the particle.

The electromagnetic theory predicts that linear and nonlinear optical effects should be enhanced in addition to the Raman effect. This prediction occurs naturally in the nonlinear expansion of the macroscopic susceptibility which describes all manner of nonlinear optical effects. The enhancement of nonlinear optical effects is more sensitive to the local field factors than linear effects, thus enhanced nonlinear optical effects can provide sensitive tests of the electromagnetic theory.

We should emphasize that there are alternative mechanisms to obtain surface enhanced Raman scattering from metal surfaces (3). However, we feel that for the surfaces discussed in this work the experimental evidence to date overwhelmingly supports the electromagnetic (enhanced local field) model. These experiments include measurement of the range of enhancement (4), the effect of particle shape (5), the dependence of the enhancement on the dielectric properties of the film and the surrounding medium and the wavelength dependence of the enhancement (6). The other enhanced linear and nonlinear optical effects described in this work which are predicted by electromagnetic theory afford further support for that model.

Some experiments may not be fully explained by the simplest models. However, experiments with enhanced linear and nonlinear optics have been performed on surfaces or structures that are much more complicated than the simple models employed by the theories. The details of the calculated local field factors depend sensitively upon the size and shape of the metal structure, the orientation and location on the structure of the molecule, on the effect of neighboring structures and molecules as well as on the distance of the molecule from the particle. Because of these complexities perfect agreement between simple models and complicated surface morphologies cannot be expected. Frequently, order of magnitude agreement between theory and experiment is found. We receive these successes enthusiastically and are not deterred by the occasional failure to find agreement as close as one would like. The ultimate experimental test of the importance of the local field enhancement must make use of structures that more closely approximate the theoretical models.

Since the majority of the experimental and theoretical work done on enhanced light·scattering from metallic surfaces has been limited to the Raman effect it is appropriate to start this review with a discussion of the rich variety of possible nonlinear light scattering processes. Nonlinear optical effects are conveniently described in the dipole limit by a power series expansion of the relationship between induced microscopic dipole moments p, and the local fields $E_\ell(\omega)$, experienced by a polarizable entity.

$$p = \alpha E_\ell + \beta E_\ell E_\ell + \gamma E_\ell E_\ell E_\ell + \ldots \tag{1}$$

The expansion coefficients α, β, and γ are the linear, the second order, and the third order generalized polarizabilities respectively (7).

With appropriate frequency arguments, equation 1 describes the polarization source terms to a wealth of nonlinear optical effects. Of course α describes the linear refractive index and linear absorption. Spontaneous Raman scattering may be viewed as the sidebands generated by a periodic change in the linear polarizability of an entity as it modulates the applied field. The modulation frequencies are the characteristic frequencies of the polarizable entity. For instance, for a molecular vibrational coordinate Q, $\alpha = \alpha_0 + (\partial\alpha_0/\partial Q)Q$, where the second term is the first order Raman polarizability. The second order polarizability β describes second harmonic generation, sumfrequency mixing, and optical rectification to name a few. The third order polarizability γ describes third harmonic generation, various four-wave mixing processes, optically induced changes to the linear refractive index, and two photon absorption. The many four wave mixing processes possible include Coherent Antistokes Raman Scattering (CARS), and stimulated Raman scattering. Higher order processes are also possible such as fifth harmonic generation but are not further discussed here. Examples of enhanced nonlinear optical effects at metallic surfaces, involving each of the first three terms of the power series expansion have been demonstrated and are reviewed in this chapter.

The task of nonlinear optics is to describe how all induced dipoles collectively interact and produce a macroscopic polarization. The electromagnetic theory shows how the presence of the metallic surface modifies this interaction. In passing from the induced microscopic dipole moment p of equation 1 to the macroscopic polarization P, local field factors $L(\omega)$ are introduced formally to account for the contribution to the local field from the applied fields E, and the near and far fields of neighboring atoms, molecules and metallic structures. The factor $L'(\omega)$ is introduced to take into account the possible amplification of outgoing fields. The macroscopic polarization P is given by:

$$P = L^{'}(\omega)\chi^{(1)}L(\omega)E(\omega) + L^{'}(\omega)\chi^{(2)}L(\omega_1)L(\omega_2)E(\omega_1)E(\omega_2)$$
$$+ L^{'}(\omega)\chi^{(3)}L(\omega_1)L(\omega_2)L(\omega_3)E(\omega_1)E(\omega_2)E(\omega_3) + \ldots \tag{2}$$

The tensor properties of the susceptibilities is suppressed in
equation 2 for the sake of simplicity. The macroscopic generalized
susceptibility χ contains the sums over all the microscopic generalized polarizabilities in the interaction volume.

The calculation of the appropriate local field factors $L(\omega)$
and $L^{'}(\omega)$ can be quite complicated since detailed microscopic
structure of the surface must be specified as well as long range
structure. The importance of microscopic surface roughness in the
enhanced Raman effect is well documented by experiment. Simple
electromagnetic models which model the rough surface as an ensemble
of spherical metallic particles having dimensions much less than the
wavelength of light can account for the enhanced Raman effect with
order of magnitude estimates. Care must be taken concerning whether
the optical process occurs outside the sphere, as for Raman scattering from a molecular layer, or from just within the sphere as in
the case of light absorption by the uncoated metal particle itself.
The boundary conditions, for example, in the case of a small sphere,
require that the field just outside the sphere be larger than the
field inside the sphere by a factor of $\varepsilon(\omega)$, the dielectric constant
of the metal. This can be important since the magnitude of the
real part of the dielectric constant of simple metals rapidly becomes
large with increasing wavelength beyond the plasma frequency.

The field experienced by a polarizable object inside the surface
of an isolated sphere of diameter much less than the wavelength is
greater than the incident field by the factor

$$L(\omega) = 3/[\varepsilon(\omega) + 2] \tag{3}$$

where $\varepsilon(\omega)$ is the complex dielectric constant of the metal. Thus
$L(\omega)$ becomes large for $\varepsilon(\omega) = -2$. This is the resonant condition
for plasma oscillation of the sphere. The dipolar approximation
used to obtain this local field factor must be extended to include
higher order electric and magnetic multipoles as the spheres approach a wavelength in diameter. Retardation effects that are not
important for spheres much smaller than a wavelength must be considered for much larger spheres.

In a more general case of an ensemble of interacting ellipsoids we have for the field inside the particle (8)

$$L(\omega) = 1/[1 + \{\varepsilon(\omega) - 1\}\{A+B\}] \tag{4}$$

where A is the depolarization factor which depends upon the shape
of the particle and the dielectric constant of the surrounding medium
and B is a factor which depends on the dipole interaction between
the particles.

It is clear from this relation that the local field enhancement
can vary widely according to the details of the optical parameters
of the metal and surroundings, particle shapes and their positions
with respect to their neighbors. In silver island films the local
field factor may be 10^2. $L(\omega)$ can be calculated exactly if all
details of the film structure are known, or approximated by an
effective medium theory (9). However, from an experimental point of
view it is often most useful to determine the value of $L(\omega)$ from the
linear optical absorption $A(\omega)$ and transmission $T(\omega)$ of the metal
surface (8). For instance for silver island films of mass thickness
d_m (6)

$$|L(\omega)|^2 = (\dot{c}/\omega d_m)\{A(\omega)/T(\omega)\}/Im[\epsilon(\omega)] \ . \tag{5}$$

$L'(\omega)$ is much more complicated (10,11). One must solve the boundary
value problem for an infinitesimal radiating dipole in the presence
of a metallic sphere of finite size. Evidently multipolar near field
solutions of the sphere must be taken into account.

An important first step in providing a sound experimental basis
for understanding the origins of the local field enhanced optical
effects is to start with a metallic surface having well defined
optical frequency dielectric properties, or better yet, having well
defined microscopic structure which corresponds to exact theoretical
analysis.

ENHANCED FIRST ORDER POLARIZATION

We turn first to discussion of the recent work done on enhanced
linear optics at metallic surfaces. The leading term in the power
series expansion of the nonlinear polarization governs linear
absorption through the imaginary part of the linear susceptibility.
Enhancement of absorption can occur when the incident photon energy
falls within the particle plasma resonance. Enhancement of fluorescence is expected to occur also when the emission wavelengths lie
within the plasma particle resonance. This situation is depicted
diagramatically in figure 1 sketch A and sketch C (a). Enhanced
elastic light scattering is also governed by the linear susceptibility.

The first experimental attempt to connect linear absorption
with enhanced Raman scattering comes from studies of sols, (12)
although some earlier work (13) had suggested a connection with
inelastic scattering from a roughened silver surface. Silver and

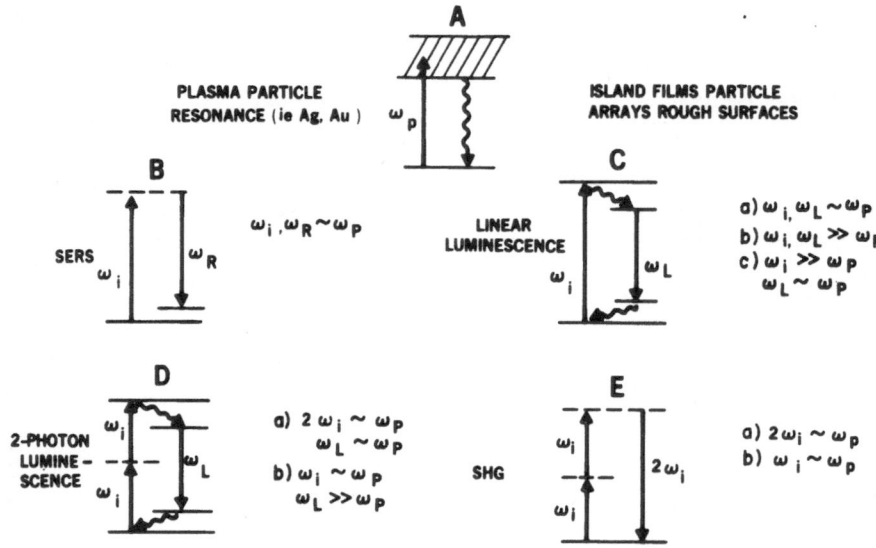

Fig. 1   Energy level diagram of four linear and nonlinear optical
processes that have been demonstrated to be enhanced when
the incident field ω_i or emitted field ω_R, ω_L overlap the
plasma particle resonance ω_P.

gold sols were laced with pyridine and the Mie extinction was compared with observed Raman enhancement as a function of wavelength.
Two peaks in the extinction spectra were found for both silver and
gold sols after adding pyridine to the sol. The longer wavelength
peaks (550nm for Ag and 750nm for Au) grew with time in both cases
indicating that clustering or chaining was occurring. The shorter
wavelength peaks appeared at 400nm for Ag and 525nm for Au. From
the Mie theory, the size of the particles corresponding to the short
wavelength peaks was estimated to be 30nm for Au. The authors
found that the Mie extinction for the long wavelength peak was
strongly correlated with the enhanced Raman excitation profile in
both silver and gold sols. Although this is gratifying evidence
for the local field enhancement mechanism, the absence of enhanced
Raman scattering in the gold sol for the short wavelength peak is
puzzling. The absence of enhanced Raman scattering for the short
wavelength absorption peaks might be explained by the smaller fields
outside the particle compared to the larger external fields at
longer wavelengths because of the extra factor of ε coming from the
boundary conditions. Studies correlating absorption and Raman
scattering supporting this local field enhancement have made use of
silver island films, microlithographic arrays and periodic grating
structures.

It is clear from eq. 2 that the local field effect should also enhance the absorption and luminescence of adsorbed molecules driven by the intense local field. An important difference, however, between the latter effect involving resonant excitation of molecules and nonresonant excitation, such as the Raman effect, is that the electronic transitions of the absorber must fall within the plasma resonance bandwidth of the metallic surface to result in enhancement. The resonant interaction between the adsorbed molecule and the metal particle can profoundly affect the optical properties of the metal particle even for very thin molecular overlayers.

The dipole radiators that are the microscopic sources of emitted fluorescence in general experience related local field factors which enhance the outgoing radiation as well. The fluorescence intensity F is proportional to the product of $|L(\omega)|^2$ and $|L'(\omega)|^2$, where $L(\omega)$ account for enhancement of the incident field which is equivalent to enhancement of the absorption by molecule, and $L'(\omega)$ accounts for the enhancement of the outgoing wave. We can then write

$$F \propto \eta |L(\omega)|^2 |L'(\omega)|^2 |E(\omega)|^2 \tag{6}$$

where the fluorescent quantum yield of the molecule η may also be strongly affected by the surface. The metal provides an alternative nonradiative decay path for the excited molecule. For optical processes such as Raman scattering, photoemission or photodissociation, this quenching by the metal surface is not important, but for molecules having long radiative lifetimes quenching becomes important. It is well known that for molecules near smooth metal surfaces the luminescence is heavily quenched by energy transfer from the molecule to the metal (14).

A convincing way of demonstrating local field enhancement of the luminescence was achieved with dye coated silver island films having continuously varying mass thickness as a function of position on a glass slide (15). These films are fabricated by evaporating silver onto a slide using a moving shadow mask, with mass thickness varying from zero to about 200 Å. This results in a continuous variation of particle size and density, as a function of position, and this in turn results in a continuous variation of plasma resonance frequency of the film as shown in figure 2. Thus with a thin layer of dye molecules spun onto the slide the enhancement of the fluorescence could be monitored as a function of the optical properties of the silver island film, and thus as a function of the local field factors.

Figure 2 shows the luminescence intensity of various dyes on the wedge silvered film using different dye molecules and different metals. Clear peaks in the fluorescence occur, which are especially

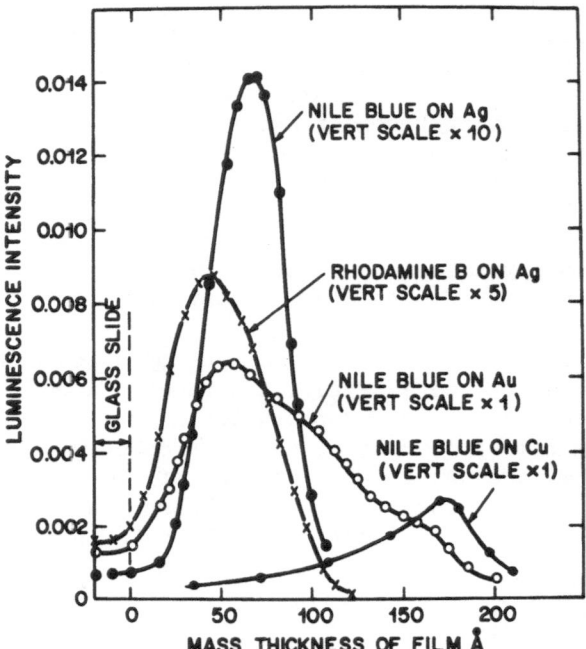

Fig. 2 Luminescence intensity of various dyes adsorbed on
metal island films as a function of film thickness.

evident on silver, when the absorption band of the dye molecules
falls within the plasma resonance of the particle. For Rhodamine B
on silver the peak occurs at a mass thickness of 50Å.

Fluorescence experiments offer the possibility of determining
the factor $L(\omega)$ and $L'(\omega)$ respectively by suitable choice of adsorbed
molecule with appropriate excitation frequency and particle plasma
resonance frequency ω_P. Referring to the energy level diagram in
figure 1 C it is clear that it is possible to achieve resonance
either between ω_i and ω_P or between ω_L and ω_P, thereby allowing
separate measurement of $L(\omega)$ and $L'(\omega)$. This is generally not
possible in Raman scattering experiments because of the small value
of the stokes shift compared with the plasma resonance bandwidth of
practical metal surfaces. Experiments of this kind (15,16) with Nile
Blue for which ω_i, $\omega_L \simeq \omega_P$ as well as $\omega_L \gg \omega_P$ and $\omega_L \sim \omega_P$ indicate
that $L(\omega) > L'(\omega)$ in this case.

Absorption spectra of silver island wedges were investigated
with and without adsorbed dyes. The absorption spectra are presented in figure 3 for varying silver island film thickness. Since
the absorption and reflection spectra of the bare silver island films
show the effect of the plasma particle resonance the magnitude of
$L(\omega)$ can be evaluated. When there is no overlap of the plasma

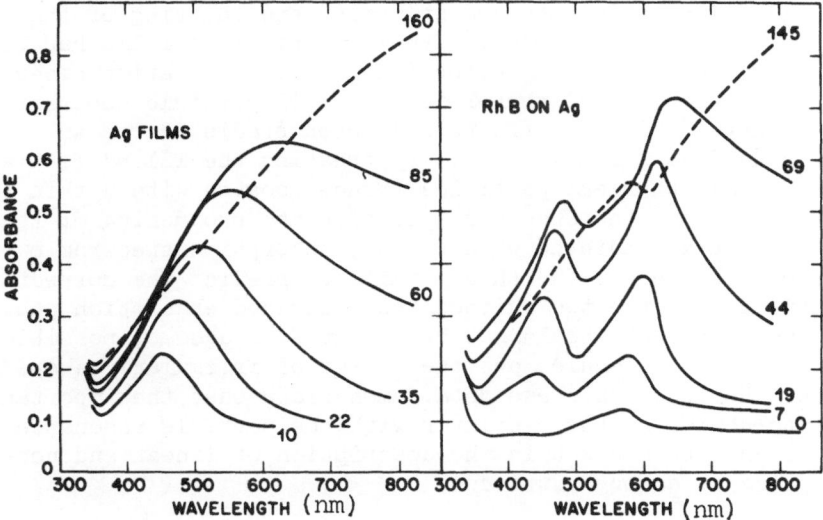

Fig. 3 Absorption spectra of bare and coated silver island films
        of differing mass thickness as a function of excitation
        wavelength.

resonance of the metal film and the dye absorption, the absorption
spectrum of the composite film looks much like the sum of the dye
absorption and the silver island film absorption. However, it is
evident from figure 3 that when the dye absorption falls within the
resonance bandwidth of the film the presence of the dye affects the
absorption of the silver particles greatly, even though the absorption of the dye alone is quite weak. As the film thickness
increases a splitting and broadening of the peaks become evident.
The dramatic change in the absorption spectrum is attributed to a
strong interaction between the dye film and the metal particles.
The particle resonance is strongly damped by the dye molecules in
their absorption band.

    Excitation spectra show the luminescence intensity to be a
convolution of the absorption spectra of the silver particle composite and the dye molecular absorption. Excitation of the silver
particles with light outside the absorption bandwidth of the dye
molecules resulted in no excitation transfer to the molecule (17).

    An interesting point concerning the dependence of local field
enhanced resonant processes on the distance of the molecule from the
particle surface has been pointed out by Nitzan and Brus (11). For
a molecule at a distance d from a metal sphere of radius a, the
local field enhancement factor varies as $(d+a)^3$; the distance of the
molecule from the center of the sphere, while the surface induced
quenching rate for $d \ll a$ depends only on d^3. Thus there should
exist a non-zero value of d for which the luminescence is greater
than that for a molecule located directly on the metal surface.

The experimentally observed splitting and shifting of the absorption spectrum of composite dye-metal particle films has been described (9) in terms of an effective-medium calculation based on the theory of the optical properties of small particle composites by Maxwell-Garnett. Similar effects have been predicted for an isolated spherical particle (18). By treating the island film as a system of metal spherical particles (Drude) coated with a thin shell of dye (Lorentzian) the observed splitting and broadening of the plasmon peaks are predicted when the dye absorption spectrum overlaps the plasmon resonance. The theory fails to predict the correct ratio of the strength of the two shifted and broadened absorption peaks, but precise agreement should not be expected because of possible clustering of dye molecules and the effect of irregularly shaped particles. Nonetheless these data demonstrate that the importance of the interaction of the overlayer with the particle resonance must be taken into account in the description of linear and non-linear optics at a rough surface.

ENHANCED SECOND ORDER POLARIZATION

We now turn our discussion to $\chi^{(2)}$, the first nonlinear contribution to total polarization. In an isotropic medium, symmetry considerations lead to the conclusion that in a dipole limit $\chi^{(2)}$ is zero. Higher order magnetic and electric multipolar contributions from within the bulk material near the surface are nonzero, however and contribute to a nonzero $\chi^{(2)}$. Furthermore, symmetry is broken for polarizable objects at an interface and their dipolar contributions can make a significant contribution to $\chi^{(2)}$ at a surface.

$\chi^{(2)}$ is the source of second harmonic generation. The SHG process is depicted schematically in figure 1 E. Theoretical accounts of SHG from a smooth surface consider the contribution of conduction and core electrons near the surface as well as surface terms due to field gradients (19). These considerations add definite angular dependence to the generated second harmonic intensity which for the case of particles is more difficult to calculate than for the planar surface. Since the local field within the particle as well as just outside are larger than the incident field, enhanced second harmonic generation from the particles as well as from overlayers is predicted. Since the reflected second harmonic is proportional to $|P(2\omega)|^2$ equation 3 predicts a second harmonic signal of

$$I(2\omega) \propto |\chi^{(2)}|^2 \quad |L'(2\omega)|^2 \quad |L(\omega)|^4 \quad |E(\omega)|^4. \tag{7}$$

The process of SHG is depicted schematically in figure 1 E. The local field factor $L(\omega)$ will be enhanced when the incident frequency

ω is resonant with ω_p, the particle plasma resonance. The factor $L(2\omega)$ will be enhanced when the generated frequency 2ω is resonant with ω_p.

C. K. Chen et al. (20) were the first to report enhanced SHG, originating from an electrochemically roughend silver surface. They employed a Q-switched Nd:YAℓG laser output at 1.06 micrometer excitation wavelength, eliminating surface damage by controlling the energy density delivered to the surface. The second harmonic signal observed at 0.53 micrometers was measured to be $\simeq$1000 times larger than the SHG signal from a smooth vacuum deposited silver film of 1000Å thickness. Enhanced second harmonic generation was also observed from gold but not copper. Electron micrographs revealed that their electrochemically etched surfaces consisted of particles about 500Å in size and occupying about 5 percent of the surface area. Noting that the SHG should not be in resonance with these particles, they took $L(2\omega) = 1$ and estimated $L(\omega) = 20$. The spectrum of emitted light from the roughend silver surface is presented in figure 4. The strong peak at the second harmonic is evident. The SHG peak is accompanied by a broad continuum extending deep in the antistokes frequencies as well as into the stokes side. The authors report that the power dependence of the continuum is strongly

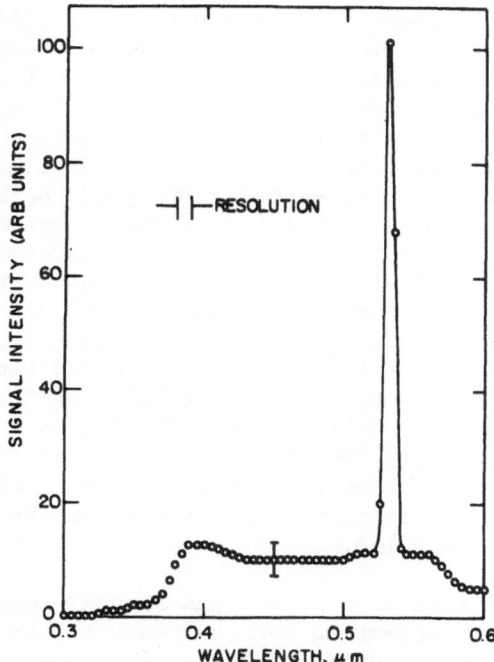

Fig. 4   Spectral distribution of the nonlinear signal from a rough bulk sample.

nonlinear. A continuum has also been observed during enhanced Raman studies and ascribed to an unknown luminescence mechanism. This is, however, the first report of a nonlinear power dependence. The origin of the continuum was not explained but the observation of a temporal decay time of several pulse widths suggests a luminescence origin, excited perhaps by the second harmonic signal.

The observation of enhanced second harmonic generation at a roughened silver surface is striking evidence in support of the local field theories. Detailed connection to the actual measured plasma resonances of controlled particle sizes, shapes and distributions is required however. Furthermore the possible contribution of planar surface plasmons launched by roughness should be determined. Second harmonic generation and enhancement by planar surface plasmons launched by prism coupling has been investigated on smooth metallic surfaces (21).

The dependence of SHG from a silver island film on the resonant excitation of a localized plasmon excitation has been explicitly demonstrated (8) with silver island films. This was accomplished by deriving the local field factors $L(\omega)$ and $L(2\omega)$ from measured effective dielectric properties of wedges of silver films, which were obtained from the measured absorption and reflection of the film at frequencies ω and 2ω as a function of position on a silver island film wedge. Although $L(2\omega)$ and $L'(2\omega)$ may differ as discussed earlier the data are presented in figure 5 as in the original report. Figure 5 shows the SHG signal as a function of location on a silver island film wedge and the derived local field factors, here labeled $f(\lambda)$. The SHG signal peaks near 40Å mass thickness which agrees with enhanced Raman scattering for silver island films on sapphire substrates. Sapphire was used here for its good thermal conductivity but the plasmon peak is shifted slightly compared to silver island films on glass. As the mass thickness increases above about 80Å the second harmonic signal increases to about 10 percent of its peak. The local field factors for incident light and the second harmonic are also plotted in figure 5. The local field factor for 1.06 microns peaks at about 60Å mass thickness and the local field factor for the second harmonic peaks near 25Å mass thickness. The appearance of the SHG peak between these values, as expected, can be considered reasonable agreement. The authors attribute the SHG signals at mass thickness greater than about 80Å to extended surface plasmons launched by scattering from roughness on the silver surface originating from the sapphire.

Similar measurements with gold films showed a much weaker peak in the second harmonic appearing near 40Å mass thickness, in excellent agreement with the location of the 1.06 micron local field factor peak near 35Å. The second harmonic local field factor was featureless and apparently does not significantly contribute to enhanced SHG. For film mass thickness greater than about 140Å the

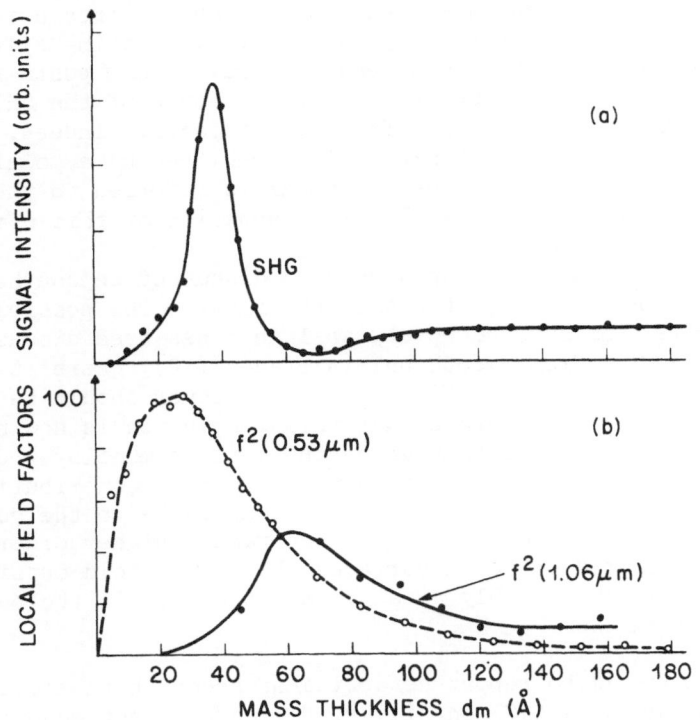

Fig. 5   Second harmonic generation from silver island films as a
         function of mass thickness. (a) Second harmonic intensity
         excited at 1.06 microns. (b) Local field enhancement factors
         for the excitation wavelength and the second harmonic. The
         local field factors are labeled f (λ) here.

extended surface plasmon contribution is actually nearly twice the
local plasmon peak strength.

     The enhancement factor for SHG on silver island films over
planar surfaces was found to be $\simeq$1000 compared with the data on
smooth films obtained by Bloembergen (19). This is comparable to
that found on electrochemically roughened silver. It is worth noting
that the uncertainty in the angular factors that enter into surface
SHG, especially on rough surfaces, make comparison of enhancement
between different laboratories somewhat uncertain. Note also that
there is a definite reduction in field strengths for polarizable
material within the metal particle as compared to material just
outside the particle. Second harmonic generation from overlayers,
rather than from the metal particles themselves should show greater
enhancement by this argument.

A unique experiment has been reported on SHG from a regular array of ellipsoidal particles produced by evaporating silver on a microlithographically produced regular array of SiO posts (8). The regularity of the array, and the coherent nature of the SHG suggest the possibility of observing diffraction effects. Indeed, collimated SHG was observed emitted at new angles predicted by a modified grating equation. The interested reader is referred to the chapter by Liao (5) in this book for further discussion of this effect.

In the previous paragraphs we have discussed second harmonic generation from sources just within the metal. The possible contribution to SHG from surface molecular layers has been discussed in early work on SHG from smooth bulk surfaces (19), where it was concluded that the contribution from a monolayer should be negligible. C. K. Chen et al. argue that this conclusion is not necessarily correct (22). They note that since the surface monolayer does not possess inversion symmetry their electric dipole contribution to the second order polarizability can be comparable to the contribution of hundreds of monolayers of the weaker higher order multipoles. Furthermore on rough surfaces the local field outside the surface can be considerably greater than that within the particle as discussed earlier.

Chen proceeds to report detection of second harmonic originating from adsorbed AgCl on silver, formed by electrochemical processing of a polished silver surface. Once again, an excitation wavelength of 1.06 micrometer generated by a Q-switched Nd:YAG laser was employed and SHG was detected with as little as 0.6 mJ/pulse over 0.2 cm surface area. The SHG originating from the AgCl was found to be as much as 25 times stronger than the AgCl free processed silver surface. Due to the surface roughness this signal was reported to be diffuse and unpolarized. Dramatic confirmation of the surface monolayer contribution to the SHG was obtained by adding pyridine to the electrochemical cell and switching the potential of the silver electrode to -1.1V SCE. The abrupt rise in the second harmonic signal is evident in figure 6a, occurring when the potential step is made. Figure 6b shows the SHG strength as a function of silver electrode potential. The behavior of the SHG from the surface without pyridine present remains nearly unchanged throughout the same potential range. The pyridine signal is reported to be nearly 50 times larger than the SHG from the pyridine free surface. The rise in strength near -0.6V SCE, with a peak near -1.1V SCE is a behavior nearly identical to the enhancement of Raman scattering from the same molecule. The authors point out that the extension of SHG to sum and difference frequency generation should permit one to perform surface spectroscopy through resonant enhancement. The detection of SHG from monolayers is a significant development; the large signals which evidently should be detectable even without the aid of enhancement, suggest a new and potentially sensitive surface spectroscopy.

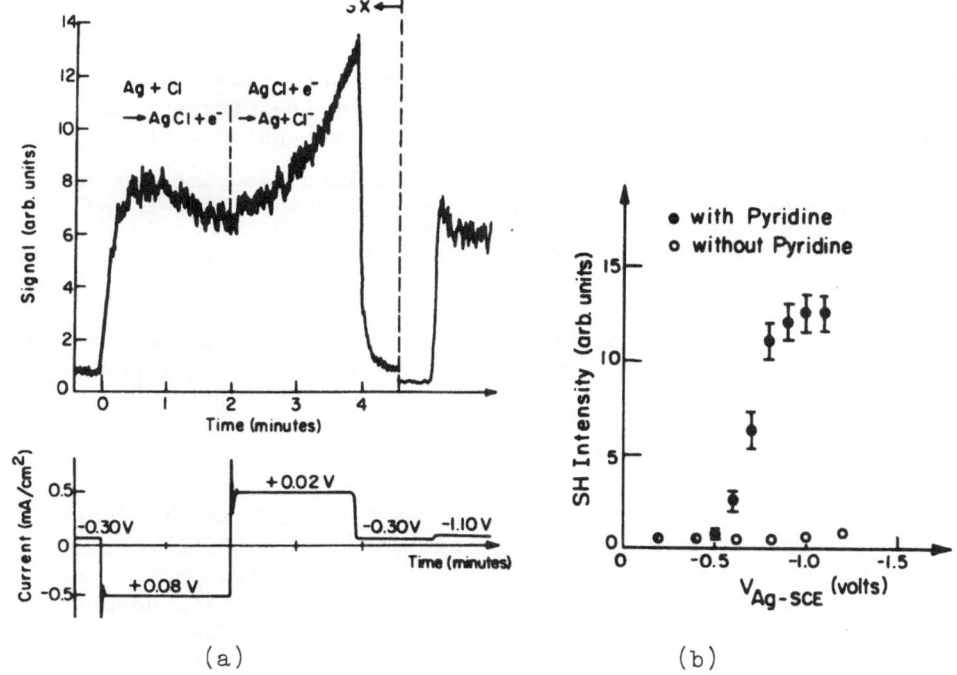

(a)   (b)

Fig. 6 a) Current and diffuse second harmonic reflection as a
           function of time during and after an electrochemical cycle.
           (volts SCE) 0.05 M pyridine was added to 0.1 M KCL after
           completion of the electrolytic cycle. b) Diffuse second
           harmonic signal vs. silver electrode potential (SCE)
           following an electrolytic cycle, with 0.05 M pyridine and
           0.1 M KCL in water.

ENHANCED THIRD ORDER POLARIZATION

     We now turn our attention to the third order nonlinear sus-
ceptibility. Of the many nonlinear optical effects that are described by $\chi^{(3)}$, perhaps the most obvious is third harmonic
generation (THG). The third harmonic signal is proportional to
$|P(3\omega)|^2 = |\chi^{(3)}|^2 |L'(3\omega)|^2 |L(\omega)|^6$. THG has been observed from a
silver surface that was not deliberately roughened (23). To date
there have been no reports of enhanced THG. Unlike SHG, dipolar
sources of SHG will originate from within the bulk since $\chi^{(3)}$ has
nonzero components in media with inversion symmetry. With THG the
wide spacing between the fundamental and generated frequencies means
that both are unlikely to resonate with a plasma resonance. However,
resonance of the fundamental frequency with the surface plasmon can
lead to enormous THG enhancements dependent on the sixth power of
the local field enhancement! However, the third harmonic of visible
or near infrared wavelengths is likely to be severely attenuated by
absorption.

Two-photon absorption is also described by a third order polarization. This can be seen by considering the definition of the generalized absorption coefficient which is the time average rate at which energy is absorbed per unit volume divided by the energy flux. With $P(\omega) = \chi^{(3)}(\omega_1\omega_2-\omega_2)E_1E_2E_2^* \exp[-i\omega_1 t]$ we have that $\alpha = \langle Re(J \cdot E^*)\rangle/2I$ where the current density is $J = \partial P/\partial t$. Taking the time derivative and dividing by $I = \{c/8\pi\} |E|^2$ one finds that $\alpha_2 = \{32\pi^2\omega_1/c^2\}\chi''^{(3)}$ where $\chi''^{(3)}$ is the imaginary part of the third order susceptibility and α_2 is defined by $\alpha = \alpha_0 + \alpha_2 I^2$.

The two-photon excitation rate for N molecules/cm^2 on a surface can then be written $R = \alpha^{(2)}N |L(\omega)|^4 |E(\omega)|^4$. The fluorescence intensity is then given by

$$F = \eta|L'(\omega_\ell)|^2|L(\omega)|^4|E(\omega)|^4 \tag{3}$$

where the quantum efficiency η depends, as in the case of linear absorption, on the coupling of the excited state with the metal particles. It is evident from this relation that two-photon absorption or two-photon luminescence studies of molecules adsorbed onto metal particles is more sensitive to the local field factor and enhancement can be much greater than for linear absorption and luminescence discussed above.

Two experiments have been performed (17) in which dye molecules adsorbed on silver island films were excited with high intensity, short duration (100ps) light pulses from continuous, modelocked lasers. The first of these used Rh B, about a monolayer thick, spun onto a silver island wedge on a sapphire substrate and excited with 1.06 micron radiation. There is no linear luminescence from the 1.06 micron excitation, but luminescence at 0.59 microns, with quadratic dependence on incident intensity, due to two photon excitation is readily observed. The energy level diagram appropriate for this experiment is shown in Fig. 1 D (a). Both the 1.06 microns and 0.59 microns radiation fall within the plasma resonance bandwidth of the silver film, so both L(1.06) and L'(0.59) can act to enhance the absorption and luminescence.

The experimental results are shown in figure 7 where the linear luminescence excited at 0.53 microns is shown for comparison on the same wedge. An intensity reference of the two-photon luminescence was obtained from a 10^{-5} M solution of Rh B in ethanol (1mm cell) measured in the same experiment. This reference is equivalent to 6×10^{14} molecules/cm^2. Thus the relative quantum efficiency per molecule of the dye on the silver film could be measured. The peak intensity observed at a silver island mass thickness of about 90Å corresponds to an enhancement factor for two-photon luminescence of about 150. This is significantly greater than the enhancement of the linear luminescence which peaks at 60Å mass thickness on the same wedge but less than the enhancement calculated from the local field

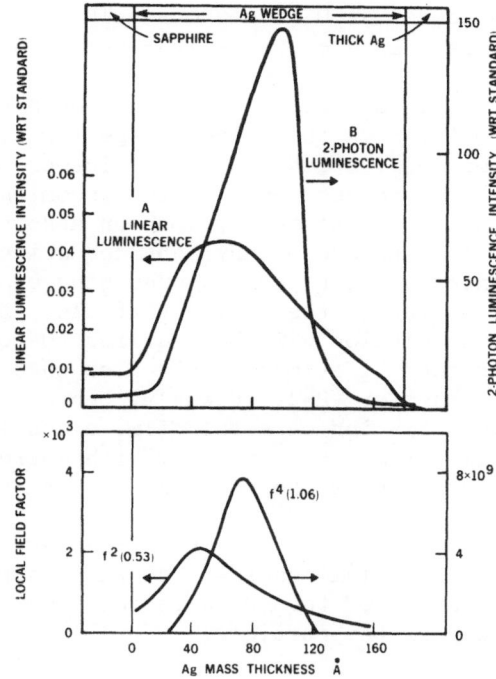

Fig. 7   a) Linear and two-photon luminescence of Rhodamine B at 0.59
         microns as a function of silver film thickness for incident
         wavelength of 0.53 microns and 1.06 microns. b) Calculated
         local field factors f (0.53) and f (1.06) for linear and two-
         photon luminescence.

factors. These local field factors were obtained from the linear
optical properties of the film. The calculated value of $L(\omega)$ and
$L'(\omega)$ at 1.06 microns and 0.53 microns are shown in figure 7 b. The
general shape of the curves and position of the predicted maxima are
in good agreement with experiment. Quantitative agreement is not
expected because of the unknown effect of quenching by the metal
particle.

     The measured second harmonic generation from the same sample
was considerably weaker than the two-photon luminescence. This shows
that the luminescence was excited by two-photon absorption and not by
second harmonic generation and linear absorption. It could not be
determined, however, whether the two-photon absorption is due to the
nonlinear susceptibility of the dye molecule or of the silver particle
followed by excitation transfer to the molecule.

     The second two-photon luminescence experiment was carried out
with Diphenylanthracene (DPA) adsorbed on a silver wedge deposited
on a sapphire substrate to avoid burning the film by the laser beam.
The energy level diagram appropriate to this case is shown in Fig. 1
D(b). DPA is excited by two-photon absorption of 0.5145 microns

radiation and luminesces at 0.43 microns. Thus the incoming radiation excites the resonance of the silver particle, but the excited electronic state of the dye molecule is well above the resonance bandwidth of the metal particle. No luminescence enhancement was observed at any position on the silver wedge for either two-photon excitation of linear excitation (at 0.265 microns) as shown in figure 8. Because of the resonance of the incoming light it is expected from equation 6 that the two-photon absorption was significantly enhanced. The absence of any enhanced emission must, within the picture presented here, be due to quenching of the excited state by energy transfer to the silver particle. Direct measurement of the two-photon absorption will be required to resolve this point. Studies of luminescence lifetimes should also shed light on the effects of energy transfer (24). However, because of the variation of molecule-particle spacing on metal films a considerable variation of measured lifetimes might be expected, with corresponding complications in interpretation of the results.

It is worth noting that in all the linear and two-photon experiments performed enhanced luminescence is only observed when the excited electronic state of the molecule falls within the plasma resonance band width of the particle.

The first measurement that demonstrated that a nonlinear optical effect was also enhanced by a roughened silver surface was performed by stimulated Raman scattering. Stimulated Raman scattering depends upon the imaginary part of the third order nonlinear susceptibility. In stimulated Raman scattering processes two optical fields must be

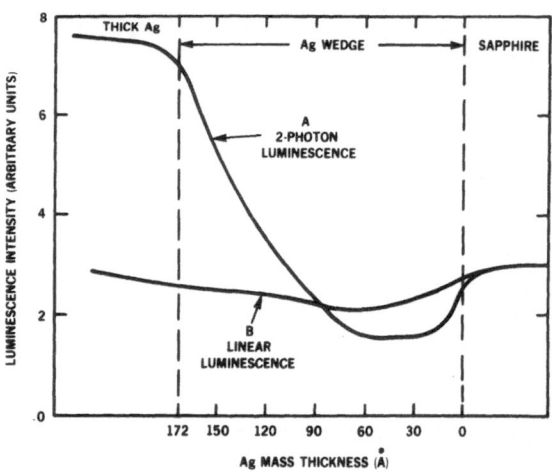

Fig. 8   Linear and two-photon luminescence of Diphenylanthracene at
         0.43 microns as a function of silver film thickness, for
         incident wavelength of 0.268 microns and 0.5145 microns.

accounted for. The strong excitation field E and the stokes shifted field E_S. When the difference between the optical frequencies equals the natural frequency of a Raman active state the Raman resonance is driven by the fields. This driven state mixes with the applied fields generating a polarization at exactly the right phase and frequency to amplify the stokes field. The growth of the stokes field is an exponential function of the intensity of the excitation field. In the small signal limit, the linear term in the expansion of the exponential dominates and this third order process responds linearly to the pump intensity.

The Raman gain experiments employed continuous trains of picosecond pulses to provide high intensity and low noise (25). Raman spectra were obtained showing the cyanide peak at 2145 cm^{-1}. In stimulated Raman gain spectroscopy the quantity that is measured is a change in the intensity of one of the two applied laser beams. The change in intensity amounts to about 0.01 percent for the enhanced Raman gain compared to a gain of about 10^{-6} percent that is expected if there is no enhancement. This signal is easily detected by modulating the pump beam and synchronous detection of the modulation impressed on the stokes beam. A background signal arises from reflectivity changes of the surface due to heating by the pump beam. This reflectivity change was found to be of the same order as the enhanced Raman signal since the absorption is enhanced by the roughness. For cyanide clear Raman spectra could be obtained even in the presence of the heating induced background.

Raman gain spectra of the broad peak near 1600 cm^{-1} that frequently appear in spontaneous Raman spectra were clearly obtained with a wavelength derivative variation of picosecond Raman gain spectroscopy (26). The important feature of the wavelength derivative technique is that wavelength independent artifacts that are stronger than the desired spectrum are effectively excluded from the derivative spectrum. By investigating the time resolved behavior of the features of the spectrum one concludes that the 1600 cm^{-1} peak is in fact a Raman transition. This is evident since, for nonresonant Raman effect, gain appears only when the pulses overlap in time and space at the surface. No Raman gain signal could be detected in the vicinity of 2000 cm^{-1} where the featureless continuum dominates surface enhanced spontaneous Raman spectra. It is worth pointing out here that the Raman gain spectrometer is insensitive to fluorescence as was easily verified by blocking the stokes laser and researching for modulated fluorescence. None was observed. It has been suggested that the origin of the continuum is an unspecified fluorescence mechanism. Although the continuum is not understood the picosecond Raman gain measurements and the nonlinear luminescence measurements (20) support this interpretation. The continuum remains a neglected subject in the topic of surface enhanced optical effects. Resolution of its nature remains both a theoretical and experimental challenge.

It is worth noting that the picosecond Raman gain technique is so sensitive that Raman spectra may be obtained on surfaces which are not expected to produce enhancement of the Raman effect. This has been demonstrated for p-nitrobenzoic acid on aluminum oxide (27). The ultrahigh sensitivity of surface picosecond Raman gain spectroscopy can be used to study surfaces or structures where the local field enhancement is small. This high sensitivity can also be used in two-photon absorption measurements which will permit separation of incoming and outgoing local field factors.

The experiments to date represent only a beginning of the study of surface enhanced nonlinear effects. Coherent antistokes Raman scattering is an example of a nonlinear optical effect that when performed at an enhancing surface might be useful for time resolved measurements of coherent dephasing of surface molecular vibrations.

A class of effects whereby photons are absorbed and a particle other than another photon is emitted exists. Enhanced photoemission of electrons, already recognized as important for photocathodes (28) warrants study. It is interesting that surface enhanced photoemission due to excitation of surface plasmons in smooth and rough metal surfaces was studied and understood long before the field of surface enhanced Raman scattering emerged. Photodissociation of adsorbed species and photoassisted chemical reactions at surfaces also offer promising areas of study. The reaction rate of a photochemical species located near the metal surface can be considerably enhanced by the local field enhanced absorption (11).

Surface enhanced nonlinear optical effects promise to elucidate the mechanisms of enhancement by providing clear experimental tests of the local field theory to determine where and to what extent it is dominant. That the local fields exist and are important is not questioned, and there is no doubt that in special cases they are very large indeed. Nonetheless, other mechanisms may well be important and in certain cases act in concert with the local field enhancement. The study of enhanced nonlinear optics on surfaces with controllable morphology promises to help unravel the apparent complexity of surface enhancement and to generate new optically based spectroscopies that will be useful for studying the physics and chemistry of metallic interfaces.

ACKNOWLEDGEMENTS

We gratefully acknowledge many helpful discussions with J. G. Bergman, L. E. Brus, P. F. Liao, A. Nitzan, A. Wokaun, T. H. Wood, and J. M. Worlock.

REFERENCES

1. R. P. Van Duyne, Laser excitation of Raman scattering from
    adsorbed molecules on electrode surfaces, in "Chemical and
    Biochemical Applications of Lasers," Vol. 4, C. B. Moore, ed.,
    Academic Press, New York (1978), p. 101.
2. The S1 photocathode consists of a silver island film coated with
    a low work function Cs-O overlayer. It is evident from early
    work on these devices [see "Photoemissive Material," edited
    by A. H. Sommer, John Wiley and Sons, New York (1968)] that
    enhanced local fields near the silver particle are of major
    importance.
3. See theoretical chapters in this volume.
4. See chapter by C. A. Murray, Molecule-silver separation
    dependence, in this volume.
5. See chapter by P. F. Liao, Silver structures produced by
    microlithography, in this volume.
6. J. G. Bergmann, D. S. Chemla, P. F. Liao, A. M. Glass, A. Pinczuk,
    R. M. Hart, and D. H. Olson, Relationship between surface-
    enhanced Raman scattering and the dielectric properties of
    aggregated silver films, Opt. Lett. 6:33 (1981).
7. J. A. Armstrong, N. Bloembergen, J. Ducuing, and P. S. Pershan,
    Interaction between light waves in a nonlinear dielectric,
    Phys. Rev. 127:1918 (1962).
8. A. Wokaun, J. G. Bergman, J. P. Heritage, A. M. Glass, P. F.
    Liao, and D. H. Olson, Surface second harmonic generation from
    metal island films and microlithographic structures, Phys.
    Rev. B, to be published.
9. See for instance the Maxwell-Garnett theory applied to silver
    films: H. G. Craighead, and A. M. Glass, Optical absorption
    of small metal particles with adsorbed dye coats, Opt. Lett.
    6:248 (1981).
10. M. Kerker, D. S. Wang, and H. Chew, Surface enhanced Raman scat-
    tering by molecules adsorbed at spherical particles, Appl.
    Opt. 19:4159 (1980).
11. A. Nitzan and L. E. Brus, Theoretical model for enhanced photo-
    chemistry on rough surfaces, to be published.
12. J. A. Creighton, C. G. Blatchford, and M. G. Albrecht, Plasma
    resonance enhancement of Raman scattering by pyridine on
    silver of gold sol particles of size comparable to the exci-
    tation wavelength, J. Chem. Soc. Faraday Trans. II, 19:4159
    (1979).
13. J. G. Bergman, J. P. Heritage, A. Pinczuk, J. M. Worlock, and
    J. H. McFee, Cyanide coverage on silver in conjunction with
    surface enhanced Raman scattering, Chem. Phys. Lett. 68:412
    (1979).
14. R. R. Chance, A. H. Miller, A. Prock, and R. Silby, Fluorescence
    and energy transfer near interfaces: the complete and quanti-
    tative description of Eu^{3+}/mirror systems, J. Chem. Phys.
    63:1589 (1975).

15. A. M. Glass, P. F. Liao, J. G. Bergman, and D. H. Olson, Inter-action of metal particles with adsorbed dye molecules: absorption and luminescence, Opt. Lett. 5:368 (1980).

16. G. Y. Ritchie, C. Y. Chen, and E. Burstein, Secondary light emission by molecules at metal surfaces, Bull. Am. Phys. Soc. 25:259 (1980).

17. A. M. Glass, A. Wokaun, J. P. Heritage, J. G. Bergman, P. F. Liao, and D. H. Olson, Enhanced two-photon fluorescence of molecules adsorbed on silver island films, Phys. Rev. B, to be published.

18. S. Garoff, D. A. Weitz, T. J. Gramila, and C. D. Hanson, Optical absorption resonances of dye-coated silver-island films, Opt. Lett. 6:245 (1981).

19. N. Bloembergen, R. K. Chang, S. S. Jha, and C. H. Lee, Optical second harmonic generation in reflection from media with inversion symmetry, Phys. Rev. 174:813 (1968).

20. C. K. Chen, A. R. B. de Castro, and Y. R. Shen, Surface enhanced second harmonic generation, Phys. Rev. Lett. 46:145 (1981).

21. H. J. Simon, D. E. Mitchell, and J. G. Watson, Optical second harmonic generation with surface plasmons in silver films, Phys. Rev. Lett. 33:1531 (1974).

22. C. K. Chen, T. F. Heinz, D. Ricard and Y. R. Shen, Detection of molecular monolayers by optical second harmonic generation, Phys. Rev. Lett. 46:1010 (1981).

23. N. Bloembergen, W. K. Burns, and M. Matsuoka, Reflected third harmonic generated by picosecond laser pulses, Opt. Commun. 1:195 (1969).

24. D. A. Weitz, S. Garoff, and C. D. Hanson, The effect of rough silver surfaces on fluorescent lifetimes, Bull. Am. Phys. Soc. 26:339 (1981).

25. J. P. Heritage, J. G. Bergman, A. Pinczuk, and J. M. Worlock, Surface picosecond Raman gain spectroscopy of a cyanide monolayer on silver, Chem. Phys. Lett. 67:229 (1979).

26. J. P. Heritage, and J. G. Bergman, Wavelength derivative surface Raman gain spectroscopy of carbonate on silver, Opt. Commun. 35:373 (1980).

27. J. P. Heritage and D. L. Allara, Surface picosecond Raman gain spectra of a molecular monolayer, Chem. Phys. Lett. 74:507 (1980).

28. J. G. Endriz and W. E. Spicer, Experimental evidence for the surface photoelectric effect in aluminum, Phys. Rev. Lett. 27:570 (1971).

CONTRIBUTORS

Billmann, J.
  Physikalisches Institut III
  Universität Düsseldorf
  D-4000 Düsseldorf
  Federal Republic of Germany

Birman, J. L.
  Physics Department
  City College, CUNY
  New York, NY 10031

Bumm, L. A.
  Chemistry Department
  Clarkson College of Technology
  Potsdam, NY 13676

Burstein, E.
  Physics Department and
    Laboratory for Research
    on the Structure of Matter
  University of Pennsylvania
  Philadelphia, PA 19104

Chen, C. Y.
  Physics Department and
    Laboratory for Research
    on the Structure of Matter
  University of Pennsylvania
  Philadelphia, PA 19104

Chew, H.
  Physics Department
  Clarkson College of Technology
  Potsdam, NY 13676

Creighton, J. Alan
  Chemical Laboratories
  University of Kent
  Canterbury, CT2 7NH, U.K.

Demuth, J. E.
  IBM Thomas J. Watson
    Research Center
  Yorktown Heights, NY 10598

DiLella, Daniel P.
  Department of Chemistry
    and Erindale College
  University of Toronto
  Toronto, M5S 1A1, Canada

Fleischmann, M.
  Chemistry Department
  University of Southampton
  Southampton, S09 5NH, U.K.

Genack, A. Z.
  Exxon Research and
    Engineering Company
  P.O. Box 45
  Linden, NJ 07036

Gersten, Joel I.
  Physics Department
  City College, CUNY
  New York, NY 10031

Glass, A. M.
  Bell Telephone Laboratories
  Holmdel, NJ 07733

Gramila, T. J.
  Exxon Research and
    Engineering Company
  P.O. Box 45
  Linden, NJ 07036

Heritage, J. P.
  Bell Telephone Laboratories
  Holmdel, NJ 07733

Hill, I. R.
  Chemistry Department
  University of Southampton
  Southampton, SO9 5NH, U.K.

Jha, Sudhanshu S.
  Tata Institute of
    Fundamental Research
  Homi Bhabha Road
  Bombay 400 005, India

Kerker, M.
  Chemistry Department
  Clarkson College of Technology
  Potsdam, NY 13676

Kirtley, J. R.
  IBM Thomas J. Watson
    Research Center
  Yorktown Heights, NY 10598

Lee, T. K.
  Institute for Theoretical
    Physics
  University of California
  Santa Barbara, CA 93106

Liao, P. F.
  Bell Telephone Laboratories
  Holmdel, NJ 07733

Lundqvist, S.
  Chalmers University
    of Technology
  S-412 96 Gothenberg, Sweden

Metiu, Horia
  Department of Chemistry
  University of California
  Santa Barbara, CA 93106

Mills, D. L.
  Department of Physics
  University of California
  Irvine, CA 92717

Moskovits, Martin
  Department of Chemistry
    and Erindale College
  University of Toronto
  Toronto, M5S 1A1, Canada

Murray, Cherry A.
  Bell Laboratories
  Murray Hill, NJ 07974

Nitzan, Abraham
  Chemistry Department
  Tel-Aviv University
  Tel-Aviv, Israel

Otto, A.
  Physikalisches Institut III
  Universität Düsseldorf
  D-4000 Düsseldorf
  Federal Republic of Germany

Pettenkofer, C.
  Physikalisches Institut III
  Universität Düsseldorf
  D-4000 Düsseldorf
  Federal Republic of Germany

Pettinger, Bruno
  Fritz-Haber-Institut
    der Max-Planck-Gesellschaft
  1000 Berlin 3
  Federal Republic of Germany

Pockrand, I.
  Physikalisches Institut III
  Universität Düsseldorf
  D-4000 Düsseldorf
  Federal Republic of Germany

Ritchie, G.
  Physics Department and
    Laboratory for Research
    on the Structure of Matter
  University of Pennsylvania
  Philadelphia, PA 19104

Sanda, P. N.
   IBM Thomas J. Watson
      Research Center
   Yorktown Heights, NY 10598

Schatz, George C.
   Department of Chemistry
   Northwestern University
   Evanston, IL 60201

Siiman, O.
   Chemistry Department
   Clarkson College of Technology
   Potsdam, NY 13676

Theis, T. N.
   IBM Thomas J. Watson
      Research Center
   Yorktown Heights, NY 10598

Tsang, J. C.
   IBM Thomas J. Watson
      Research Center
   Yorktown Heights, NY 10598

Ueba, Hiromu
   Department of Electronics
   Toyama University
   Takaoka, Toyama, Japan

Wang, D.-S.
   Chemistry Department
   Clarkson College of Technology
   Potsdam, NY 13676

Warlaumont, J. M.
   IBM Thomas J. Watson
      Research Center
   Yorktown Heights, NY 10598

Weitz, D. A.
   Exxon Research and
      Engineering Company
   P.O. Box 45
   Linden, NJ 07036

Wetzel, Herbert
   Fritz-Haber-Institut
      der Max-Planck-Gesellschaft
   1000 Berlin 3
   Federal Republic of Germany

INDEX